DESALINATION UPDATES

Edited by **Robert Y. Ning**

Desalination Updates
http://dx.doi.org/10.5772/59351
Edited by Robert Y. Ning

Contributors

Azzam Abuhabib, Mohamed Darwish, Hassan Abdulrahim, Abdel Nasser Mabrouk, Ashraf Hassan, Emrah Deniz, Yansheng Li, Robert Y. Ning, Ayman Elgendi, Evgeny Kharin, Hassan Fath, Muhammad Wakil Shahzad, Kyaw Thu, Li Ang, Azhar Bin Ismail, Kim Choon Ng, Dilek Duranoglu, Ulker Beker

Published by InTech

Janeza Trdine 9, 51000 Rijeka, Croatia

Publishing Process Manager Ivona Lovric

Technical Editor InTech DTP team

Cover InTech Design team

First published October, 2015

Printed in Croatia

Additional hard copies can be obtained from orders@intechopen.com

Desalination Updates, Edited by Robert Y. Ning

p. cm.

ISBN 978-953-51-2189-3

For EU product safety concerns:
IN TECH d.o.o., Prolaz Marije Krucifikse Kozulić 3, 51000 Rijeka, Croatia,
info@intechopen.com or visit our website at intechopen.com.

Meet the editor

Dr. Robert Ning is the vice president of the Science and Business Development of King Lee Technologies based in San Diego, CA, USA. This company was founded in 1977 and specializes in chemistry and chemicals used in reverse osmosis (RO) membrane plants. Since 1995, Dr. Ning is responsible for science and product development, as well as in the establishment of a network of process design for engineers and service providers around the world to design and service RO plant processes.
He has a BS degree in chemistry from the Rochester Institute of Technology, a PhD in organic chemistry from the University of Illinois, an MBA from the Fairleigh Dickinson University and was a postdoctoral fellow at the California Institute of Technology. He has 25 years of process chemistry experience in the pharmaceutical and biotech industries, prior to his current specialization in water and wastewater treatments.

Contents

Preface

The Intech book series of *Desalination, Trends and Technologies* (2011), *Expanding Issues in Desalination* (2011), *Advancing Desalination* (2012), and this volume of *Desalination Update* illustrates the growing research and development activities in the field of desalination of water. The chapters in this book also show the close link in the supply of water and supply of power. Power is needed to desalinate water, and water is needed to produce power via steam and cooling water. As the world is becoming increasingly in need of water and power, the education of generations of new workers in these technologies makes the publications of these books of rising importance.

Students and specialists alike will find branching strands in this field of development worthy of dedication of careers. Never has shrinking essential resources and exploding needs confront mankind as much as water. Excellent reviews in this book provide keywords, concepts, and current knowledge and status of practice useful for teaching and continued evolution.

Dr. Robert Y. Ning
Vice President, Science and Business Development,
King Lee Technologies, USA

Section 1

Reverse Osmosis, Nanofiltration, Membrane Distillation

Chapter 1

Reverse Osmosis Chemistry — Basics, Barriers and Breakthroughs

Robert Y. Ning

Additional information is available at the end of the chapter

http://dx.doi.org/10.5772/60208

Abstract

While reverse osmosis (RO) for desalination of brackish water, seawater and wastewater is a most economical and powerful method, its sensitivity to fouling points to the importance of understanding the water chemistry involved and methods of fouling control and system maintenance. As a chemical developer of antiscalants, antifoulants, and operation and maintenance chemicals needed for RO systems, we present here a basic understanding of RO chemistry, the challenges of scaling and colloidal fouling that limits % recovery of permeate and some breakthroughs we have attained.

Keywords: Reverse osmosis chemistry, scaling, colloidal fouling, antiscalant, antifoulant, tandem RO system

1. Introduction

The rapidly increasing introduction of reverse osmosis (RO) membrane plants around the world for treatment of water challenges the process of training professionals and technicians needed to design, operate and maintain such systems. The systems vary in size from 100 million gallons per day (15,800 m3/hour) municipal systems for municipal wastewater, brackish water and seawater desalination, down to 10 gallons per minute (38 liters per minute) used in kidney dialysis clinics. The sensitivities of RO membranes toward fouling and challenges in sustaining operation highlight the need for understanding the chemistry that impacts on the design, performance and maintenance of RO systems. In this chapter, the basics, barriers and breakthroughs in RO chemistry are briefly reviewed.

2. The water cycle

Water covers three quarters of the surface of our blue planet. It is the most powerful and essential solvent for life as we know it. The Water Cycle as depicted in Figure 1 is instructive for tracing the chemistry that occurs in water as it circulates in our environment and that which impacts the reverse osmosis process in water treatment.

Water evaporates from the oceans and from land as pure water vapor, then condenses in cooler atmosphere as clouds of minute droplets before falling as rain or snow. Streams form rivers and lakes before returning to the oceans, both above and below the surface of land. Of note is the portion of water that seeps deep underground and forms aquifers, from which we retrieve brackish well water. Summarized in Table 1 are the distinct stages of the water cycle in which we can discern unique chemistries that occur that will have impact on the RO system.

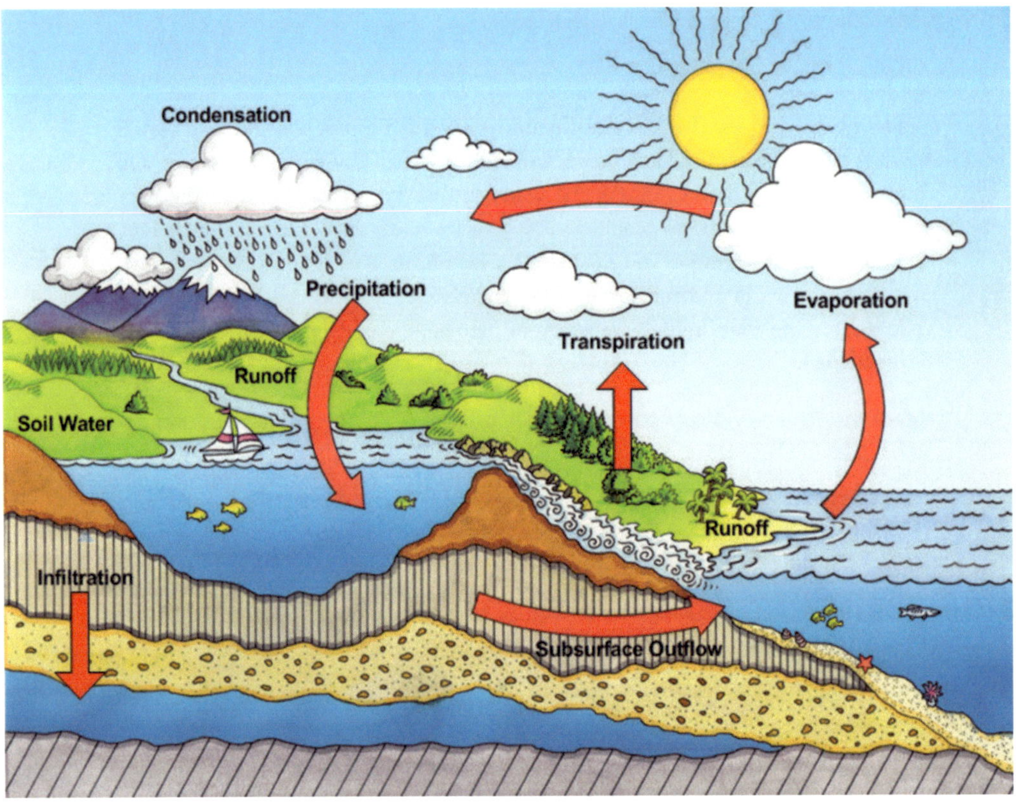

Figure 1. The Water Cycle.

2.1. Rain

As rain falls through air, dissolution of oxygen and nitric oxides will lead to oxidation-reduction reactions in the ground, and more significantly will lead to formation of carbon

http://dx.doi.org/10.5772/60208

Source of Feedwaters and Significant Chemistry that Impact RO

Source Waters	Significant Chemistry
1. Rain	- Oxidation reduction (eg. O_2+NO_x), acid base (eg. CO_2+H_2O)
2. Springs	- Dissolution, acid base (acid rain+carbonate rocks)
3. Ponds, Lakes	- Complex interactions of living and nonliving matter
4. Wells, Aquifers	- Dissolution-precipitaion (eg. $CaCO_3$,$CaSO_4$, $SrSO_4$, $BaSO_4$, silicates, iron and aluminum salts, sulfides), relatively low in organic matter
5. Rivers, municipal and industrial wastewater	- Complexation interations, precipitation
6. Ocean (99.4% of earth's water including 2% as ice)	- Biotic life, complexation interations, precipitation

Table 1. Sources of feed water and significant chemistry that impact RO

dioxide (subsequently forming carbonic acid) and sulfur dioxide (subsequently forming sulfurous acid), leaching lime stones and other alkaline rocks in the ground.

2.2. Springs

As soon as rain touches the ground, available water-soluble salts dissolve. As a general rule, it is good to remember that all salts of sodium and potassium paired with mono or divalent anions are soluble, and all salts of chlorides and nitrates paired with mono or divalent cations are soluble in water. Calcium, strontium and barium carbonates and sulfates have low solubility. Water dissolves the soluble sodium, potassium salts and chlorides and nitrates of calcium, strontium and barium from minerals, and solvates and separates them as a mixture of freely mixing cations and anions. This allows ion paring of the less soluble divalent salts like the carbonates and sulfates of calcium, strontium and barium to reach or exceed their solubilities in water, forming sediments.

2.3. Ponds and lakes

Accumulation of nutrients in ponds and lakes brings together complex interactions of living (bacteria, algae, diverse organisms) and non-living matter- organic (carbon-based) and inorganic (noncarbon-based) matter.

2.4. Wells and aquifers

Water that seeps deep into the ground, lacking air and light has relatively low organic activities and content. Chemistry is more limited to dissolution of rocks, which consist of largely silica and silicate compounds of minerals. Hydrolysis of these rocks mobilizes the various ions we find in natural waters. Paring together of insoluble salts such as described in 2.2 above form deposits of pure compounds such as calcium carbonate (lime stone), calcium sulfate (gypsum), etc.

2.5. Rivers and municipal and industrial wastewater

Rivers are highly contaminated with natural plant and animal debris, along with municipal and industrial discharges. In addition, finely dispersed inorganic particles of eroded rocks and soil form complex sediments, silting up the rivers and causing rivers to meander.

2.6. Oceans

Oceans contain 99.4% of the water on earth, including 2% present as ice. Chemistries in seawater as far as it impacts seawater ROs is predominantly colloidal organic particles excreted by the abundance of algae and planktons in the sunlit upper regions of the sea. It has been observed, that in the deep ocean, coagulated organic mass fall like snow-flakes. Seawater RO with high salinity from sodium chloride and soluble salts in the feedwater seldom scales at 50% recovery. Extensive removal of colloidal organic matter is needed to avoid fouling, while avoiding carryover of coagulants used in pretreatment.

3. Deployment of RO system

Typical RO processes are depicted in Figure 2. Pretreatment of raw water before the RO has been described extensively in open access literature [1,2], and by searching the subject "RO pretreatment" with www.googlescholar.com. The basic recommended requirements for RO feedwater quality is turbidity of less than 1 NTU, and Silt Density Index of less than 3. Examples of the use of RO permeate is for drinking, boiler feedwater, cleaning processes in microelectronic manufacturing and in pharmaceutical industry. The brine (concentrated reject) has been used in cooling towers, evaporators or simply discharged.

When pretreated water sufficiently devoid of suspended particles is fed into the RO system through a cartridge guard filter, recoveries of 50–90% are typically attained (see Figure 3). Corresponding to these recoveries, the impurities in the RO feedwater are concentrated by a factor of 2–10 fold. Physical separation of any solids from the concentrated streams that clog the fine passages of the membrane elements constitute system fouling, lowering the productivity of the system, and requiring cleaning to restore performance. Clean-in-place equipment in each RO plant provides the ability to clean the system by recirculating appropriate cleaning solutions through the membrane elements. To avoid channeling of cleaning solutions during cleaning by excessive foulants, the system performance must be carefully monitored by trending normalized permeate flow rate, differential pressure and salt passage [3,4]. Effective cleaning is needed when performance by these parameters drop by 10–15% to fully restore performance. The need to replace hundreds or thousands of membrane elements would be extremely expensive. This is the reason that RO is commonly recognized as a very sensitive

http://dx.doi.org/10.5772/60208

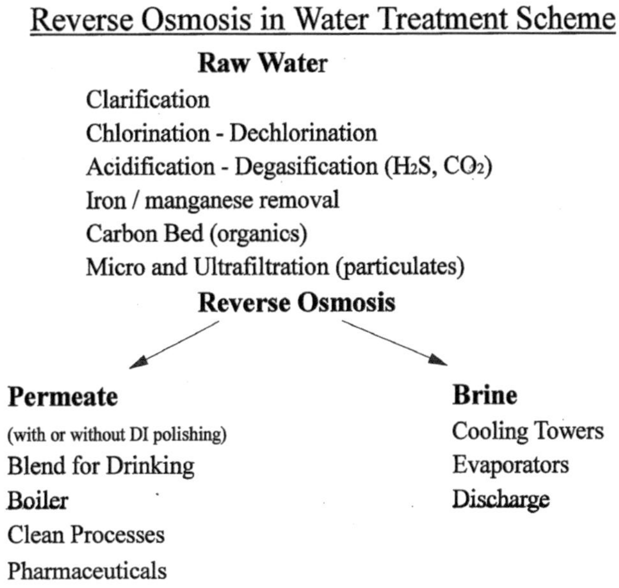

Figure 2. Reverse osmosis in water treatment scheme

unit operation. The tendencies to foul represent barriers to efficient operation and reliability of RO systems.

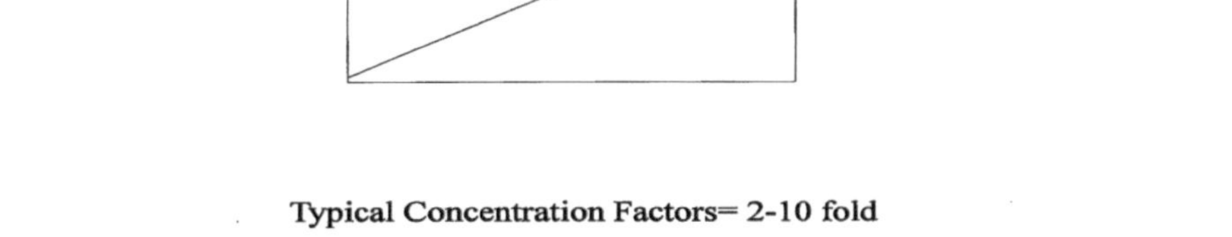

Figure 3. Concentration factor in reverse smosis system

4. Barriers presented by membrane fouling

Membrane fouling mechanisms can be classified into three categories: crystallization of insoluble salts commonly termed scaling, coagulation of colloidal particles and polymers known as colloidal fouling and microbial growth forming biofilm. We will examine each category in greater detail.

4.1. Scaling

Salts consist of pairs of positively charged cations and negatively charged anions forming neutral molecules. Sodium chloride (NaCl) is an example. Salt molecules of the same composition pack in regular repeated patterns in three-dimensional forms to form crystals of different shapes. Sodium chloride crystals are cubic when viewed under the microscope, or grown, as some hobbyists do, as large single cubic crystals. When crystalline salts dissolve in water, the ions dissociate into freely mobile cations (e.g., Na^+) and anions (e.g., Cl^-) each weakly bonded and stabilized (solvated) by water molecules (H_2O). Likewise, potassium nitrate (KNO_3) ionizes in water to form K^+ and NO_3^- ions each solvated by water molecules.

It is useful to remember that monovalent (singly charged) ions are easier for water to ionize (pull apart and solvate) than divalent (doubly charged) ions. Thus, it can be stated that in common water treatment, all salts of sodium, potassium, chloride, bicarbonate and nitrate are relatively soluble, and those of calcium, strontium, barium, carbonate and sulfate are much less soluble when paired. In the laboratory, it can be simply demonstrated that when perfectly clear water solutions of calcium chloride and sodium sulfate are mixed, crystals of calcium sulfate (gypsum, see Figure 4) will form. Such a demonstration illustrates two phenomena, firstly, ions freely mix in water solution, and secondly, that the doubly charged ions (Ca^{+2}) and (SO_4^{-2}) pair up to be less soluble in water. For this reason, most commonly observed scaling in brackish water RO systems are $CaCO_3$, $CaSO_4$, $SrSO_4$ and $BaSO_4$. In seawater, due to the high concentration of sodium chloride and a variety of competing ions, the same pairing of divalent ions to initiate regular stacking of crystal forms is greatly suppressed. Seawater ROs do not normally scale at 50% recovery. We will consider below the concentration of total dissolved solids in RO concentrate as one of the many critical parameters that affect scaling potentials. Other important parameters are degrees of supersaturation, presence of seed crystals, nature of the solutes, interfering impurities, pH and temperature.

4.1.1. Supersaturation

Different salts have different solubilities in water. When their natural solubilities are exceeded, given time, they will crystallize forming scales. When ionized by water and freely mixing in solution with ions from other salts, an important law determining the limits of solubility is a term called Solubility Product Constant. This law states that each combination of cations and anions in solution reaches a saturation value when the product of the concentration of the cation and the anion, whether they are the same concentration or not, cannot exceed a certain constant value. For instance, Table 2 shows the solubility product constants of four types of scales in terms of solubility products as mg/L concentrations.

Although when crystallized, $CaSO_4$ is a 1:1 pairing of Ca^{+2} and SO_4^{-2} ions, in solution for the calculation of solubility product constant, their concentrations do not have to be equal. Thus 100 mg/L Ca x 963 mg/L SO_4 = 96,300, so is 10 mg/L Ca x 9630 mg/L SO_4 = 96,300. Both solutions

http://dx.doi.org/10.5772/60208

Figure 4. Gypsum seed crystals

have reached solubility product constant limit. In the case of CaF2, three ions are involved per molecule, thus 1 mg/L Ca^{+2} x 27 mg/L F^{-} x 27 mg/L F^{-} = 723 solubility limit. Barium sulfate is theoretically so insoluble that 1 mg/L Ba x 1.64 mg/L SO4 = 2.64 limit.

Salt	Mg/L Solubility Product
CaSO4	96,300
SrSO4	5,300
BaSO4	2.64
CaF2*	723

*The fluoride concentration is squared when calculating the solubility product. (taken from Stumm and Morgan, 1981)

Table 2. Solubility product of low solubility salts (concentrations expressed in mg/L as ions)

4.1.2. Seed crystals

Scaling begins with the formation of seed crystals, which provides the molecular scaffold to which additional molecules attach themselves to form unique three-dimensional structures. Figure 4 shows the needle-like crystals of gypsum ($CaSO_4.2H_2O$). Shown in Figure 4 are crystals in the tens of microns size range. This is intended to show the peculiar shapes of these crystals. Seed crystals that we refer to are much smaller than these, as configuration of small number of molecules. In solutions of low supersaturation devoid of seed crystals that are allowed to stand without agitation, it is possible to obtain large single crystals as the water is allowed to evaporate slowly. Conversely, in the crystallization processes for the production of crystalline products, addition of crushed crystals as seeds will accelerate the completion of crystallization. Repeated dissolution and crystallization of both inorganic and organic compounds is a powerful method for the preparation of high purity compounds.

4.1.3. Nature of solutes

Most commonly encountered scales in RO systems are calcium carbonate, calcium sulfate, strontium sulfate, barium sulfate, calcium fluoride, silica and silicates. True salts (pairs of ionizable cations and anions in water) crystallize in distinct shapes with gritty textures, whereas silica and silicates are not true salts, and appear in amorphous forms. They are dehydration polymers of silicic acid [Si(OH)4] with hydroxide forms of metals, most commonly those of aluminum, iron, magnesium and calcium [5,6], and they appear on membranes as a thin film of colorless or slightly yellow gel.

4.1.4. Interfering impurities – Antiscalant actions

The water solution matrix from which scales form has a decided effect on the solubility and the rates of crystallization of different scales. Freely moving cations and anions have to pair themselves before stacking on crystalline lattices in predetermined geometries. When the concentration of total dissolved solids (TDS) increases, the ease by which the pairing of the ions of the insoluble compound is reduced. In addition, different ions of similar charges, sizes and shapes can occupy unintended atomic positions on the surface of a growing crystal, covering up a growing surface stopping crystal growth, or in cases of partial inhibition, causing the resulting crystal to be deformed, a phenomenon called crystal modification. Effective antiscalants generally contain anions with multiple negative charges that have high affinities for the growing surfaces of seed crystals, preventing scaling and stabilizing the supersaturated condition. This mechanism of antiscalant action is referred in the literature as "threshold inhibition". The efficiency of this inhibition mechanism allows for the economic use of low dosages (< 10 mg/L) of continuously injected antiscalants into the RO feedwater.

4.1.5. Solution pH

The main effect of pH on scaling is the reversible formation of carbonate ions from bicarbonate. Calcium carbonate scaling occurs at higher pH due to the presence of higher concentrations of carbonate ions. The effect of pH can be seen in the reversible formation of CO_2 gas or

http://dx.doi.org/10.5772/60208

carbonate ions in natural water. Acid addition drives the equilibrium to the left (toward CO_2 formation). Caustic addition drives the equilibrium to the right (toward carbonate ion formation as shown below). A third and important phenomenon should be noted. Since CO_2 gas is easily degassed from water, its loss amounts to losing acidity, thereby increasing the alkalinity of the system, causing an increase in the carbonate ion concentration. This degassing of CO_2 in the presence of sufficient concentration of calcium can cause calcium carbonate scaling.

$$CO_2 + H_2O \rightarrow H_2CO_3$$
$$H_2CO_3 + OH^- \rightarrow HCO_3^- + H_2O$$
$$HCO_3^- + OH^- \rightarrow CO_3^{-2} + H2O$$

Natural waters are buffered by bicarbonate ions at pH 5.35–7.35. The pKa of the conversion of carbonic acid to bicarbonate, when titrated with alkalinity, is 6.35. At this pH, the concentration of carbonic acid equals that of bicarbonate ions. At pH=7.35, one log value above the pKa value, the ratio of molar concentration of $HCO_3^-/H_2CO_3 = 10$, rising to a ratio of 100 at pH=8.35. The pKa for the conversion of bicarbonate to carbonate is 10.33. When raising the pH of RO feedwater with caustic injection to help ionize the weak boric acid and silicic acid for rejection by membranes in the region of pH= 11, the use of antiscalant to control calcium carbonate scaling is needed due to increased carbonate ion concentration.

4.1.6. Temperature

Generally, higher temperatures increase the solubility of salts. In situations where higher temperatures cause degassing of acids like CO_2 and H_2S or bases like NH_3, membrane scaling chemistry may be affected.

4.2. Colloidal fouling

When fouled membrane elements are opened during autopsies for examination, more frequently than not, the foulants appear to be amorphous pastes or gels, with no crystalline shapes under the microscope. This is the result of colloidal fouling.

Colloidal particles by definition are pre-existing particles in water that are smaller than 0.1 micron in size. They are not visible to the naked eyes. They escape filtrations in RO pretreatment processes, and do not register in turbidity and Silt Density Index measurements. The latter records plugging rates while passing sample water through 0.4 micron filters. In municipal wastewater treatment processes, colloidal fouling of RO is commonly seen following microfiltration and ultrafiltration pretreatment, because fractions of smaller colloidal particles escape filtration.

An educational demonstration of colloidal particles in water can be easily done. Put perfectly clear distilled water and water from the tap in two separate clear glass beakers. Turn off all lights in the room. Shine a beam of laser light from a laser pointer through the beakers. The beaker with tap water will show a line through the water, traced by colloidal particles. The beaker with distilled water will show no line through the water.

What characterizes the behaviors of colloidal particles is that they have huge surface areas per unit weight. Consider taking a brick of hard cheese, and cutting it in two. The cut exposes a new surface area. Cut each part into two again, gaining new surface areas. Imagine repeating this innumerable times. The particles each get smaller until they are too small to see, while the total surface area grows exponentially. Surface characteristics of colloidal particles become dominant factors in their inter-particle interactions, in which they can approach each other in very close distances. Controlled production of colloidal particles of uniform structures and coating them on prepared surfaces now constitute our newly named field, nanotechnology.

Naturally formed nano particles in water are impossible to speciate and quantitate. As foulants in RO systems, we can only hope to deal with them empirically with continuously injected antifoulants as anti-coagulants, dispersants and antideposition agents. It is assumed that by modifying the surface properties of the colloidal particles and the membrane surface, colloidal fouling can be and has been controlled in an increasing number of cases [7,8].

We consider below some main factors that affect the rates of coagulation and deposition of colloidal particles in RO systems.

4.2.1. *Surface interactions*

Hydrophilic, hydrophobic, ionic attraction/repulsion, hydrogen-bonding and van der Waals forces are terms that characterize complex surface interactions between particles, keeping them separated or causing them to coalesce. These forces typically increase by factors inversely proportional to the square of distances that separate them. Thus the close distances that these minute particles can approach each other greatly magnify their interactions.

4.2.2. *Charge neutralization*

Acidity or alkalinity (pH) changes in water can alter ionizable groups on the surfaces of particles to become more or less charged. Natural organic particles are predominantly characterized by surface ionization of carboxylic acids (pKa 4–5), ammonium groups (pKa 9) and phenols (pKa 10). Inorganic particle surface charges such as calcium carbonate, calcium phosphate, ferric and aluminum hydroxides are sensitive to changes in the pH 5–7 range. Particles in natural water at neutral pH generally carry net negative charges on their surfaces. In this state of mutual repulsion, they are stably suspended. When highly positively charged coagulants like aluminum and ferric salts, or cationic polymers are added to neutralize the surface charges, the colloidal dispersion become destabilized. Particles coagulate and flocculate, facilitating removal by filtration.

A good demonstration of charge neutralization causing coagulation is the curdling of milk by adding vinegar or lemon juice. Milk is a highly concentrated stable suspension of colloidal protein, fat and calcium phosphate, all of which are nutrients needed by the baby. When the negatively charged surfaces of the dispersed particles are protonated (H^+ added) by the added acid, the stable suspension collapses and curdles.

4.2.3. Concentration

While large sticky particles that pre-existed in the RO feedwater are retained by the first membrane element of an RO system, colloidal fouling and scaling occurs first in the last element, spreading forward, as the loss of membrane productivity also moves forward in the system. Colloidal fouling can be correlated to two factors: increased concentrations of the foulants as well as the rapid rise of concentration of total dissolved solids (TDS). The latter will be considered next as a very important factor.

4.2.4. TDS/ionic strength

The total dissolved solids (TDS) in the RO concentrate express the concentration by weight of all the dissolved salts. Collectively as a solution, the ionic strength takes into consideration the stronger effects of di and trivalent ions compared to the monovalent ions. Ionic strength greatly affects the stability of equilibrated colloidal dispersions.

As a model for understanding the colloidal fouling in RO systems, one can look at the formation of deltas at the mouth of large river systems. The Mississippi, Nile, Ganges and Yangtze rivers all have prominent delta regions. Deposits in the delta result from coagulation of colloidal particles, both organic and inorganic, when the low salinity river water suddenly meets with the high ionic strength seawater. This is equivalent to the rapid concentration of RO feedwater to high salinity in a few seconds during passage through the RO system. Coagulation of colloidal particles begins in the second and third stages of the RO system, and continues in the concentrate discharge system. It takes only a thin layer of coagulated colloidal matter to cause noticeable declines in membrane flux initially, before registering as increasing differential pressures through the membrane elements.

4.2.5. Organic coating

Colloidal particles can adsorb surface-modifying natural contaminants, or deliberately added coagulants or antifoulants that will affect their precipitation or dispersion, respectively. In pretreatment with low-pressure microfiltration and ultrafiltration membranes, coagulants can be used advantageously to form cake-layers to protect pores from plugging. Antifoulants can be found to control cake-layer fouling on RO membranes.

4.2.6. Flocculants

Large molecular weight polymers of natural or synthetic origin can gather colloidal particles, flocculating them into filterable masses, or causing membrane fouling.

4.2.7. Particle size distribution

Waters feeding RO systems contain a wide range of particle sizes. Currently, colloidal particle distributions cannot be measured by the most sensitive, commercial instruments usable in the field. Lacking instruments for process control, colloidal fouling by foulants derived from RO pretreatments is an ongoing challenge. Colloidal fouling is routinely seen in RO feedwater with <1 NTU turbidity, and Silt Density Index of <2, which is an index derived from plugging rate of 0.45 micron filter, and even after microfiltration and ultrafiltration pretreatment [7,8].

4.3. Microbial fouling

Membrane fouling by colloids that originate from microbes in natural and wastewater invariably occurs. Dead or alive, microbes – viruses, sperm/eggs, bacteria, algae, fungi, diatoms, flagellates, amoebas, ciliates, planktons, worms, crustaceans, etc are physical particles ranging in size from 0.05 microns into the visible range. For purposes of procreation, colonization or defense, they also excrete huge volumes of biopolymers – mostly polysaccharides – into the aqueous environment. Polysaccharides are high molecular weight polymers composed of, almost exclusively, abundant carbon, hydrogen and oxygen elements from the environment, biologically synthesized by plant and animals in the water and on land. As a mechanism of membrane fouling, all this can be included in the category of colloidal fouling. However, in RO systems, microbial fouling is unique in that bacteria can grow explosively, and the biofilms they generate on the wetted surfaces of the equipment become operational and create maintenance challenges. We will touch on three aspects of microbial fouling.

4.3.1. Exponential growth

Operation of RO systems is most affected by bacteria. In natural bodies of water, bacteria exist in planktonic (moving) state or in sessile (anchored) state. It is estimated that 99% of all bacteria are anchored to surfaces and protected in a covering of polysaccharides that they create known as biofilm [9].

Bacteria that enter the RO system with the feedwater are mainly in the planktonic state. In relatively stagnant regions, and especially during system shutdowns without preservatives, they are encouraged to anchor to some surface and proliferate. When conditions are favorable, *Escherichia coli* (E. coli), a very common bacterium will undergo cell division once every twenty minutes. This means during an 8 hour shift in an RO plant, one bacterium at the start can potentially become 16,777,216 bacteria (2^{24}) at the end of one shift! Bacteria and microorganisms basically fix nutrients from the air (CO_2, O_2, N_2, H_2O) and water (sugars, amino acids, fats, nitrate, phosphate, calcium, potassium, sodium, trace minerals, water, digestible biomass, etc) to form a new biomass that physically separates from the water phase and becomes mechanical obstructions in the operation of the RO system. Even minute traces of biomass in high purity product water are objectionable to applications in microelectronics, pharmaceutical, and power production plants.

4.3.2. Biofilm and slime

In the literature, polysaccharides excreted by bacteria and planktons are referred to as exopolymers, transparent exopolymers (TEP), extracellular polymeric substances (EPS) or exopolysaccharide matrix. They form intricately defined habitat structures we call biofilm [9, 10]. Biofilms can literally be called bacterial cities, with high rises, subterranean compartments, roadways, and tunnels, occupied by communities of single or multiple species. Bacteria respond to sensor molecules from the environment and neighbors. Some species of bacteria common in water treatment are strong producers of slime, such as *Pseudomonas aeruginosa, Aerobacter cloasae,* Enterobacteriaceae, iron bacteria (Gallionella) and sulfate-reducing bacteria (Desulfovibrio). Besides coating membrane surfaces with biofilm that reduces membrane permeability, free-floating slime from them block water flow-channels in the RO membrane elements, increasing differential pressures.

4.3.3. Colloidal particles and biopolymers in feedwater

Raw water from rivers, lakes, shallow wells and the sea contains an abundance of organic and inorganic colloidal particles that escape filtrations in RO pretreatment steps. Colloidal fouling of ROs are often seen even when the RO feedwater has a turbidity of <1 NTU, and Silt Density Index of <4 [7,8]. Exopolymers in seawater, especially during periods of high activities in the sea of picophytoplanktons present major challenges for designing pretreatment systems for seawater ROs (SWROs) [11,12].

4.4. Disposal of RO concentrate

For RO plant far from the sea, disposal of RO concentrates is a costly barrier to overcome [13, 14]. Conventional concentrate disposal methods involve discharge to surface water bodies or to a municipal sewer system. Deepwell injections require great depths and distances to avoid contamination of aquifers. Evaporation ponds are not practical for large plants with available land, evaporation rate and environment to permit treatment of million gallons a day discharge. All these discharge methods are increasingly being regulated and taxed for concerns with environmental impacts. Zero liquid discharge for maximum water recovery and having to only handle solid mineral recovery or waste has become an active field of research and development [15].

5. Breakthroughs on overcoming scaling and fouling barriers

Due to the limitation of time and knowledge, breakthroughs we discuss here will be limited to largely our own published data.

5.1. Advances with antiscalants and antifoulant development

A wide variety of antiscalants are now available for controlling all scaling seen in RO systems using the threshold mechanism for controlling seed crystals as described in 4.1.4 above. Full control is accomplished with continuous injection of an appropriate antiscalant at less than 10 mg/L dosage. The antiscalant and the dosage are selected based on the feedwater analysis, to control all scaling potentials for the specific water quality. For off-the-shelf small to medium-sized RO systems, or where water quality analysis is incomplete or not available, effective general-purpose antiscalants at a safe dosage can be prescribed, which will perform satisfactorily in almost every case. By carefully monitoring with input of daily sets of operating data into personal computer programs that can generate trend-charts of normalized permeate flow rate, differential pressure and salt passage, RO system performance can be kept within 10–15% of the performance at startup [4]. With appropriate maintenance cleaning method, and timely adjustments in the choice of antiscalant and dosage if scaling appears, cleaning frequency can be kept to a minimum, and membrane service life can be extended to > 15 years.

The barrier of colloidal fouling and microbial fouling is the frontier for research and development today. We learn from every RO plant problem solved, and pilot plant study completed. The impossibility of speciating and quantitating colloidal particles within the cost constraints today blocks the first step necessary to proceed with developing control methods. Short of doing projections of colloidal fouling based on water analyses, we have made progress with addressing colloidal fouling in individual RO systems after assuring that scaling is fully controlled. A series of antifoulant chemicals have found success in empirically showing efficacy in controlling fouling by colloidal sulfur, silica, calcium phosphate, oil and grease, humic matter and exopolymers [7,8 and unpublished data]. Progress on controlling microbial fouling resulted from focus on not killing planktonic bacteria, but focusing on stripping biofilms with multistep cleaning methods.

Many fouling problems we encountered in RO plants were resolved by the elimination of some pretreatment steps, which actually contributed to colloidal fouling, and significantly to the cost of pretreatment and maintenance. Since colloidal contaminants in raw waters are impossible to remove completely, and since we do not care what goes into the RO system, as long as they pass through the system, helping them to get through the system without sticking is the guiding light for the development of continuously injected antiscalants and antifoulants. With the dual approach of simplifying existing pretreatment and maximizing the effectiveness of antiscalants and antifoulants, much value has been delivered to existing RO plants and to owners of future plants. Improvements made in RO processes and current capabilities of our antiscalants and antifoulants are summarized in Table 3, and briefly discussed below.

1. **Calcium Carbonate**: Many plants operate on using both acid and antiscalants to control calcium carbonate scaling. To overcome pH buffering capacities of natural waters, large amounts of concentrated sulfuric acid, and even hydrochloric acid are used to lower the Langelier Saturation Index of the RO concentrate. Since the new antiscalants alone can control Langelier Saturation Index of 3.3, and Stiff Davis Stability Index of >4.5 for high concentrations of total dissolved solids, we have found that acid injection simply for controlling calcium carbonate scaling can be universally eliminated.

2. **Iron and Manganese**: Soluble reduced iron and manganese in well waters quickly precipitate when exposed to oxygen in the air to form insoluble oxy-hydroxides. Traditional pretreatment involves accelerated oxidation followed by multimedia filtration or oxidation and simultaneous filtration with Greensand coated with black manganese dioxide. Manganese dioxide is consumed as oxidizing agent, and has to be periodically regenerated with chlorine or potassium permanganate. Such pretreatment method we have found to shed colloidal iron and manganese particles into the RO membranes, fouling them as well as catalyzing oxidative damages on the membranes when exposed

RO Foulant	Traditional Pretreatment	Current Capability
1. Calcium Carbonate	Acidification to lower LSI,	Antiscalant alone:
	then with antiscalant to	LSI=3.2;S&DSI=>4.5
		maximum LSI=2.5
2. Iron, manganese	Oxidation/filtration: Greensand,	Antiscalant alone:
	manganese dioxide, catalytic ox.	Fe and Mn at > 8ppm
3. Silica: Reactive	Lime, Ion-Exchange (OH)	Antiscalant: >320ppm
Nonreactive	None	Antifoulant: variable
4. Calcium Sulfate	Lime, Ion-Exchange	Antiscalant: >400x sat'n
5. Strontium Sulfate	Ion-Exchange	Antiscalant: >43x sat'n
6. Barium Sulfate	Ion-Exchange	Antiscalant: >51x sat'n
7. Calcium Fluoride	Lime, Ion-Exchange	Antiscalant: >16,000x sat'n
8. Colloidal Organic	UF, MF, coagulation/	Antifoulant alone
Matter and Sulfur	clarification/ MMF.	
	slow sand and carbon filter	
slow sand and carbon filter		

Table 3. Summary of improvements in pretreatment and current capabilities

to traces of oxidants carried over into the system. Using appropriate antiscalants alone, iron and manganese can be kept soluble in the RO feedwater, and safely pass through the system with the concentrate. Systems with > 8 mg/L of iron in feedwater containing the antiscalant alone, can be fed directly to the RO system through a cartridge filter without fouling events.

3. **Silica – Reactive and Non-reactive**: Reactive silica measured by colorimetric assay using ammonium molybdate reagent is essentially silicic acid monomer [$Si(OH)_4$]. It spontaneously polymerizes by dehydration reaction to higher and higher molecular weights to form nonreactive silica, and together with hydroxides of iron, aluminum, magnesium etc, to form silicates during concentration in the RO system [5,6]. Traditional pretreatment to reduce silica concentrations involves lime softening or ion-exchange with quaternary ammonium resins in the hydroxide form to adsorb the weak silicic acid (pKa = 9.9, 11.8, 12, 12). Alternatively, appropriate antiscalants can inhibit the rate of polymerization to the extent that up to 320 mg/L of reactive silica can be tolerated in the RO concentrate without deposition of some polymeric hydrated silica and silicate gels. Feedwaters also contain nonreactive silica and silicates in the soluble colloidal forms. Silica in the non-reactive (towards molybdate reagent in colorimetric assay) form can be quantitated with emission spectroscopy as total silica, then subtracting the reactive component measured separately with colorimetric assay. Nonreactive silica and silicates of sufficient chain size,

shape and composition can stick and coat the surfaces of RO membrane causing severe drop in productivity. This type of colloidal fouling has to be controlled with antifoulant injection.

4. **Calcium, Strontium and Barium Sulfate**: Traditionally, reductions of concentrations of calcium, strontium and barium are accomplished by softening with lime or with ion-exchange resins. More efficiently, these sulfate scales can be controlled to high levels shown in the Table 2, expressed as multiples of saturation at Solubility Product Constants shown in Table 2.

5. **Calcium Fluoride**: It commonly appears in well waters as a milky suspension. It can be removed from RO feedwaters by softening with lime or ion-exchange resins, and by filtration of the larger preexisting particles. Colloidal particles of CaF_2 can be controlled with an antifoulant. Antiscalants can control very high scaling potentials of this salt.

6. **Colloidal Organic Matter and Sulfur**: Removal of preexisting particles involves coagulation/filtration with multimedia filters, slowsand filters, activated carbon towers and more recently, microfiltration and ultrafiltration, with or without assistance of coagulant addition. For clarified water feeding RO systems, we can, or try to, control colloidal fouling with antifoulants.

5.2. Tandem RO process- (Maximum water recovery at 1,000 psi pressure)

The physical limitation of operational pressure of RO systems is about 1,000 psi (68 atmospheres), since in maximizing water recovery, the total dissolved solids concentration in the RO concentrate is also maximized. The natural osmotic pressure created between the RO concentrate and the permeate across the membrane is in opposition to the mechanical pressure applied. Assuming osmotic pressure increases by a factor of 0.0115 psi/ppm of TDS [16], then we would run out of net driving pressure of a 1000 psi pump at a TDS value of about 87,000 ppm of TDS, about twice the salinity of seawater.

Maximized water recoveries in large municipal RO plants are in the region of 70–87% [17]. Addressing the barrier of concentrate disposal (4.4), and the need for minimization of concentrate volume, we demonstrated the possibilities of using a low pressure RO in tandem with a high pressure RO in continuous operation, to reach overall recoveries of brackish water in the region of 90–97% [18; 19; 20; 21; 22]. Beyond the limits of the RO system, thermal evaporation of the RO concentrates would be necessary for water recovery, solids recovery, or for use in cooling towers before final evaporation.

6. Conclusions

This brief overview of basics, barriers and breakthroughs in RO chemistry serves as an introduction for nonchemists in the practice of RO application in desalination.

For those pursuing process development, colloidal fouling of RO systems presents continued challenges, especially with the rising demands to treat municipal and industrial wastewaters

http://dx.doi.org/10.5772/60208

before discharge. The need in oil and gas exploration in preparing injection water and reusing produced water has become obvious. Expert attention to pilot studies becomes difficult due to the varieties of wastewater qualities and rapid changes and turn-over of raw water qualities and projects.

For large stable plants, continuous improvements are in progress with the following objectives:

a. Increasing overall water recoveries with tandem RO process.

b. Continuous RO system operation with no down time for cleaning.

c. Reliable control of microbial growth.

d. Long membrane service life of >15 years.

e. Zero liquid discharge of waste.

f. Economic recovery of minerals from water (water mining).

g. Training of ever increasing number of people needed to design, build, operate and maintain RO plants.

We write this chapter hoping very much that it will be found useful for training.

Author details

Robert Y. Ning

Address all correspondence to: www.kingleetech.com

King Lee Technologies, San Diego, CA, USA

References

[1] Gaid, K. (2011). A large review of the pretreatment, Chap. 1, Expanding Issues in Desalination, Open Source Book, intechweb.org, pp 3-56.

[2] Ning, R.Y. (2011). Pretreatment for reverse osmosis systems, Chap. 2, Expanding Issues in Desalination, Open Source Book, intechweb.org, pp 57-62.

[3] Ning, R.Y. (2012). Chemistry in the operation and maintenance of reverse osmosis systems, Chap. 4, Advancing Desalination, Open Source Book, intechweb.org, pp 85-95.

[4] Troyer, T.L., Tominello, R.S., and Ning, R.Y. (2012). Standardized data and trending for RO plant operators, Chap. 5, Advancing Desalination, Open Source Book, intechweb.org, pp 97-110.

[5] Ning, R.Y. (2002). Discussion of silica speciation, fouling, control and maximum reduction, Desalination, 151, 67-73.

[6] Ning, R.Y. (2010). Reactive silica in natural waters- a review, Desalination and Water Treatment, 21, 79-86.

[7] Ning, R.Y. (2005). Chemical control of colloidal fouling of reverse osmosis systems, Desalination, 172, 1-6.

[8] Ning, R.Y. (2007). Colloidal fouling of RO membranes following MF/UF in the reclamation of municipal wastewater, Desalination, 208, 232-237.

[9] Prakash, B., Veeregowda, B.M. and Krishnappa, G. (2003). Biofilms: A survival strategy of bacteria, Current Science, 85, 1299-1307.

[10] Lewis, K. (2001). Minireview: Riddle of biofilm resistance, Antimicribial Agents and Chemotherapy, 45 (4), 999-1007.

[11] Choules, P., Schrotter, J.-C., Leparc, J., Gaid, K. and Lafon, D. (2008). Improved operation through experience with SWRO plants, Ultrapure Water, October, pp. 13-22.

[12] Villacorte, L.O., Schurer, R., Kennedy, M.D., Amy, G.L. and Schippers, J.C. (2010). Removal and deposition of transparent exopolymer particles in a seawater UF-RO system, IDA Journal, First Quarter, 45-55.

[13] Howe, K.J. and Ahuja, N. (2005). Strategies for concentrate management from inland desalination, American Water Works Association- Membrane Technology Conference, 2005.

[14] Mickley, M. (2010). Overview of global inland desalination concentrate management: solutions, challenges and technologies, IDA Journal, 2 (3), 48-52. (International Desalination Association)

[15] Ning, R.Y., Tarquin, A.J. and Balliew, J.E. (2010). Seawater RO treatment of RO concentrate to extreme silica concentrations, Desalination and Water Treatment, 22, 286-291.

[16] Byrne, W. (1995). Reverse Osmosis- a Practical Guide for Industrial Users, Tall Oaks Publishing Inc., Littleton, CO, USA, p.76.

[17] Ning, R.Y. and Troyer, T.L. (2008). Minimizing liquid waste discharge from municipal reverse osmosis plants, Water Practice, 2(3), 1-5.

[18] Ning, R.Y., Tarquin, A., Trzcinski, M.C., and Patwardhan, G. (2006). Recovery optimization of RO concentrate from desert wells, Desalination, 201, 315-322.

[19] Ning, R.Y. and Troyer, T. (2009). Tandem reverse osmosis process for zero-liquid discharge, Desalination, 237, 238-242.

[20] Ning, R.Y., Troyer, T.L. and Tominello, R.S. (2009). Antiscalants for near complete recovery of water with tandem RO process, Desalination and Water Treatment, 9, 92-95.

[21] Ning, R.Y. and Tarquin, A.J. (2010). Crystallization of salts from super-concentrate produced by tandem RO process, Desalination and Water Treatment, 16, 1-5.

[22] Subramani, A., Cryer, E., Liu, L., Lehman, S., Ning, R.Y. and Jacangelo, J.G. (2012). Impact of intermediate concentrate softening on feed water recovery of reverse osmosis process during treatment of mining contaminated ground water, Separation and Purification Technology, 88, 138-145.

Chapter 2

Modified Nanofiltration Membranes Performance Improvement for Desalination Applications

A. A. Abuhabib

Additional information is available at the end of the chapter

http://dx.doi.org/10.5772/60212

Abstract

The study focuses on NF membranes modification and performance improvement while desalinating brackish water. The study provides valuable information about flux and rejection changes and relationship with pressure changing before and after modification. Experimental works included in the study investigate modified and unmodified NF membranes performance while filtering synthesized single salt and mixture salt solution at various concentrations (ranged from 1000 ppm to 4000 ppm) and various pressure magnitudes (pressure ranged from 2 to 10 bars). The rejection rates witnessed an increase after membrane modification took place with about 11–30% for magnesium sulfate and sodium sulfate, and 50–60% for sodium chloride and potassium chloride.

Keywords: Nanofiltration, Membrane, Desalination, Modification, UV graft

1. Introduction

Nanofiltration membranes have made noticeable establishment and found a way into many industries since their first introduction in the early 1990s. The major industries in which these membranes are variously applied and served are water and wastewater industries. The characteristics of these membranes determined by high flux, high rejection of salts, and low energy consumption associated with low pressure requirements enabled these membranes to apply significantly and perfectly [1-5]. However, obtaining an improved flux and rejection as well as fouling resistance of NF membranes for various applications are of major interest for researchers [6]. Technically, surface modification is considerably applied to improve membrane properties in terms of flux, salt rejection, and fouling resistance [6, 7]. In addition, micropollutant removal could also be improved by surface modification [8]. Various modification techniques have been applied for NF membranes by researchers including radical polymerization [9], low temperature plasma [10], pre-oxidation [11], layer-by-layer alternating polyelectrolyte deposition (APD) [12], ionizing radiation [13], and photochemical techniques [14]. However, photochemical grafting techniques (mainly UV-initiated grafting) have been widely used due to their low cost of operation, mild reaction conditions, selectivity to absorb UV light without affecting the bulk polymer, and the possibility of easy incorporation into the end stages of a membrane manufacturing process [15].

Several hydrophilic monomers are commonly used including N-vinyl-2-pyrrolidinone (NVP), 2-hydroxylethyl methacrylate (HEMA), acrylic acid (AA), acrylamide (AAM), and 2-acrylamidoglycolic acid (AAG) for membrane surface modification of which such monomer would be grafted on the membrane surface by UV-initiated graft polymerization [7, 16-19]. Generally, modification process sufficiency is measured by measuring membrane properties and performance after modification. Two common methods are considered and followed for the UV-initiated grafting of membranes: the dip method and the immersion method. For the same support, with the same the monomer concentration and irradiation time, the degree of grafting achieved using the dip method is two to three times higher than with the immersion method. However, in certain cases, membranes modified by the dip method showed lower rejection factors compared to the membrane support and membranes modified by the immersion method [19].

Normally, UV-initiated grafting of polyethersulfone membranes involves two parallel competitive processes, crosslinking and chain scission, that determine the final membrane transport properties [20]. Both mechanisms are very important for any modified membrane where crosslinking and chain scission may affect hydrodynamic resistance and membrane selectivity, respectively. Kaeselev et al. [21] illustrated that the membrane hydrodynamic resistance is increased relatively with crosslinking while the membrane selectivity loss is affected by chain scission in a direct manner. This study focuses on NF membrane performance improvement in terms of flux and rejection for desalination applications via surface modification.

2. Materials and methods

2.1. Materials

NF membrane used manufacturing properties, salts (NaCl, $MgSO_4$, Na_2SO_4, and KCl) used for filtration experiments, chemicals used for modification (acrylic acid and ethylenediamine dihydrochloride), and experiment set-up. One commercial NF membrane denoted as NF-1 was purchased from Amfor Inc. China. A summary of membrane information is given in Table 1. The monomers used in this study were acrylic acid (99% purity, purchased from Sigma-Aldrich, USA) and ethylenediamine dihydrochloride (99% purity, purchased from Frinde-

http://dx.doi.org/10.5772/60212

mann Schmidt chemicals, Germany). In addition, high purity salts including MgSO4 and NaCl, 99% purity, were purchased from John Kollin, UK.

Membrane	Material	Manufacturer operational data
NF-1	Polyethersulfone (PES)	Operational pressure is 150 psi, temperature is 25°C, and water flux is 100 $L.m^{-2}.h^{-1}$ and rejection rate is 98% for 2000 ppm of $MgSO_4$ solution.

Table 1. NF membranes manufacturer characteristics

2.2. Methods

2.2.1. Membrane modification

The membrane was modified using a monomer solution of 4% acrylic acid and 1% ethylene-diamine dihydrochloride (w/v) following the immersion method [7, 22]. The monomer was grafted on the membrane surface using a commercial UV device (LC5) supplied by Hamamatsu, Japan. The membrane was exposed to UV radiation for 5 minutes. The unmodified membrane was marked with a UV time of 0 min. The modification procedure is detailed as followed:

1. Membrane was soaked in ultra-pure water for 24 hours and dried on room temperature for 3 hours.

2. The membrane was placed in a membrane holder where only the membrane surface was exposed to the monomer.

3. The monomer was placed on the membrane surface and left for 15 minutes; then the remaining unabsorbed monomer was removed.

4. The membrane was exposed to the UV light for 5 minutes.

5. The membrane was soaked in NaOH solution (concentration: 1 M) for 5 minutes then washed out with ultra-pure water for 1 minute.

2.2.2. Membrane permeation and rejection

Filtration experiments were performed in a stainless steel cylindrical batch cell (HP 4750), served as a dead-end filtration system (cell volume: 300 cm^3), supplied by Sterlitech (UK). The working pressure in the cell was applied by a nitrogen gas cylinder in the range of 2 to 10 bar for unmodified and modified membranes; the experiments were conducted at room temperature. The membrane active area was 14.6 cm^2. Membranes were washed with ultrapure water and compacted at 10 bar of pressure for 20 minutes prior to use. No further pretreatment was performed on the membranes, bearing in mind that the manufacturer did not provide certain instructions for preparing the commercial membrane prior to use. A new filter was used for each experiment. Ultrapure water was used with conductivity below 1 μs/cm. The system was flushed with ultrapure water before and after use. Solutions of $MgSO_4$, NaCl, Na_2SO_4, and KCl at four various concentrations each (1000, 2000, 3000, and 4000 ppm) were used as the feed for unmodified and modified membranes to measure their rejection and determine the best performing membranes in terms of rejection. The concentrations of the feed and permeate were measured depending on solution conductivity measurements using a commercial conductivity meter supplied by Martini instruments (Romania).

3. Results

The major findings illustrated and analyzed in this section are concluded in the following two sections.

3.1. Flux

Pure water flux and solution flux (both single and mixture salt solutions) at various pressure magnitudes are highlighted and discussed in this section. The flux observed to be reduced after modification but with less than 30% compared to the unmodified membrane flux.

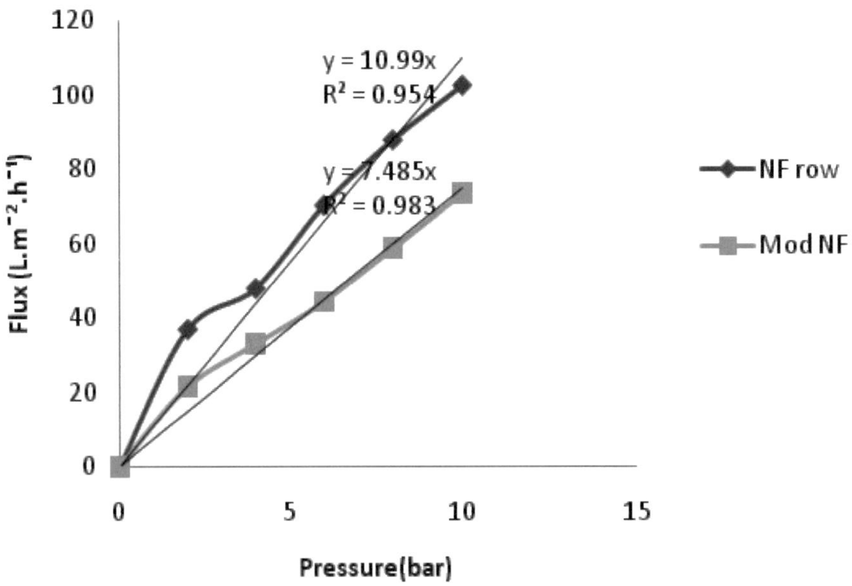

Figure 1. Pure water flux of unmodified and modified membranes

As illustrated in Figure 1, pure water flux for both unmodified and modified membranes was obtained at different applied pressures ranging from 2 to 10 bar. Membrane permeability was defined as the slope of pure water flux versus pressure. Apparently, the pure water flux of the membrane decreased after modification by about 30%. Accordingly, the permeability had shown some 24% decreasing after modification. Both flux and permeability decreasing is evidently confirming pore size decreasing following the modification process. Membrane permeation decreasing and increasing with membranes modified by UV grafting can be found in the literature. The increase in permeation of modified membranes was observed by Puro et al. [23] when modifying commercial polyethersulfone membranes (NTR7450 Nitto Denko) following the immersion method. The study also demonstrated pore size increasing in some of the modified membranes. In contrast, UV-initiated grafting of membrane pore walls may reduce pore size according to Yu et al. [24]. In their work, they stated that for membranes with small pores, most of the polyacrylic acid may be grafted on the membrane surface, not on the pore walls. Abu Seman et al. [7] observed both mechanisms depending on the degree of grafting, which was related to UV irradiation time and monomer concentration.

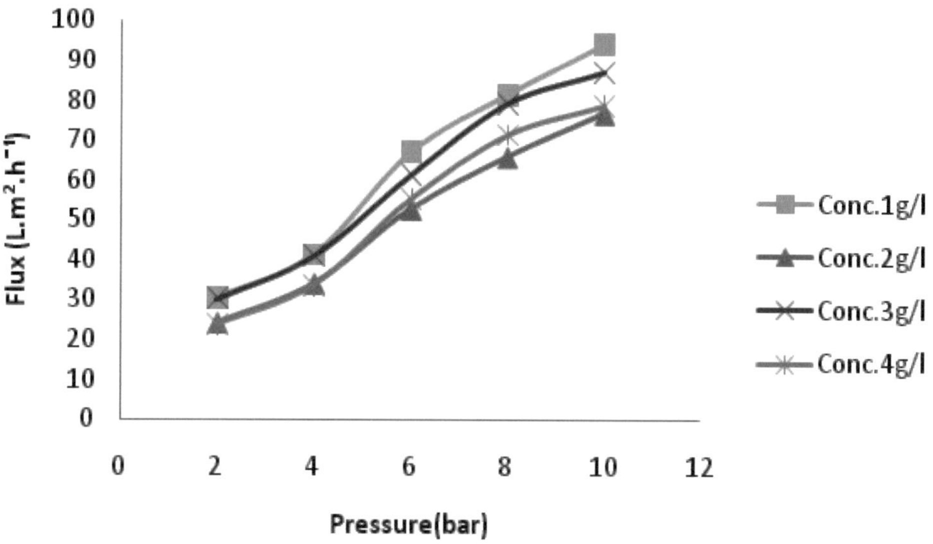

Figure 2. $MgSO_4$ solution flux for unmodified membrane at various concentrations

Figures 2 and 3 showed the effect of $MgSO_4$ concentration increasing on unmodified and modified membranes flux. For unmodified membrane the flux decreased with concentration. However, the decreasing for unmodified and modified membranes was 11% and 17%, respectively (flux decreasing from 1 g/L to 4 g/L). The same observations noticed for $Na2SO_4$ solution (Figures 4 and 5) but with different percentage (30% and 7% for unmodified and modified membranes, respectively).

For NaCl (Figures 6 and 7), the observations were different, as the flux at 2 and 4 bar pressure for both unmodified and modified membranes decreased with concentration but increased at higher pressure magnitude with concentration.

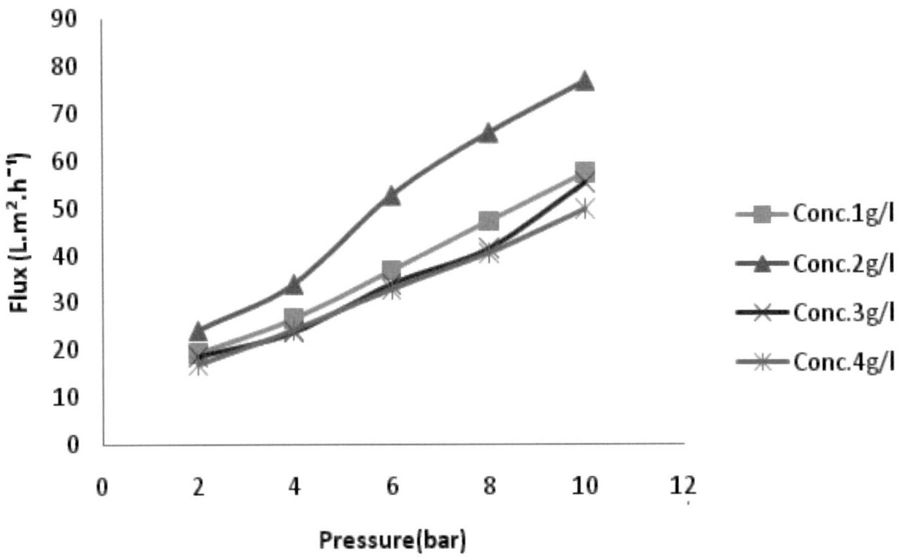

Figure 3. $MgSO_4$ solution flux for modified membrane at various concentrations

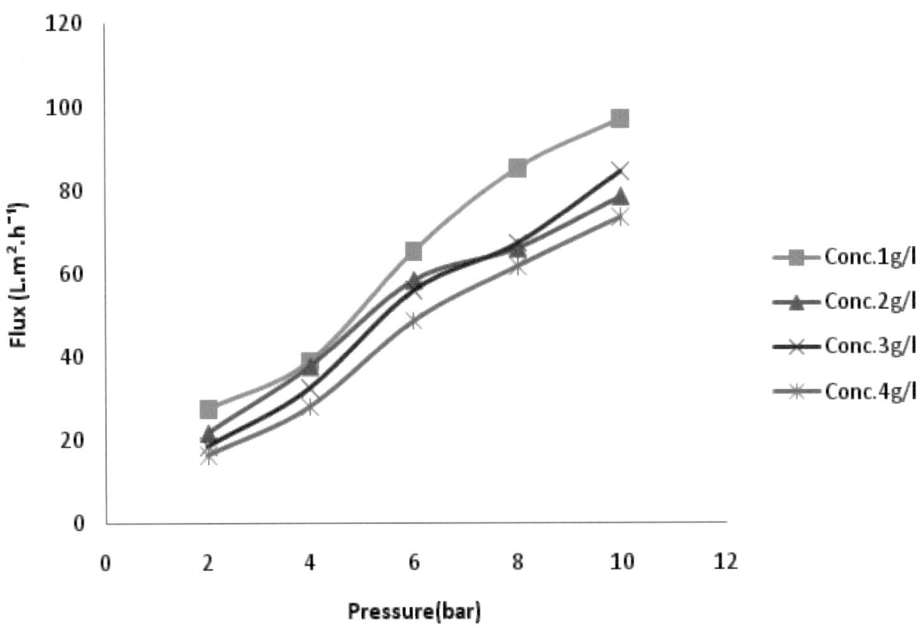

Figure 4. Na_2SO_4 solution flux for unmodified membrane at various concentrations

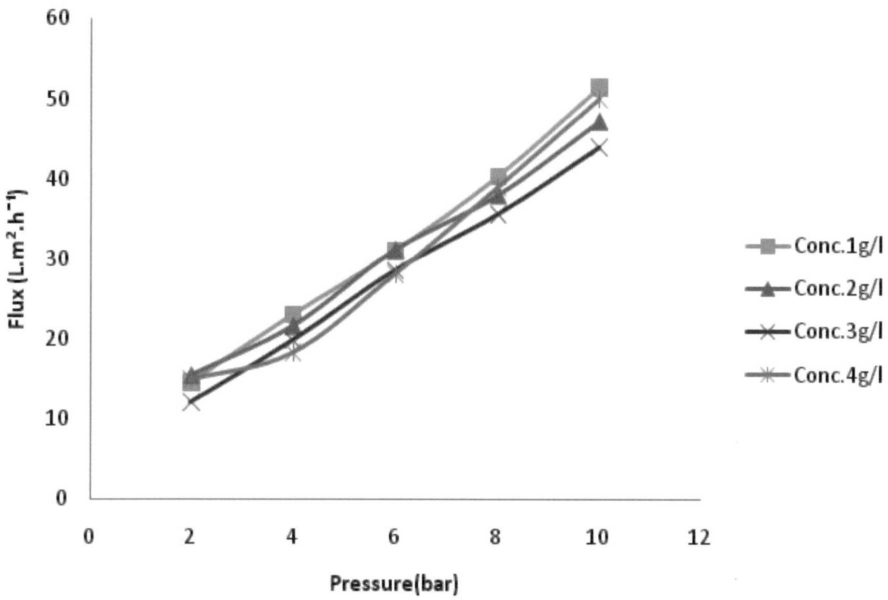

Figure 5. Na_2SO_4 solution flux for modified membrane at various concentrations

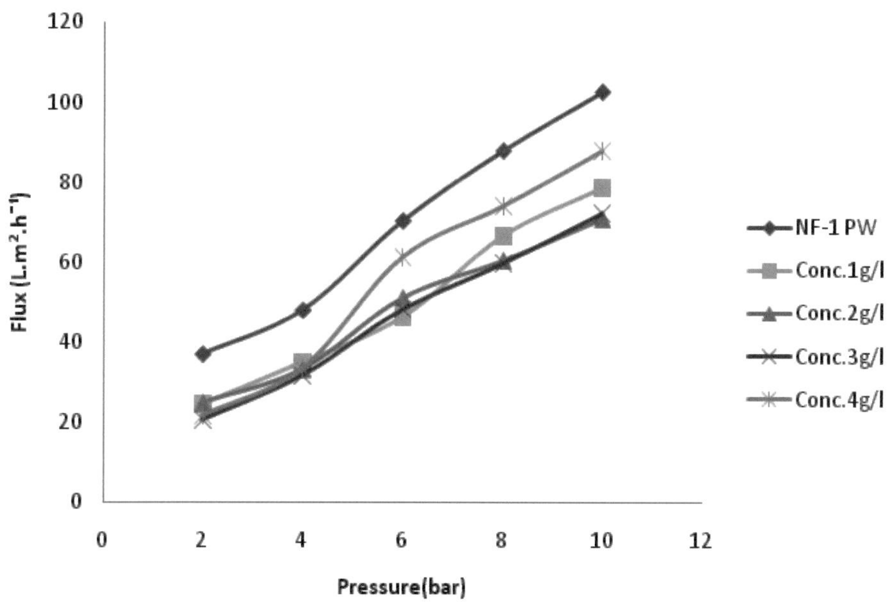

Figure 6. NaCl solution flux for unmodified membrane at various concentrations

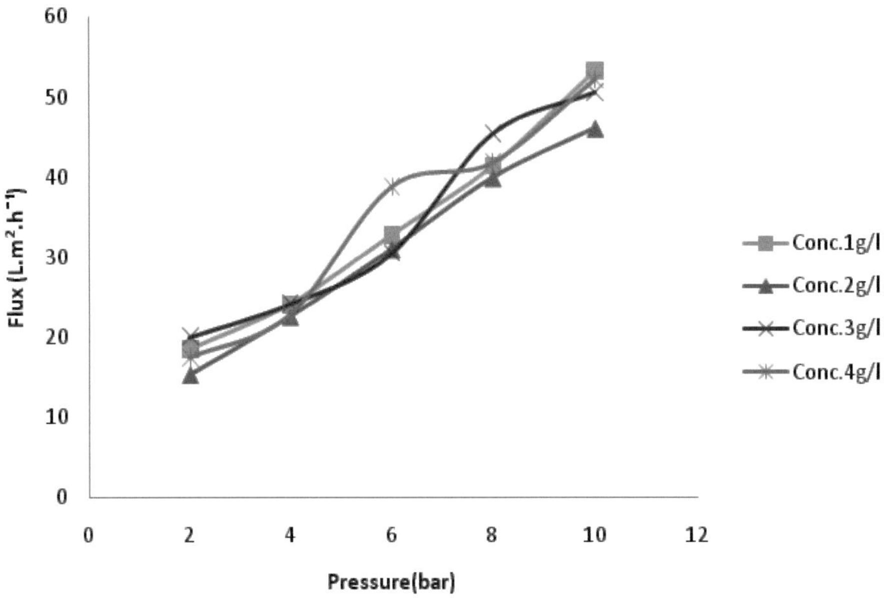

Figure 7. NaCl solution flux for modified membrane at various concentrations

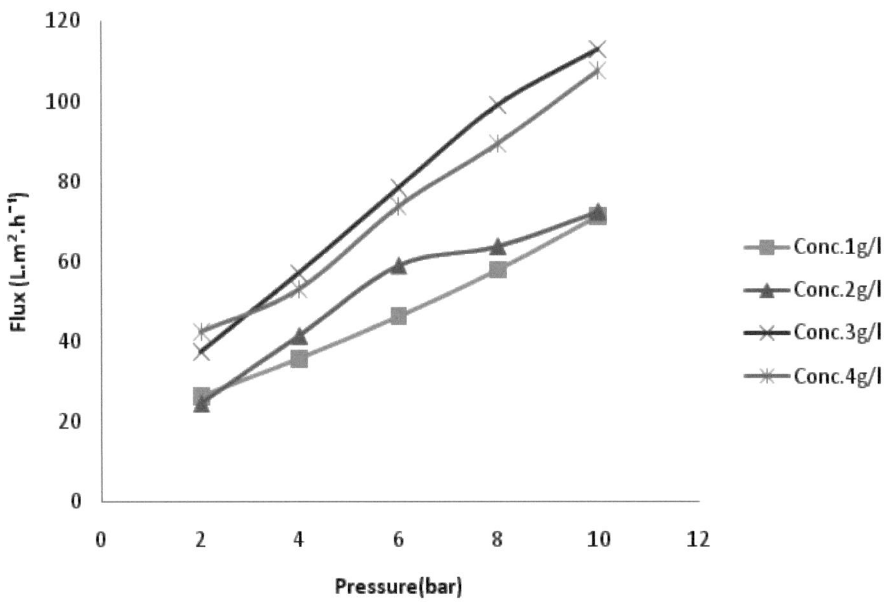

Figure 8. KCl solution flux for unmodified membrane at various concentrations

The same observations noticed for KCl solution (Figures 8 and 9) but with different percentage (55% and 1% for unmodified and modified membranes, respectively). It is worth mentioning that increasing concentrations have much more influence on unmodified membrane than the modified one for all solutions.

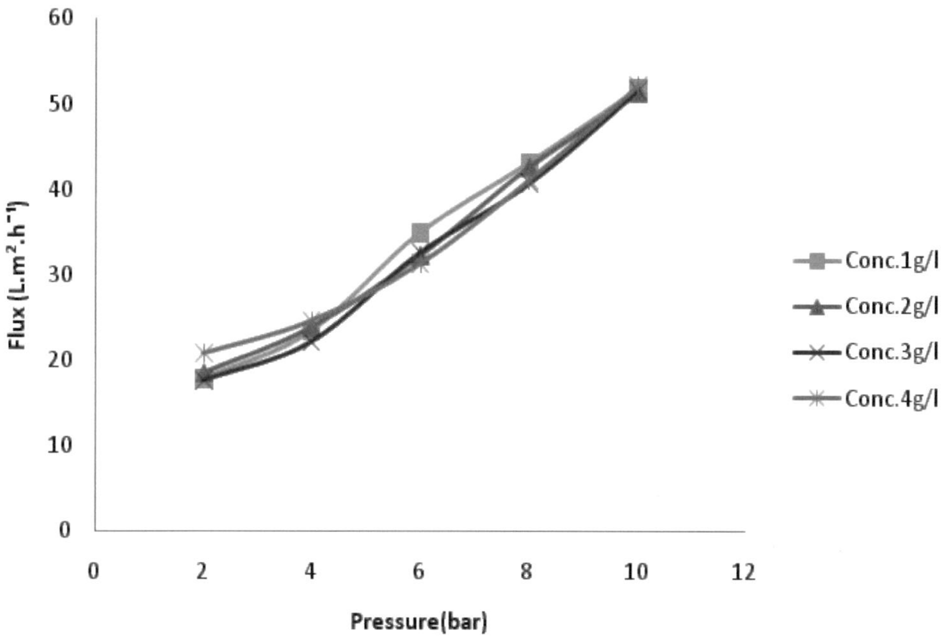

Figure 9. KCl solution flux for modified membrane at various concentrations

It is worth mentioning that flux may decline with time and concentration increasing, as osmotic pressure increased leading to reducing net driving pressure. Such observation is potentially associated with a dead-end filtration system, while for cross-flow system, flux decline may be lesser or may occur with longer time running.

3.2. Rejection

Rejection rates at various pressure magnitudes and for different concentrations are highlighted and discussed in this section. In addition, rejection of both modified and unmodified NF membranes for solutions consisting of a mixture of salts is also included. The rejection rates witnessed an increase after membrane modification took place with about 11–30% for magnesium sulfate and sodium sulfate, and 50–60% for sodium chloride and potassium chloride.

Figures 10 and 11 illustrate membrane rejection for both $MgSO_4$ and Na_2SO_4 at various concentrations, respectively. $MgSO_4$ and Na_2SO_4 rejection of modified and unmodified membranes decreased with concentration increase. The rejection observed for modified

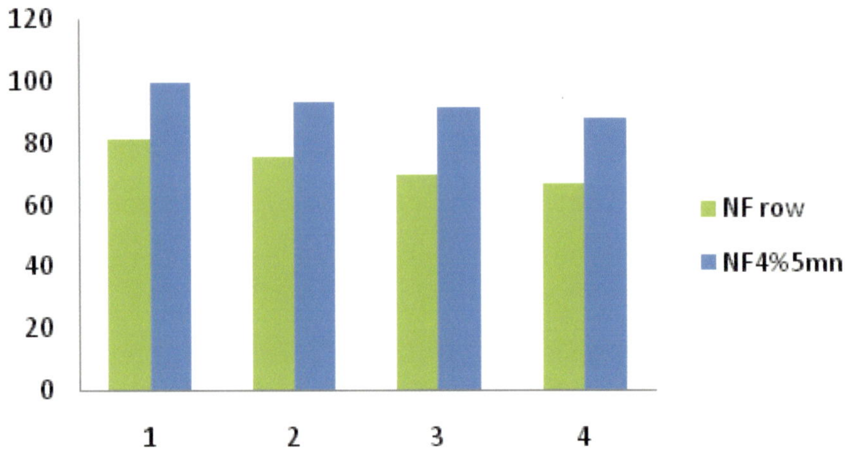

Figure 10. Membrane rejection (unmodified and modified) for $MgSO_4$ at various concentrations

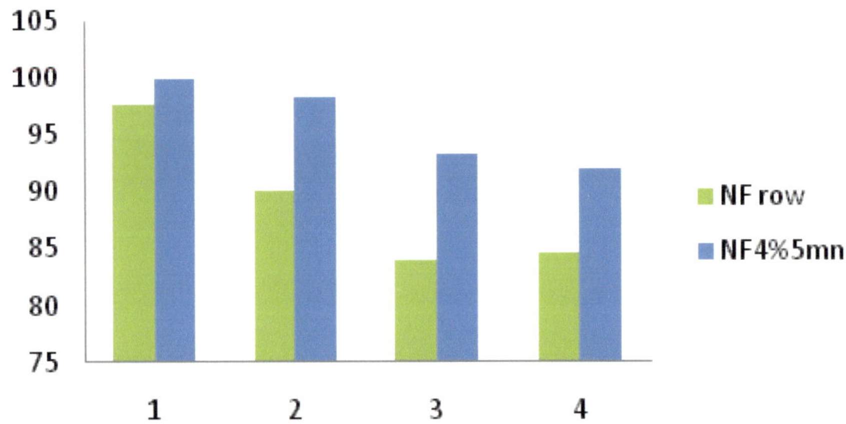

Figure 11. Membrane rejection (unmodified and modified) for Na_2SO_4 at various concentrations

membrane was higher than that observed for unmodified membrane at all concentrations for both $MgSO_4$ and Na_2SO_4. However, concentration increasing found to have more influence on unmodified membrane than the modified one. For the modified membrane, $MgSO_4$ rejection decreased by 12% from 1g/L to 4 g/L while rejection decreased for unmodified membrane with nearly 20%. For Na_2SO_4, rejection was reasonably decreased for unmodified membrane from 98% to 85% (difference is 13%) while for the modified membrane, the rejection decreased from 100% to 92% (difference is 8% only). This attributes to the fact that pore geometry and size have reasonably changed leading to better rejection as well as lowering concentration increasing effect on the membrane at this concentration range (1–4 g/L) [6, 25].

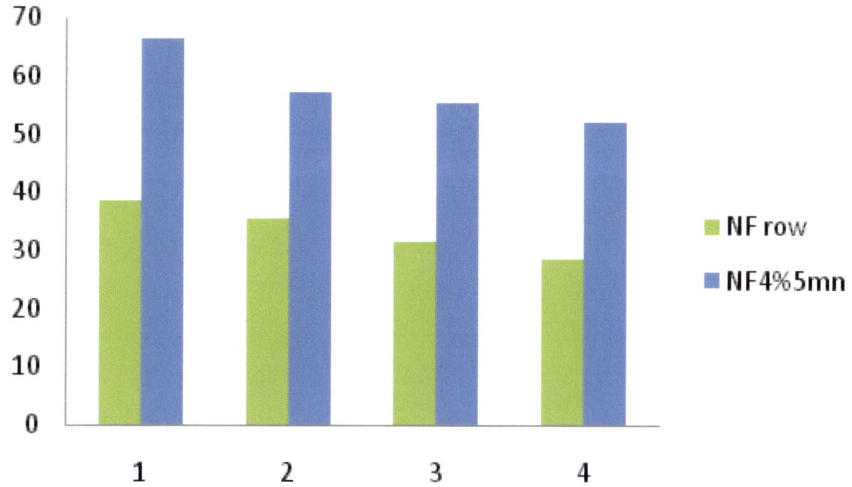

Figure 12. Membrane rejection (unmodified and modified) for NaCl at various concentrations

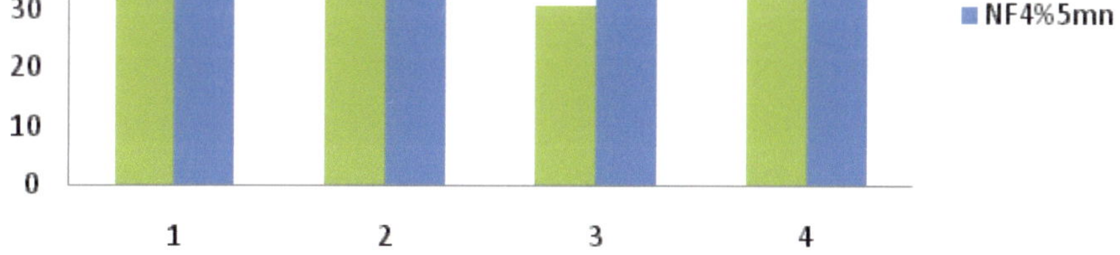

Figure 13. Membrane rejection (unmodified and modified) for KCl at various concentrations

Figures 12 and 13 illustrate membranes rejection for both NaCl and KCl at various concentrations, respectively (concentration range: 1–4 g/L). Two main observations worth concluding: firstly, the significance increase of both NaCl and KCl rejection at all concentrations for modified membrane over unmodified ones (for NaCl, modified membrane had rejection increasing from 39% to 72% in average while for KCl rejection increased from 59% to 75%); secondly, concentration increase had lower influence on both NaCl and KCl rejection for the modified membrane over the unmodified membrane (rejection decreased with concentration for NaCl and KCl by 26% and 44 % for unmodified and 25% and 22% for modified membrane). Although NF membranes are more vulnerable to chloride ions than sulfate ions, modified membrane seemed to have more consistent performance and less concentration increasing influence in terms of rejection than unmodified membrane. Generally, salt rejection values may suffer some decline with time as applied dissolved solids concentrations increase.

4. Conclusions

A commercial NF membrane was modified via UV-grafted surface modification method to obtain better salt rejection and reasonable flux while desalting brackish water. The study provides valuable information about flux and rejection changes and the relationship with pressure changing before and after modification. Experimental works included in the study investigate modified and unmodified NF membranes performance while filtering synthesized single salt solution at various concentrations (ranged from 1000 ppm to 4000 ppm) and various pressure magnitudes (pressure ranged from 2 to 10 bars). Following the modification, the rejection rates showed an increase with about 11–30% for magnesium sulfate and sodium sulfate, and 50–60% for sodium chloride and potassium chloride. It is worth mentioning that concentration increase was found to have lower effect on membrane rejection after modification.

Author details

A. A. Abuhabib[1,2]

Address all correspondence to: azz200@hotmail.com

1 Environmental Engineering Department, Engineering Faculty, Islamic University of Gaza, Gaza, Palestine

2 Development planning Department, University College of Applied Sciences (UCAS), Gaza, Palestine

References

[1] N. Hilal, H. Al-Zoubi, N.A. Darwish, A.W. Mohammad, M. Abu Arabi, A comprehensive review of nanofiltration membranes: Treatment, pretreatment, modelling, and atomic force microscopy, Desalination, 170 (2004) 281-308.

[2] A. Mohammad, Y. Teow, W. Ang, Y. Chung, D. Oatley-Radcliffe, N. Hilal, Nanofiltration membranes review: Recent advances and future prospects, Desalination, (2014).

[3] S. Subramanian, R. Seeram, New directions in nanofiltration applications — Are nanofibers the right materials as membranes in desalination? Desalination, 308 (2013) 198-208.

[4] R. Haddada, E. Ferjani, M.S. Roudesli, A. Deratani, Properties of cellulose acetate nanofiltration membranes. Application to brackish water desalination, Desalination, 167 (2004) 403-409.

[5] N. Hilal, V. Kochkodan, H. Al Abdulgader, D. Johnson, A combined ion exchange–nanofiltration process for water desalination: II. Membrane Selection, Desalination.

[6] A. Abuhabib, A. Mohammad, N. Hilal, R.A. Rahman, A.H. Shafie, Nanofiltration membrane modification by UV grafting for salt rejection and fouling resistance improvement for brackish water desalination, Desalination, (2012).

[7] M.N. Abu Seman, M. Khayet, Z.I. Bin Ali, N. Hilal, Reduction of nanofiltration membrane fouling by UV-initiated graft polymerization technique, Journal of Membrane Science, 355 (2010) 133-141.

[8] J.-H. Kim, P.-K. Park, C.-H. Lee, H.-H. Kwon, Surface modification of nanofiltration membranes to improve the removal of organic micro-pollutants (EDCs and PhACs) in drinking water treatment: Graft polymerization and cross-linking followed by functional group substitution, Journal of Membrane Science, 321 (2008) 190-198.

[9] S. Belfer, J. Gilron, Y. Purinson, R. Fainshtain, N. Daltrophe, M. Priel, B. Tenzer, A. Toma, Effect of surface modification in preventing fouling of commercial SWRO membranes at the Eilat seawater desalination pilot plant, Desalination, 139 (2001) 169-176.

[10] A. Ang, E. Kang, K. Neoh, K. Tan, C. Cui, T. Lim, Low-temperature graft copolymerization of 1-vinyl imidazole on polyimide films with simultaneous lamination to copper foils—effect of crosslinking agents, Polymer, 41 (2000) 489-498.

[11] B. Zhang, D. Wang, Y. Wu, Z. Wang, T. Wang, J. Qiu, Modification of the desalination property of PAN-based nanofiltration membranes by a preoxidation method, Desalination, 357 (2015) 208-214.

[12] G. Greene, R. Tannenbaum, Adsorption of polyelectrolyte multilayers on plasma-modified porous polyethylene, Applied Surface Science, 233 (2004) 336-342.

[13] M. Kim, K. Saito, S. Furusaki, T. Sugo, J. Okamoto, Water flux and protein adsorption of a hollow fiber modified with hydroxyl groups, Journal of Membrane Science, 56 (1991) 289-302.

[14] N. Hilal, L. Al-Khatib, B.P. Atkin, V. Kochkodan, N. Potapchenko, Photochemical modification of membrane surfaces for (bio)fouling reduction: a nano-scale study using AFM, Desalination, 158 (2003) 65-72.

[15] M.N. Abu Seman, D. Johnson, S. Al-Malek, N. Hilal, Surface modification of nanofiltration membrane for reduction of membrane fouling, Desalination and Water Treatment, 10 (2009) 298-305.

[16] A.C. Talla-Nwafo, Modification of NF membranes for improved salt rejection, Howard University, 2009.

[17] A.H. Mohd Yusof, M. Ulbricht, Polypropylene-based membrane adsorbers via photo-initiated graft copolymerization: Optimizing separation performance by preparation conditions, Journal of Membrane Science, 311 (2008) 294-305.

[18] M. Taniguchi, G. Belfort, Low protein fouling synthetic membranes by UV-assisted surface grafting modification: varying monomer type, Journal of Membrane Science, 231 (2004) 147-157.

[19] J. Pieracci, J.V. Crivello, G. Belfort, UV-Assisted graft polymerization of N-vinyl-2-pyrrolidinone onto poly(ether sulfone) ultrafiltration membranes using selective UV wavelengths, Chemistry of Materials, 14 (2001) 256-265.

[20] S.-i. Kuroda, I. Mita, K. Obata, S. Tanaka, Degradation of aromatic polymers: Part IV--Effect of temperature and light intensity on the photodegradation of polyethersulfone, Polymer Degradation and Stability, 27 (1990) 257-270.

[21] B. Kaeselev, P. Kingshott, G. Jonsson, Influence of the surface structure on the filtration performance of UV-modified PES membranes, Desalination, 146 (2002) 265-271.

[22] B. Van der Bruggen, Chemical modification of polyethersulfone nanofiltration membranes: A review, Journal of Applied Polymer Science, 114 (2009) 630-642.

[23] L. Puro, M. Manttari, A. Pihlajamaki, M. Nystrom, Characterization of modified nanofiltration membranes by octanoic acid permeation and FTIR analysis, Chemical Engineering Research & Design, 84 (2006) 87-96.

[24] V. Yangali-Quintanilla, A. Sadmani, M. McConville, M. Kennedy, G. Amy, A QSAR model for predicting rejection of emerging contaminants (pharmaceuticals, endocrine disruptors) by nanofiltration membranes, Water Research, 44 (2010) 373-384.

[25] R. Malaisamy, A. Talla-Nwafo, K.L. Jones, Polyelectrolyte modification of nanofiltration membrane for selective removal of monovalent anions, Separation and Purification Technology, 77 (2011) 367-374.

Chapter 3

Phase Diagram and Membrane Desalination

Ayman Taha Abd El-aziem El-gendi

Additional information is available at the end of the chapter

http://dx.doi.org/10.5772/60419

Abstract

Desalination technologies have made a significant impact in seawater and brackish water desalination. Recently, the evolution of membrane development has improved performance to lower operating costs and membranes have become the preferred technology for water desalination. Fortunately, different raw materials can be used for preparing membrane sheets which include either organic or inorganic materials, such as cellulose acetate, polyamide, polyimide, ceramic, natural, or artificial polymers. On the one hand, as a result of the variety of the raw materials which already exist in the entire world, different membrane separation processes might be applied dependent on the nature of the membrane sheet and the requirements of treatment process. On the other hand, there are different types of membranes can be used for membrane desalination by using different technologies such as reverse osmosis (RO), membrane distillation (MD), and forward osmosis (FO). The ternary phase diagram for membrane casting solution has an important role to get the required membranes.

Keywords: Desalination technologies, membrane, phase diagram, membrane separation processes

1. Introduction

This chapter intends to focus on using membrane separation processes for desalination based on the selectivity of the membrane, such as the preferential permeation of water (classical selectivity) or the solute (reverse selectivity). Phase inversion process is the most important technique used to prepare asymmetric polymeric membranes. In addition, the morphology and performance of membranes depend on their effects, which also depend on the membrane preparation process parameters. From a thermodynamic point of view, study on polymer–solvent– nonsolvent system can be well depicted in a ternary phase diagram. The Flory-Huggins theory was found to be a convenient and useful framework for the thermodynamic analysis of component mixing in a membrane preparation system. As a result of the variety of the raw materials which already exist in the entire world, different membrane separation processes might be applied dependent on the nature of the membrane sheet and the requirements of treatment process. In addition, there are different engineering forms which have been established for membranes that include flat, tubular, spiral wound or hollow fiber. Generally, the membrane preparation techniques are classified as a function of the raw materials used, the engineering forms and the characteristics of the required separation process. Over the past decades, the polymeric membranes have achieved commercial importance in a variety of separations applications in the chemical, food, pharmaceutical, and biotechnology industries. Today, the membrane industry is faced with the challenge of inventing new membrane materials.

In this chapter, we will focus particularly on membrane desalination techniques, such as reverse osmosis (RO), forward osmosis (RO) and membrane distillation for desalination. For these methods, as with any membrane process, the membrane is one of the most important characteristics which determines the usefulness and effectiveness of the entire process. For this reason, several targets of research have shifted in recent years towards developing new and more efficient materials that allow for a compromise between two fundamental properties of the separation, often antagonistic, namely the selectivity and permeability.

However, membrane processes are becoming economically competitive after the development of highly permeable polymer membranes. These membranes are less expensive than inorganic membranes and their implementation is much easier. Several types of polymers can be used such as cellulose acetate, polysulfones, polyamides and polyimides.

This chapter aims at preparing desalination membranes in order to get water selective membranes suitable for the retention of salts water mixtures. Hence two objectives must be reached: first, the selection of water selective materials well resistant in almost pure water and second, the preparation of high flux membranes needed for the recovery of water.

As a matter of fact, the preparation of polymeric membranes usually involves the phase inversion process, in which a homogeneous casting solution induces phase separation into a polymer-rich phase and a polymer-poor phase by the exchange of solvent with nonsolvent in an immersion bath (i.e: as water bath). Phase separation would continue to form the membrane structure until the polymer rich phase is solidified. Solidification during phase inversion could be induced by gelation and/or crystallization of the casted polymer solution. The equilibrium ternary phase diagram system is still a good tool for controlling the morphology and interpreting the membrane structure. Significantly, knowledge of phase equilibria (cloud points, binodals, spinodals, and critical compositions) enables one to change the conditions for the preparation of membranes such as the compositions of the casting solution, the temperature and of the coagulation bath type to obtain an optimum membrane structure.

http://dx.doi.org/10.5772/60419

The phase diagram has an important role to report the agreement between experimental work in order to get the required membranes, and the ternary phase diagram miscibility gaps for the evaluations of membrane-forming system. In the ternary phase diagram (polymer (p)/ solvent (s)/nonsolvent (ns)) a miscibility gap with metastable regions exists. According to the theory of phase separation three modes of phase separation can take place in such ternary system: nucleation and growth of the polymer lean phase, spinodal phase separation and nucleation and growth of the polymer rich phase. Since polymer is one of the components of the ternary system, solidification of a part of the system can take place.

2. General introduction on membrane science and technology

A general introduction on membrane science and technology is given in this section. It begins with the definition of terms and provides a description of membranes structures and membrane separation process. A short overview over the developments of membrane science and technology is given.

2.1. Membrane definition

A. Zydney et.al [1] defined membrane as a flexible layer surrounding a cell, organelle (such as the nucleus), or other bodily structure. The movement of molecules across a membrane is strictly regulated in both directions. Therefore, a membrane is a layer of material which serves as a selective barrier between two phases and remains impermeable to specific particles, molecules, or substances when exposed to the action of a driving force. Some components are allowed to pass through the membrane into a permeate stream, whereas others are retained by it and accumulated in the retentate stream (Fig. 1).

C.P Mark [2] defined a membrane as follows: What is a membrane? Possibly the broadest definition is a region of discontinuity interposed between two phases. This statement implies that membranes can be gaseous, liquid, or solid, or combinations of these phases. The term "region" in the definition is used to eliminate ordinary interfaces. Thus, the interfaces of two immiscible liquids, of a gas and a liquid, or of a gas and a solid, would not ordinarily be considered as membrane structures.

From Wikipedia [3]: membranes can be of various thicknesses, with homogeneous or heterogeneous structure. Membranes can also be classified according to their pore diameter. There are three different types of pore size classifications: microporous (dp < 2 nm), mesoporous (2 nm < dp < 50 nm), and macroporous (dp > 50 nm). Membranes can be neutral or charged, and particles transport can be active or passive. The latter can be facilitated by pressure, concentration, chemical, or electrical gradients of the membrane process. Membranes can be generally classified into three groups: inorganic, polymeric, or biological membranes. These three types of membranes differ significantly in their structure and functionality.

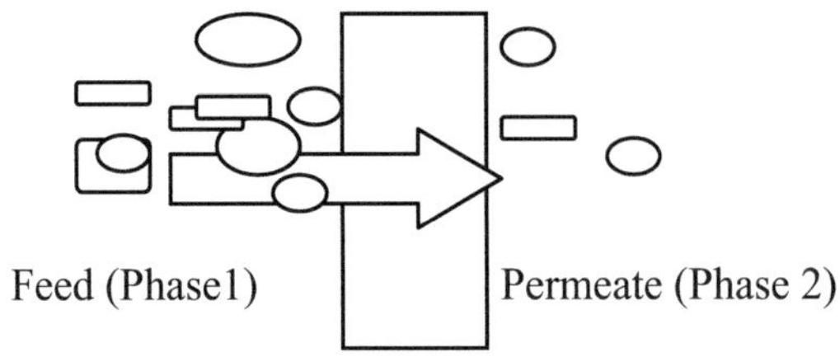

Figure 1. Membrane module.

2.2. Classification of membranes [1, 2, 6]

The membranes can be classified as follows:

a. Classification by nature of membrane

b. Classification by structure of membrane (Fig. 2)

c. Classification by application of membrane

d. Classification by mechanism of membrane function

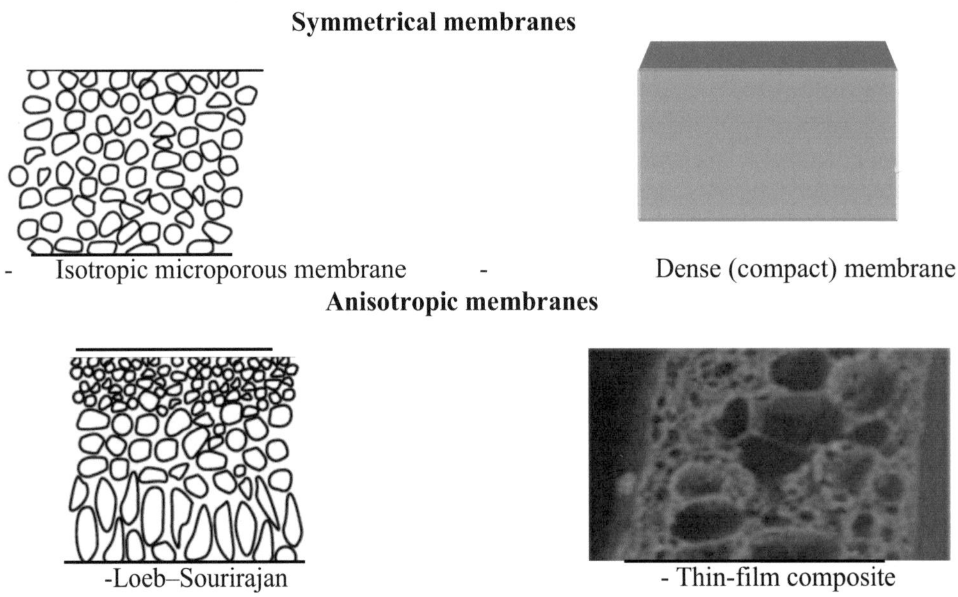

Figure 2. The main types of membranes based on structure [4].

2.3. Membrane types based on morphology [1, 2, 6]

The membrane types can be divided into isotropic membranes, anisotropic membranes (polymeric, ceramic, metal), and liquid membranes. These types will be outlined below.

http://dx.doi.org/10.5772/60419

2.3.1. Isotropic membranes

An isotropic membrane is divided into three types: Microporous membranes, nonporous membranes, and electrically charged membranes.

2.3.1.1. Microporous membranes (ultrafiltration, microfiltration)

In fact, the microporous membrane is very similar in function and morphology to a traditional liquid filtration. In addition, it has a rigid, highly voided structure with unorganized, distributed, and interconnected pores. Moreover, these pores differ from those in a conventional filter by being smaller, on the order of 0.01–10 μm in diameter. Therefore, all particles larger than the largest pores will be completely rejected by the membrane. While particles smaller than the largest pores, but larger than the smallest pores could be partially rejected, according to the pore size distribution of the membrane. However, particles much smaller than the smallest pores could pass through the membrane. Thus, the separation of solutes by microporous membranes is mainly dependent on the molecular size and pore size distribution. In general, only molecules that differ considerably in size could be rejected effectively by microporous membranes, for example, in ultrafiltration and microfiltration.

2.3.1.2. Nonporous, dense membranes

Nonporous, dense membranes composed of a dense film through which material can pass on by diffusion under the driving force of a pressure, concentration, or electrical potential gradient. However, the separation efficiency of various solutes of a mixture is dependent on their transport rate through the membrane, which is determined by their solubility and diffusivity in the membrane material. In addition, nonporous, dense membranes could reject charged or uncharged solutes of similar size if their concentration in the membrane material differs significantly. Typically, pervaporation, most gas separation, and reverse osmosis membranes use nonporous membranes to do the separation process.

2.3.1.3. Electrically charged membranes

Electrically charged membranes are dense or microporous. A membrane with fixed positively charged ions is called an anion-exchange membrane. Similarly, a membrane containing fixed negatively charged ions is referred to as a cation-exchange membrane. Separation with charged membranes is achieved mainly by the exclusion of ions of the same charge as the fixed ions of the membrane structure. The separation is affected by the charge and concentration of the ions in solution. For example, monovalent ions are excluded less effectively than divalent ions and, in solutions of high ionic strength, selectivity decreases. Electrically charged membranes are used for processing electrolyte solutions in electrodialysis.

2.3.2. Anisotropic membranes (asymmetric membrane)

The permeate flux through a membrane is inversely proportional to the membrane thickness. Really, high permeability is desirable in membrane separation processes for economic reasons; therefore, the membrane thickness should be as thin as possible. Conventional film fabrication technology limits the manufacture of mechanically strong, defect-free films to about 20 μm thickness. Moreover, the development of a new membrane fabrication technique to produce asymmetric membrane was one of the major breakthroughs of membrane fabrication technology during the last years. Asymmetric membranes consist of an extremely thin surface layer supported on a much thicker surface layer, porous substructure. The surface layer and its substructure may be formed in a single operation or separately. In composite membranes, the layers are usually made from different polymers. The rejection properties and flow rates of the membrane are determined exclusively by the surface layer; the substructure functions as a mechanical support. The advantages of the higher fluxes provided by anisotropic membranes are so great that almost all commercial processes use such membranes.

2.3.3. Ceramic, metal, and liquid membranes

Ceramic membranes are a special class of microporous membranes. They are being used in ultrafiltration and microfiltration applications for which solvent resistance and thermal stability are required.

3. The basic membrane separation processes [5]

The membrane separation processes have been used in several applications for biological, industrial, drinking water, and wastewater treatment purposes. Moreover, the advantages of this separation system are their powerful separation capacity of small and large molecules and compounds from mixtures, organic and inorganic solutions. In addition, they are characterized by their low-power consumption and area required comparing to other separation techniques. Furthermore, membrane separation technologies have some advantages over other mass transfer processes, including high rejection, low-energy consumption, low cost-to-performance ratio, and compact and modular designs. Thus, they are generally considered to be "clean technologies." Fig. 3. shows the basic membrane separation process.

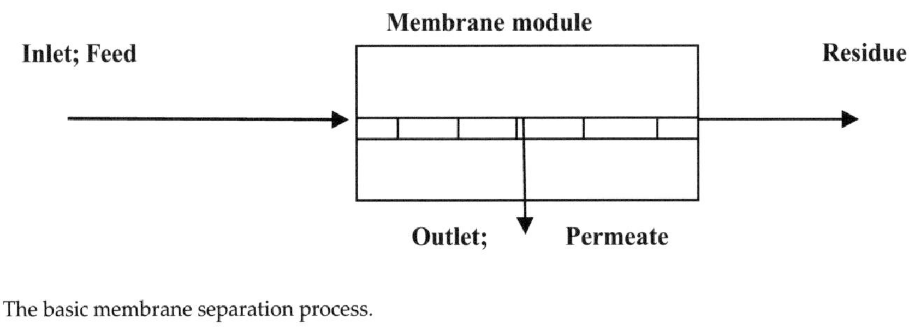

Figure 3. The basic membrane separation process.

Membrane separation processes are often used since these applications realize high removal of constituents such as charged solutes, uncharged solutes, and organic molecules. In the recent years, membrane separation techniques have grown from a simple laboratory tool to an industrial process with considerable technical and commercial impact. Really, membrane separation processes are easier, more efficient, and more economical than conventional separation techniques. The separation is usually performed at ambient temperature by using membrane process; this allows temperature-sensitive solutions to be produced without the constituents being damaged [2]. In fact, this is very important where temperature-sensitive products have to be processed (i.e., food, drug industry, biotechnology). The main membrane separation processes are membrane distillation, pervaporation, and pressure-driven membrane processes, and these processes are outlined below in detail.

3.1. Pressure-driven membrane process

Pressure-driven membrane process techniques have become an alternative to conventional water treatment methods, as well as in the treatment and concentration of wastewater of both industrial and municipal origin [6]. The occurrence and prevalence of organic micro-pollutants, such as pesticides, and other organic micro-pollutants are becoming a concern in drinking water, wastewater, and water reuse applications due to potential adverse health effects associated with these compounds. Thus, membrane processes are becoming increasingly widespread in water treatment and wastewater reclamation/reuse applications where a high product quality is desired [7, 8, 9].

3.1.1. Pressure-driven membrane separation processes [1, 2, 10]

The pressure-driven membrane separation processes are as follows:

- Microfiltration (MF)
- Ultrafiltration (UF)
- Nanofiltration (NF)
- Reverse osmosis (RO)

Literally, all RO, NF, and UF membranes are asymmetric. This differentiates easily most membranes from traditional filters, which are symmetric or, in other words, identical on the both sides of the common filter. In addition, membrane has a tight top layer facing the pollutant solution to be treated. This layer, also called the skin layer, is thin typically <<0.1 μm. The membrane thickness is about 150–250 μm, and the bulk of the membrane simply provides structural support for the skin layer. The asymmetric structure means that the pores are wider, which prevents the pores from being plugged. This provides good fouling resistance, since foulants have a tendency to either be totally rejected or pass all the way through a membrane. Pressure-driven membrane filtration processes are classified by pressure into two categories:

- High-pressure membrane processes, such as reverse osmosis (RO) and nanofiltration (NF), normally have a relatively small pore size compared to low-pressure membranes, since small pore size membranes require high-driving force to drive fluid through the membrane compared to big pore size membranes. Further, these types of membranes normally require a relatively high pressure from 4 to 70 bar. In addition, these high-pressure processes primarily remove contaminants through chemical diffusion.

- And low-driven pressure membrane, such as microfiltration (MF) and ultrafiltration (UF), normally requires a low pressure from 0.4 to 7 bar. These membranes primarily remove contaminants through physical sieving. More details about these types are outlined below.

i. Microfiltration (MF)

Microfiltration (MF) is characterized by a membrane pore size between 0.05 and 10 μm and operating pressures below 2 bar. MF is primarily used to separate particles and bacteria from other smaller solutes. Further, MF is a low-pressure technique that removes particles with a molecular weight above 50, 000 or a particle size over 0.05 μm. In addition, MF is a process where ideally only suspended solids are rejected, while even proteins pass the membrane freely.

ii. Ultrafiltration (UF)

Ultrafiltration (UF) refers to membranes having smaller pore size compared to microfiltration membranes. UF is characterized by a membrane pore size between 2 nm and 0.05 μm and operating pressures between 1 and 10 bar. UF is a low-pressure technique where protein and suspended solids are rejected freely. There is consequently no rejection of mono- and di-saccharides, salts, amino acids, organics, inorganic acids, or sodium hydroxide.

iii. Nanofiltration (NF)

Nanofiltration (NF) is characterized by a membrane pore size between 0.5 and 2 nm and operating pressures between 5 and 40 bar. NF is used to achieve a separation between sugars, other organic molecules, and multivalent salts on one hand, and monovalent salts and water on the other. True NF rejects only ions with more than one negative charge, such as phosphate, while passing single charged ions. NF also rejects uncharged, dissolved materials and positively charged ions according to the size and shape of the molecule in question.

iv. Reverse Osmosis (RO)

Reverse osmosis (RO) is the tightest possible membrane process. Water the only material passing through it while all the dissolved and suspended materials are rejected. RO as a desalination technology has led to a reduction in the desalinate cost in the last decade. RO is now the potential competitor to the world-leading desalination technology, multistage flash desalination (MSF). Desalination by RO uses less energy than the other desalination technologies, which makes RO more attractive, as RO uses a semipermeable membrane under high pressure to separate salts. In principle, the performance of the RO desalination depends on the used membrane. Table 1 represents cost for desalting m^3 of potable water using different desalination technology.

It clearly illustrates that RO has a significant economic advantage for treating brackish water, while price quotes for ED are not readily available. For desalination of seawater, RO has an economic advantage over other thermal technology. The comparison between the four membrane processes is summarized in Fig. 4. Table 2 illustrates some of the characteristics with NF and RO membranes. Total worldwide use of membranes, based on membrane surface area, is approximately presented in Fig. 5.

MSF	MEE	VC	Seawater RO	Brackish RO	Brackish ED	Reference
1.10–1.50	0.46–85	0.87–0.92	0.45–0.92	0.20–0.35		[11]
0.80	0.45		0.72–0.93			[12]
0.89	0.27–0.56		0.68			[13]
			1.50	0.37–0.70	0.58	[14]

Table 1. Desalination costs ($/m³ fresh water – multiply by 3.8 for $/1000 gal).

Membrane type	RO (Dow Filmtec BW30-4040)	NF (Dow Filmtec NF90-4040)
Operating pressure, bar	9.04	5
Power consumption, kW	0.37	0.22
Specific energy costs, kWh m^{-3}	1.47	0.89
Permeate TDS, ppm	65	318

Table 2. Operating pressures, power consumption, and permeate quality with NF and RO membranes [15].

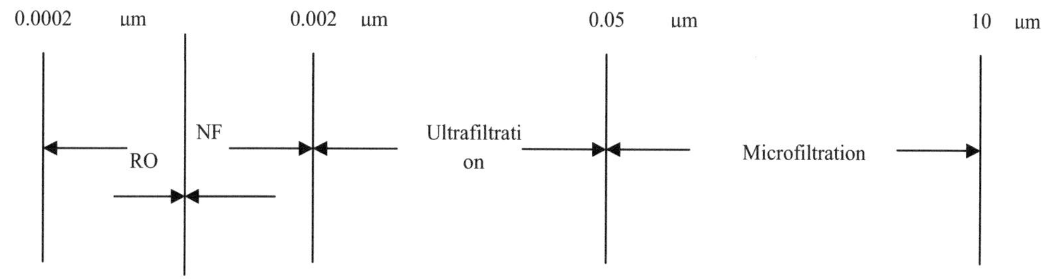

Figure 4. Pressure-driven membrane processes based on the smallest particle [2, 16].

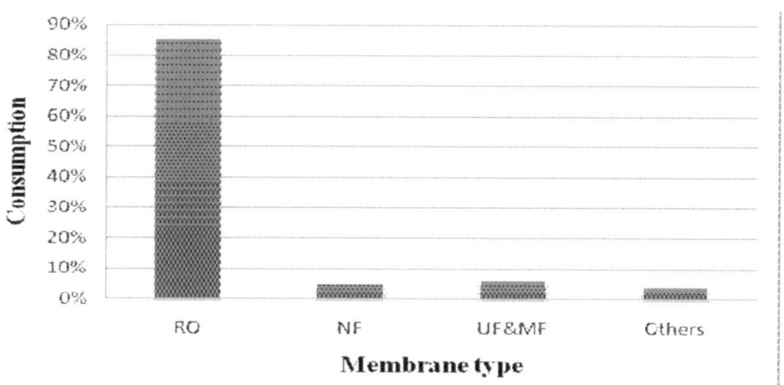

Figure 5. Membrane use in the world based on surface area [16, 10].

3.2. Membrane material

Membrane processes are becoming economically competitive after the development of highly permeable polymer membranes. These membranes are less expensive than inorganic membranes and their implementation is much easier. Several types of polymers can be used such as cellulose acetate, polysulfones, polyamides, polyvinylchloride, and polyimide. The suitable polymer material for membrane and its membrane-forming technique should be first found out to study water desalting with membrane process. Really, selection of the polymer membrane materials for separation process is based mainly on some of the important features such as good chemical resistance and high mechanical strength of the polymer film in the mixture solution. Therefore, it should have good interaction preferably with one of the components of the mixture for separation. The main polymeric materials used for membrane processes have been depicted in Table 3.

Membrane material	Membrane process							
	MF	UF	NF	RO	PV	MD	GS	FO
Cellulose acetate	•	•	•	•	•		•	•
Cellulose nitrate	•							
Cellulose esters	•							
Polyacrylonitrile		•			•			
Aromatic polyamide	•	•	•	•				•
Aliphatic polyamide	•	•	•	•	•			•
Polyimide		•	•	•	•		•	•
Polybenzimidazole			•	•	•			•
Polybenzimidazolone			•	•	•			
Polycarbonate	•							
Polysulfone	•	•			•			
Polyethersulfone	•	•				•	•	
Polyetheretherketone	•	•						
Polyydimethylsiloxane(PDMS)	•				•			
Poly(vinyl alcohol)	•							
Poly(vinyl chloride)	•							
Polyacrylonitrile	•	•						
PVC copolymer	•	•			•			
Polysulfone	•	•						
Sulfonated-PSF		•		•	•			
Polytetrafluoroethylene(PTFE)	•				•			
Polyvinylidenefluoride(PVDF)	•	•			•	•		
Polycarbonate	•							
polyester	•							
Polypropylene	•				•	•		
Polyester	•							
Polyethylene	•				•	•		
Chitosan		•	•	•	•			

Note: PV, pervaporation; MD, membrane distillation; FO, forward osmosis; GS, gas separation

Table 3. Typical applications of various polymeric materials in selected membrane processes [1–67].

http://dx.doi.org/10.5772/60419

- As concluded in Table 3, the more useable membrane materials for RO are as follows:
  - Cellulose acetate; Aromatic and aliphatic polyamide; Polyimide
  - Polybenzimidazole; Polybenzimidazolone; Sulfonated-PSF; Chitosan

Thus, one or more of these polymers will be selected for membrane desalination using RO.

- Also, as concluded in Table 3, the more useable membrane materials for MD are as follows:
  - Polyvinylidenefluoride (PVDF); Polytetrafluoroethylene (PTFE); Polypropylene
  - Polyethylene; Polyethersulfone

Thus, one or more of these polymers will be selected for membrane desalination using MD.

4. Preparing desalination membranes

This section aims at preparing desalination membranes in order to get water selective membranes suitable for the retention of salts from water mixtures. Hence, two objectives must be reached: first, the selection of water-selective materials well resistant in almost pure water, and second, the preparation of high flux membranes needed for the recovery of water. The preparation of symmetric (such as dense membrane), asymmetric, and composite membranes made from polymeric materials, to be used in the different membrane processes and applications, are reported also in this part [1, 2].

Typically, the preparation of polymeric membranes, with different physical structures, takes place by two methods [2]:

1. Physical treatment of a certain polymer membrane (such as annealing, elongation, etc.)
2. Preparation of the membranes from a certain polymer by several membrane formation processes (casting process such as wet- and dry-phase inversion)

There are differences between two methods. On the one hand, the physical structure is created by the treatment after preparation of the membranes. On the other hand, in the second case, the structure is created at the same time as membrane formation.

Fortunately, all kinds of different synthetic materials can be used for membrane formation process. In addition, there are a number of different techniques to prepare the required membrane; some of these methods can be used to prepare polymeric as well as inorganic membranes. The equilibrium thermodynamic properties of the ternary system polymer/solvent/nonsolvent play an important role in the description of membrane preparation process. Thus, depending on the choice of the ternary system, three possible results can occur during membrane formation process, which are as follows [17, 18]:

i. Nonporous film (symmetrical dense membrane)

ii. Porous film (symmetrical porous membrane)

iii. Asymmetric membrane which has a thin dense top layer on a porous supporting layer

The asymmetric membrane is the desired structure in the study of desalination. An explanation of these results could be dependent on two sets of factors:

i. The ternary system (polymer/solvent/nonsolvent) and its equilibrium thermodynamic properties (such as liquid–liquid phase separation and solidification)

ii. The effect on the kinetics of the above-mentioned demixing phenomena and the exchange of solvent and nonsolvent during membrane preparation process

The preparation process of asymmetric membranes could include two sets of factors. The preparations of symmetric, asymmetric, and composite membranes are reported in this section.

4.1. Symmetrical membrane [19]

- Dense symmetrical membranes (nonporous membrane)

These types of membranes are capable of separating molecules of approximately the same size from each other (such as toluene/heptane). On the one hand, dense symmetrical membranes are widely applied in research, development, and other laboratory studies for characterizing membrane properties (such as swelling, mechanical properties). On the other hand, these types of membranes are, however, rarely used commercially, because the membranes permeate flux is very low for practical separation processes. Moreover, such membranes are used in some applications such as pervaporation, vapor permeation, gas separation, and dialysis. The dense membranes are prepared by two-step mechanism, solution casting and thermal melt-pressing process as described below.

- Two-step mechanism

The dense membranes, such as polyimide membranes, were obtained by addition mixing reaction of a dianhydride and diamine in the presence of suitable solvent. This reaction involves two consecutive steps, which are as follows:

1. The dianhydride and diamine react at room temperature to give polyamic acid (PAA).
2. The PAA is casted on a glass plate or Teflon to give PAA membrane, after the casting has been drawn; the casted film is dried at high temperature, thus the solvent will evaporate to leave a thin film. The PAA membrane is converted into polyimide (PI) membrane by thermal treatment under vacuum.
3. Solution casting (dry-phase inversion)

Solution casting uses a casting knife to cast polymer solution across a casting plate. After the casting has taken place, the solvent evaporates to leave a thin film. The best casting solution concentrations are in the range 15–20 wt% polymer in order to prevent the spreading of polymer over the casting plate. Solvents having high boiling points are unsuitable for this method, because their low volatility needs too long evaporation times.

- Melt pressing (melt forming)

Some polymers do not dissolve in suitable casting solvents. For example, polyethylene, polypropylene, and nylons, which have to be formed into membranes by melt pressing. This process occurs by sandwiching the polymer at high pressure between two heated plates. A pressure of 136–340 bar is used for 0.5–5 minutes, while holding the plates almost above the melting point of the polymer.

4.2. Preparation of porous membrane

In fact, this class of membrane induces separation mechanism by discriminating between solute sizes. Typically, porous polymeric films can be prepared by several ways, such as sintering, stretching, track etching, and phase inversion processes. Really, the final morphology of the prepared membranes is dependent on the properties of the polymer solution and the process conditions. Furthermore, the majority of membranes are formed by controlled phase separation of polymer solutions into two phases: in the former one with a high polymer concentration (rich polymer phase), and in the latter one with a low polymer concentration (poor polymer phase). The rich polymer phase solidifies immediately after phase separation and forms the membrane. The membrane performance is dependent on the morphology of the prepared membrane.

4.2.1. Porous membranes preparation by phase inversion process

Phase inversion can be defined as a process whereby a polymer is changed in a controlled manner from a liquid state to a solid state [1, 2]. Phase inversion of polymer solutions can be induced in several ways [20]. The four main techniques for the preparation of polymeric membranes by phase inversion are as follows [21, 22, 23]:

i. Immersion precipitation

Most commercially available membranes are formed by immersion precipitation. The phase inversion process induced by immersion precipitation is a well-known technique to get asymmetric polymeric membranes. A polymer solution is casted as a thin film on a plate support (as glass) and is immediately immersed in a coagulant bath. Precipitation can take place because the good solubility between solvent and nonsolvent in the polymer solution.

ii. Thermally induced phase inversion (TIPS)

The TIPS is dependent on the theory that solubility of the polymer in a solvent usually decreases when the temperature is decreased. After the demixing process is occurred, the solvent is removed by extraction, evaporation, or freeze drying.

iii. Precipitation from the vapor phase

This process was used as early as 1918 by Zsigmondy. Phase separation of the polymer solution is induced, during this method, by a nonsolvent vapor which penetrates in the polymer solution.

iv. Air-casting of a polymer solution (precipitation by solvent evaporation) [24, 25]

The simplest technique to form phase inversion membranes is precipitation by solvent evaporation. This process was already used in the early years of this century. In this technique, the polymer is dissolved in a mixture of solvent (volatile) and nonsolvent (less volatile). During the evaporation of the solvent, the solubility of the polymer decreases, and then the composition of mixture will shift to a higher nonsolvent and polymer content. This leads eventually to the polymer precipitation, and then phase inversion can takes place leading to the formation of skinned membrane.

It was clear that the differences between the four techniques originate from the differences in desolvation mechanisms.

4.3. Phase inversion membranes

Most of the membranes used today are phase inversion membranes (asymmetric structure membranes as RO) obtained by immersion precipitation [1, 26]. These membranes are asymmetric since their cross section reveals an asymmetric structure: dense skin layer (dense permselective layer <0.1 μm) supported by a porous sublayer (more open porous substrate ≥15 μm). In industrial applications, symmetrical microporous membranes, which induced by precipitation from the vapor phase, have been almost completely displaced by asymmetric membranes, which have much higher liquid fluxes with a good quality.

The asymmetric membrane structure was not recognized until Loeb and Sourirajan [27] prepared the first asymmetric membranes according to the wet-phase inversion process (Loeb–Sourirajan technique) in 1962. The Loeb–Sourirajan technique made a critical breakthrough in membrane process technology. Since the asymmetric membranes were prepared by Loeb and Sourirajan, the research and developments in asymmetric membrane preparation methods and properties were accelerated by using scanning electron microscopes, which enabled the effects of structural modifications to be easily assessed. More importantly, the demonstration of the advantages of the asymmetric membrane structure opens the way to improve the membrane separation processes. In addition, the development of the phase inversion process has preceded many systematic studies on the effects of membrane formation parameters and mechanism.

Figure 6 shows a schematic of the phase inversion process induced by immersion precipitation. This process consists of four main steps:

i. Casting step: the polymer solution is casted by using a film applicator to form a thin film of a polymer solution on a suitable substrate (i.e., glass support).

ii. Preconcentration step: the casted film is left for a certain evaporation time at room temperature (preconcentration time).

iii. Immersion step (phase inversion step): the casted film is immersed into an appropriate coagulation bath (i.e., water). During this step, the solvent will pass into the coagulation bath while nonsolvent will penetrate the polymer solution. The exchange of diffusion of a solvent and a nonsolvent through the interface in a polymer casting solution and a nonsolvent will yield the phase inversion process for polymer casting solution with different exchange rates which induce membranes with symmetric or asymmetric structures (the interchange of solvent and nonsolvent coagulant due to the diffusion form membrane). The exchange rate is induced by the difference between solubility parameters, the diffusion force of nonsolvent and physical factors of the phase inversion process [28]. Furthermore, the properties of the polymer casting solution are presumed to affect the structure of the resulting membrane [29, 30].

iv. Drying step: this step *removes traces of water and solvent* from the prepared membranes.

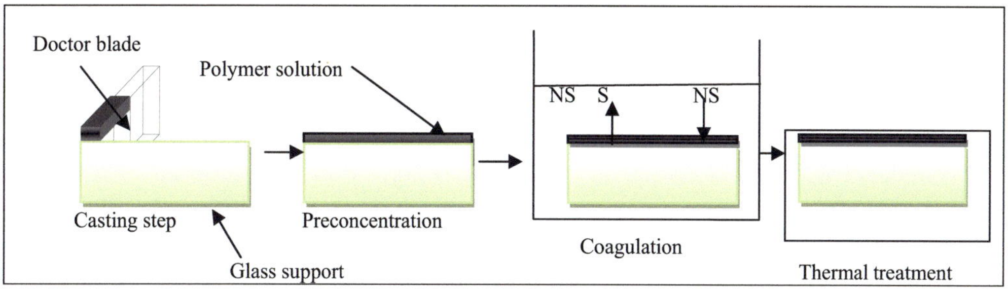

Figure 6. Membrane preparation steps via immersion precipitation. S, solvent; NS, nonsolvent [31].

4.4. Mathematical description of phase inversion process

It known that the state of any system, open, closed, or isolated can be described by state functions: entropy, enthalpy, and free energy. The thermodynamic state of a system of binary, ternary, or more components with limited miscibility can be described in terms of the free energy of mixing. At constant pressure and temperature, three different states can be distinguished [2]:

1. A stable state (homogeneous solution): in this state, the free energy of mixing is positive

$$\Delta G > 0 \ (P, T = const) \tag{1}$$

2. An unstable state: the homogeneous solution separates spontaneously into two phases, which is located within the miscibility gap, the free energy of mixing is negative

$$\Delta G < 0 \tag{2}$$

3. An equilibrium state: given by the phase boundary composition, the free energy of mixing is zero

$$\Delta G=0 \tag{3}$$

4.5. Microporous formation mechanism [2, 32]

The phase inversion process induced by several techniques in a homogeneous solution to form microporous membranes can be related to three basic procedures [2, 32]:

1. Thermogelation of a homogeneous solution of a binary system, ternary system, or more components

2. Evaporation of a volatile solvent from a homogeneous solution

3. Addition of a nonsolvent or nonsolvent mixture to a homogeneous solution

All three procedures may result in symmetric microporous structures or in asymmetric structures with a more or less dense skin at one or both surfaces suitable for nanofiltration, reverse osmosis, ultrafiltration, or microfiltration. The free energy of polymer homogeneous mixing under certain conditions of temperature and composition is negative (an unstable state) that is the only thermodynamic presumption for all three basic preparation procedures; thus the system must have a miscibility gap over a defined concentration and temperature range.

Typically, a binary mixture of a polymer and a solvent should be assumed for understanding the mechanism of liquid–liquid demixing. Phase separation induced by thermal gelation, evaporation of solvent, and addition of nonsolvent can be illustrated with the aid of the phase diagram of a polymer mixture solution. On the one hand, the starting point for preparing phase inversion porous membranes is in a thermodynamically stable state for homogeneous solution, for example the point A at temperature T_1 (Fig. 7). On the other hand, the simplest procedure to yield a microporous system is by thermogelation of a binary (two component) mixture. At high temperature, thermogelation of a binary system forms a homogeneous solution for all compositions, but a miscibility gap over a wide range of compositions is shown at a lower temperature. This behavior is illustrated schematically in Fig. 7, which shows a phase diagram of a binary mixture of a polymer and a solvent as a function of temperature.

From Fig. 7, the points P and S indicate the pure components of polymer and solvent respectively, while points on the line P–S indicate mixtures of these two components. When the temperature of the system decreases, demixing of the homogeneous solution will take place when the binodal is reached. Furthermore, the homogeneous solution will demix into two phases, and this is referred to as liquid–liquid demixing. Thus, if a homogeneous solution mixture of the composition Xp, at a temperature T_1, as represented by the point A (Fig. 7), is cooled to the temperature T_2, as represented by point B, and lies inside demixing (miscibility) gap, it is not stable thermodynamically, and will separate into two different phases, the composition of which are represented by the points B′ and B′′. The point B′ shows the (polymer-rich, solvent-poor) solid phase and the point B′′ shows the (solvent-rich, polymer-poor) liquid phase. The lines B′–B and B′′–B represent the ratio of the amounts of the two phases in the mixture, that is, the overall porosity of the obtained microporous system.

http://dx.doi.org/10.5772/60419

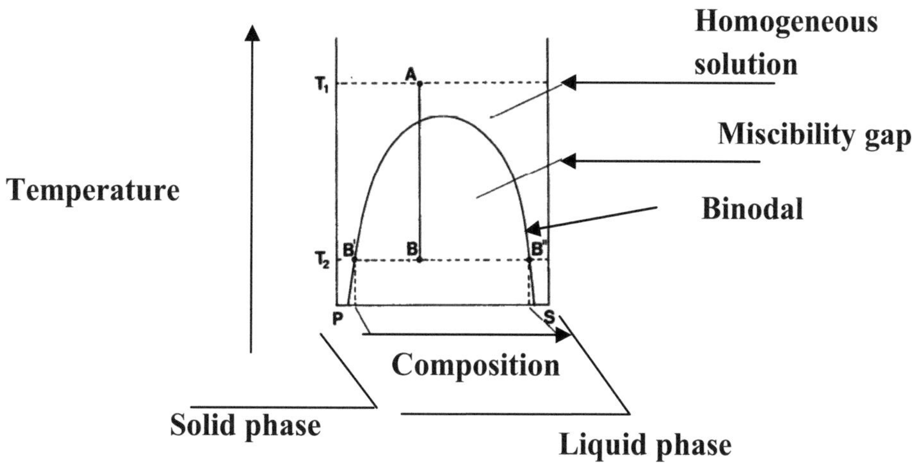

Figure 7. Formation of a microporous system by thermal gelation of a binary (two-component) mixture exhibiting a miscibility gap at certain conditions of temperature and composition, adapted from [2, 32].

4.6. Phase inversion process and ternary phase diagram

In fact, the polymeric membrane preparation usually involves the phase inversion process, in which a homogeneous casting polymer solution induces phase separation into a polymer-rich phase and a polymer-poor phase by the exchange of solvent with nonsolvent in a coagulation bath (i.e, as water bath). Phase separation would continue to form the membrane structure until the polymer-rich phase is solidified. Solidification during phase inversion could be induced by gelation and/or crystallization of the casted polymer solution. The equilibrium ternary phase diagram system is still a good tool for controlling the morphology and interpreting the membrane structure. Significantly, knowledge of phase equilibria (cloud points, binodals, spinodals, and critical compositions) enables one to change the conditions for the preparation of membranes, such as the compositions of the casting solution, temperature, and coagulation bath type to obtain an optimum membrane structure. The phase diagram has an important role to report the agreement between experimental work in order to get the required membranes, and the ternary phase diagram miscibility gaps for the evaluations of membrane-forming system. In the ternary phase diagram (polymer (p)/solvent (s)/nonsolvent (ns)), a miscibility gap with metastable regions exists. According to the theory of phase separation, three modes of phase separation can take place in such ternary system: nucleation and growth of the polymer-lean phase, spinodal phase separation, and nucleation and growth of the polymer-rich phase. Since polymer is one of the components of the ternary system, solidification of a part of the system can take place. Phase inversion process (phase separation process) is the most important technique used to prepare both asymmetric and symmetric polymeric membranes. In addition, the morphology and performance of membranes depend on their effects [33], which also depend on the membrane preparation process parameters. From a thermodynamic point of view, study on polymer–solvent–nonsolvent system can be well depicted in a ternary phase diagram, as illustrated in Fig. 8. The Flory-Huggins theory [34] was found to be a convenient and useful framework for the thermodynamic analysis of component mixing in a membrane preparation system.

The systems include a mixture of ternary components: a polymer, a solvent, and a third component, which may be a nonsolvent. This ternary system is completely miscible over a certain composition range but shows a miscibility gap over another composition range, as shown in Fig. 8, which shows an isothermal phase diagram of the ternary components. The pure components are represented at the corners of the triangle, while the boundary lines between any two corners of triangle represent mixtures of two components, and any point inside the triangle diagram represents a mixture of all three components. The main elements of ternary phase diagram are binodal and spinodal curves, a critical point, tie lines, and a glassy region shown in Fig. 8. For the studied ternary system, the interaction parameters evaluated for the binary mixture allowed a remarkably accurate prediction of the ternary phase diagram.

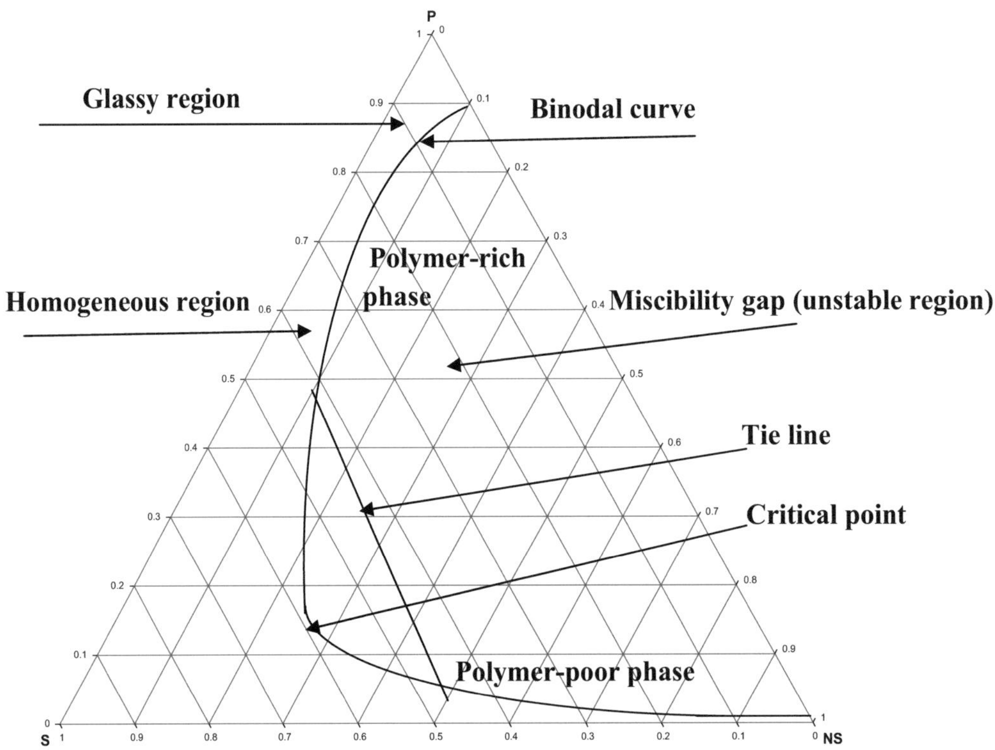

Figure 8. Ternary-phase diagram of polymer (P)–solvent(S)–nonsolvent (NS) system.

If a nonsolvent (such as water) is added to a homogeneous solution consisting of binary mixture of polymer and solvent [2, 32], the membrane casting solution composition is represented by the point A as shown in Fig. 9. Thus, if the solvent is removed from the polymer solution (this takes place by immersion precipitation), the composition of the casting solution mixture will change to follow the line A–B. At point C, the composition of the yield system will reach the demixing gap, and therefore two produced phases will start to form a polymer-rich phase at the upper boundary of the demixing gap and a polymer-poor phase at the lower boundary of the demixing gap. The polymer concentration in the polymer-rich phase will be high enough to be considered as solid; this occurs at a certain composition of the ternary mixtures, which is represented by point D in Fig. 9. At this point, the membrane structure is more or less formed. Further exchange of solvent and nonsolvent will lead to the final composition of the membrane, the porosity of which is determined by point B (pore fraction = (56/90) = 59.8%). Point B represents a mixture of the solid polymer-rich phase and the liquid solvent-rich phase as represented by points B1 and B2, respectively.

The main important step for membrane formation from such ternary systems is always to prepare a homogeneous polymer solution (thermodynamically stable). This will often represent a point on the polymer/solvent axis as point A in Fig. 9. However, it is also possible to add nonsolvent to such a binary mixture that all the polymer solution components are still miscible (homogeneous region) before the binodal curve is reached.

4.7. Demixing types in a ternary system [35]

In a ternary system consisting of a polymer/solvent/nonsolvent, changes in composition can cause demixing (Fig. 9). The demixing types that can be occurring are as follows:

a. Liquid–liquid phase separation

The addition of such an amount of nonsolvent can induce demixing process when the polymer solution becomes thermodynamically unstable (Fig. 9). When the binodal is reached, liquid–liquid demixing will take place. Free enthalpy of the polymer solution decreases by separating into two liquid equilibrium phases. Demixing can take place by two ways: (1) nucleation and growth of droplets of the second phase and (2) instantaneous spinodal demixing. The composition area surrounds the spinodal demixing gap where phase separation (inversion) by nucleation and growth occurs.

b. Crystallization or gelation

As a matter of fact, many polymers are partially crystalline. Crystallization can take place if the temperature of the solution is lower than the melting point of the polymer. Gelation is a considerable important phenomenon during membrane preparation to form the skin layer (top layer) [1, 2, 32]. When gelation takes place, a dilute or more viscous polymer solution is transformed into a system of infinite viscosity, i.e. a gel. As demixing starts, the free enthalpy of the solution is decreased by polymer molecules to form ordered structures. On the one hand, single crystals can be formed at low polymer concentrations. On the other hand, at higher concentrations of polymer, very small crystalline regions could work as physical crosslink between the polymer molecules and the result is a thermoreversible gel.

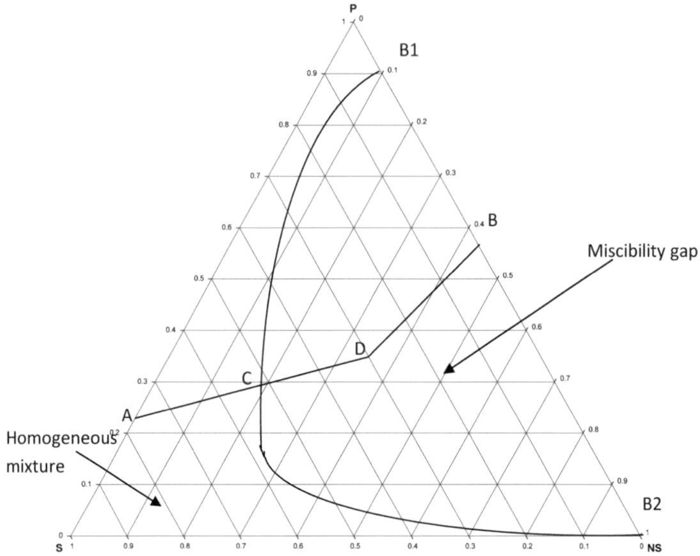

Figure 9. Formation of a microporous system by the addition of a nonsolvent to a homogeneous polymer solution in a ternary mixture exhibiting a miscibility gap at certain conditions of temperature and composition (P, polymer; S, solvent; NS, nonsolvent) [2, 32].

4.8. Mechanism of formation skin top layer and porous sublayer

4.8.1. Mechanism of formation asymmetric membrane [36, 37]

In the phase inversion process induced by immersion precipitation, a polymer casting solution, cast on a support (i.e, glass, Teflon), is immersed in a coagulation bath containing a nonsolvent (as water). The solvent moves from polymer solution to nonsolvent while the nonsolvent penetrates the casted polymer solution.

The mechanism of membrane formation has been reported by some researchers. On the one hand, C. Smolders et al. [38] suggested that the porosity of the sublayer (bottom layer) proves to be liquid–liquid phase separation while the formation of pores may be ascribed to nucleation and growth of the dilute polymer phase. On the other hand, Koenhen et al. [39] suggested that the mechanism for the formation of the sponge-like structure can be ascribed to liquid–liquid phase separation with nucleation and growth of the diluted polymer phase. Furthermore, Bokhorst et al. [40] studied the formation of asymmetric cellulose acetate membranes and reported that the top layer is formed by gelation while liquid–liquid phase separation forms the sublayer of the membrane.

Hence, the skin layer (top layer) is formed by gelation and the porous sublayer (bottom layer) is the result of liquid–liquid phase separation by nucleation and growth. On the one hand, the factor which is determining the type of phase inversion at any point in the cast sheet is the local polymer concentration at the moment of precipitation. On the other hand, in the first split second after immersion in the coagulation bath, there is a fast depletion of solvent from the casted film and a relatively small penetration of nonsolvent. Thus, this means that the concentration of polymer in the film/bath interface increases, which crosses the gel boundary (transition H to G in Fig. 10). The thin and dense gel-layer which is induced by this way will work as a resistance to solvent out-diffusion and at positions beneath the skin layer (top layer). At lower polymer and higher nonsolvent concentrations, demixing process will take place. Thus, the type of demixing process in that case will be liquid–liquid phase separation (transition H to L in Fig. 10). At the end of the process, the demixing gap is entered at the polymer-rich side of the critical point (CP), so the nuclei consist of the polymer-poor phase, and a porous sublayer structure is formed, the pores of which are filled with the dilute solvent/nonsolvent phase.

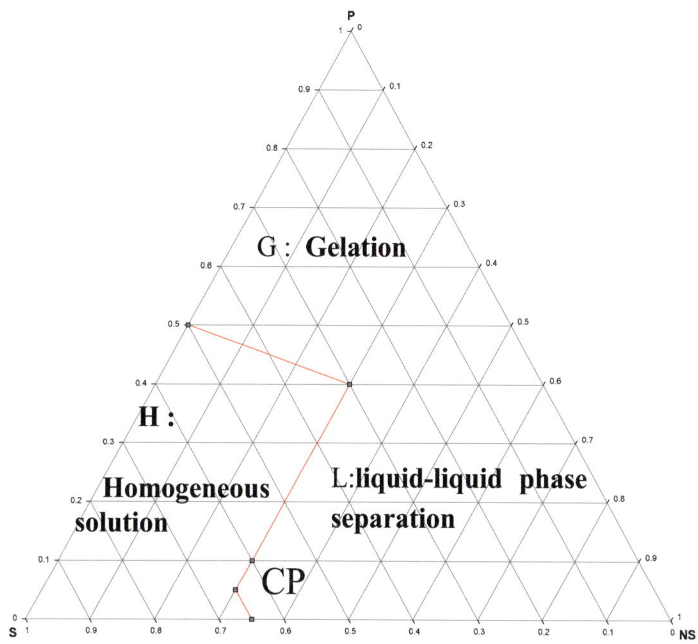

Figure 10. Mechanism of formation of asymmetric membrane, ternary phase diagram containing P, polymer; S, solvent; NS, nonsolvent; CP, critical point; G, gelation (gel region); L, liquid–liquid phase separation(two phase region); H, homogeneous solution (one phase region) [41, 42].

4.8.2. Mechanism of formation symmetric microporous membrane

Conversely to the case of precipitation from the vapor phase, the casted polymer solution, in this case, is in contact directly with a nonsolvent vapor phase. In addition, the nonsolvent vapor phase is saturated with the solvent used. On the one hand, in this process, there is no solvent outflow from polymer casted solution but only a nonsolvent inflow to polymer casted solution. On the other hand, the only possible demixing mechanism is liquid–liquid phase separation to form symmetric microporous membrane without a dense top layer. The change of composition in the casted film is illustrated in Fig. 11 (transition D to E).

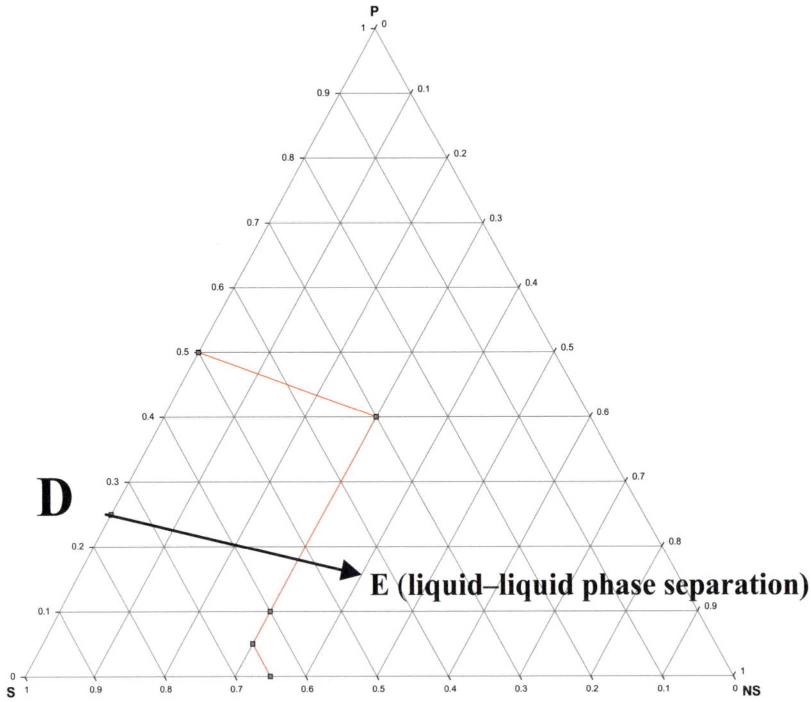

Figure 11. Formation symmetric microporous membrane of a polymer film with initial composition D when brought in contact with a nonsolvent vapor phase saturated with solvent [39–41].

4.9. Effect of some parameters on membrane morphology [43]

In this section, the influence of some parameters on membrane morphology will be described. It is known that there are two different types of membranes can be obtained, asymmetric membranes (i.e., reverse osmosis, nanofiltration, and ultrafiltration) and symmetrical membranes (i.e., dense membrane), depending on the type of formation mechanism, i.e. instantaneous demixing or delayed onset of demixing.

It is obvious that the kinetics of phase separation during the immersion of a polymer/solvent mixture in a nonsolvent bath plays a major role in the control of membrane morphology [44]. The accelerated rate of phase separation due to the addition of nonsolvent additive may be restricted in the rearrangement of polymer aggregates, resulting in the formation of membrane with small macrovoids. Surface properties of membranes can be controlled depending on the casting conditions, polymer solution, and coagulation bath temperature and compositions [2, 32].

The following factors will be described in this section:

- Choice of solvent/nonsolvent (coagulant)systems
- Polymer concentration

http://dx.doi.org/10.5772/60419

- Effect of bath temperature

There are a number of other parameters, in addition to those listed, which will not be considered here.

4.9.1. Solvent/nonsolvent effect

In order to obtain an optimal membrane structure, an additive or cosolvent is frequently used. The addition of a fourth component to a casting solution or a nonsolvent coagulant bath is an important method used in membrane modification. Many researchers studied the effects of additive materials in a coagulation bath with other factors, which can influence membrane formation.

The choice of solvent/nonsolvent is one of the main variables in the wet-phase inversion process which is influenced by the membrane formation. Generally, the membrane morphology can be characterized by scanning electron microscopy (SEM). In the case of delayed liquid–liquid demixing, a sponge-like structure can be formed, and in the case of instantaneous demixing, finger-like structure can be formed [1, 2, 32]. Generally, in the case of higher affinity between the solvent and coagulant, however, instantaneous demixing process takes place, but for a lower affinity between the solvent and coagulant, the delayed demixing process takes place [2, 32].

By addition of cosolvent to polymer/solvent solution with low-affinity coagulant, the polymer solution system will be shifted from instantaneous demixing to delayed demixing process. This is due to the lower affinity of cosolvent with coagulant than solvent. In the case of delayed demixing, a dense skin layer is formed; this inhibits the inflow of nonsolvent (water) into the polymer solution. Further, in the case of delayed demixing, the number and the size of nuclei of polymer-poor phase become smaller. Therefore, macrovoids significantly disappear. By increasing the amount of cosolvent, the numbers of macrovoids become smaller, and the top layer shows a very packed and dense shape. When the ratio of cosolvent to solvent is increased, almost all the macrovoids disappear and the spongy structure starts formation. Without the addition of cosolvent, large fingers are formed because of instantaneous demixing and the upper part of the membrane is porous.

Water-miscible solvent (wms) and water-immiscible solvents (wis) can be used in the casting solution in order to prepare asymmetric membranes. Because (wis) is immiscible with water, it is left in the membrane after coagulation and can further modify the membrane structure through its solvency power.

Kim et al. [45] developed the formation of integrally skinned asymmetric polyetherimide nanofiltration membranes by phase inversion process and showed that the polymer solution system is shifted from instantaneous demixing to delayed-demixing process. This is due to the lower affinity of 1, 4-dioxane (cosolvent) with water (nonsolvent) than DMF (solvent). By increasing the amount of 1, 4-dioxane, the number of macrovoids became smaller and the top layer shows a very packed and dense shape. Without the addition of 1, 4-dioxane, large fingers were formed because of instantaneous demixing and the upper part of the membrane was porous.

Shieh et al. [46] studied the preparation of polyetherimide membranes from water-miscible/immiscible mixture solvents and reported that the PEI/NMP/H_2O system exhibits an instantaneous liquid–liquid phase separation property, but the phase separation rate reduces with the addition of methlyene chloride (MC) into the casting solution. The decrease in the phase separation rate may be mainly due to the difference in solubility parameters; water is 47.9 $(MPa)^{1/2}$, while MC is 19.8 $(MPa)^{1/2}$. The elongated macrovoids diminish gradually as the MC content increases, and no macrovoids are formed when the MC content is 30 wt% in the casting solution.

4.9.2. Effect of Immersion Bath Temperature (IBT)

Immersion bath temperature is an important parameter in membrane morphology and performance. It was found that with asymmetric membranes, the porosity increases with the increase of the immersion bath temperature [47]. Raising the temperature increases mutual diffusivities between the nonsolvent (as water) and the solvent (as DMF) in the casting solution during demixing process. In addition, in the former case, the increase of coagulation bath temperature [48] rises the speed of nucleus growth of polymer-poor phase, that is, increases the formation of porous structure in the membrane. Conversely, in the latter case, at the reduction of the temperature, the speed of nucleus growth is limited and the formation of a large number of small nucleuses is increased, so a great number of pores/voids are formed in the bottom layer of the asymmetric membranes. Therefore, the formation of macrovoids is prevented and denser membrane structures are obtained.

Kim et al. [42] prepared skinned asymmetric polyetherimide nanofiltration membranes by phase inversion process and reported that light transmittance of experiment showed an increase with coagulation bath temperature; demixing of the casting solution was gradually shifted from delayed demixing to instantaneous demixing. When the casting solution [PEI/DMF/1, 4-dioxane (16/28/56)] was coagulated at the temperature of 25°C, porous spongy-like structure was formed. However, lower coagulation temperature (<10 °C) makes the membrane structure dense spongy.

4.9.3. Effect of polymer concentration

Generally, increasing polymer concentrations in the casting solution slows down the demixing process, leads to a much higher polymer concentration at the interface [49], and increases the thickness of the top layer and decreases the porosity of the membrane. The macrovoids formation will be diminished, and the pore size will decrease. Conversely, the reduction of the polymer concentration in the casting solution indeed leads to a more open and porous structure with large pores. Such a phenomenon has been reported in cellulose acetate/dimethyl sulfoxide/water [50] and Nomex/*N*, *N*-dimethylacetamide/water systems [51]. Furthermore, mechanical properties of the membranes prepared with a low polymer concentration are lower than for a membrane prepared with a high polymer concentration.

http://dx.doi.org/10.5772/60419

5. Process block flow diagram for membrane desalination

Fortunately, there are different types of membranes which can be used for membrane desalination by using different technologies such as (RO), (MD), and (FO). In this section, major emphasis is placed on the preparation of membrane distillation (MD) and reverse osmosis (RO) via phase inversion process.

5.1. Membrane Distillation (MD)

Membrane distillation is a thermal, vapor-driven transportation process through microporous and hydrophobic membranes. It is a low-cost membrane separation technology and energy-saving process. Membrane distillation (MD) is generally used to remove water from aqueous solutions of inorganic solutes. In this field, desalination is the best-known MD application. The main membrane materials for membrane distillation are typically prepared from polytetrafluoroethylene (PTFE), polyvinylidene fluoride (PVDF), polyethylene, polypropylene (PP), and polyether sulphone (PES). The process block flow diagrams for the preparation of MD from these materials are presented in Figs. 12–18.

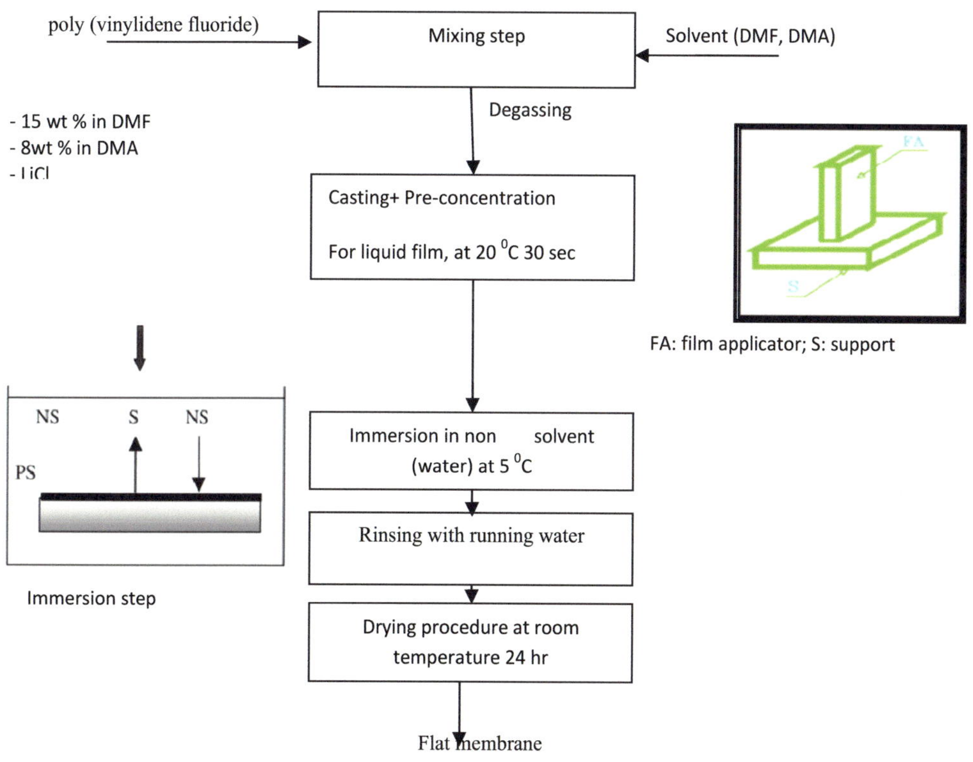

Figure 12. Process flow sheet for the preparation of PVDF membranes for desalination by direct contact membrane distillation [52].

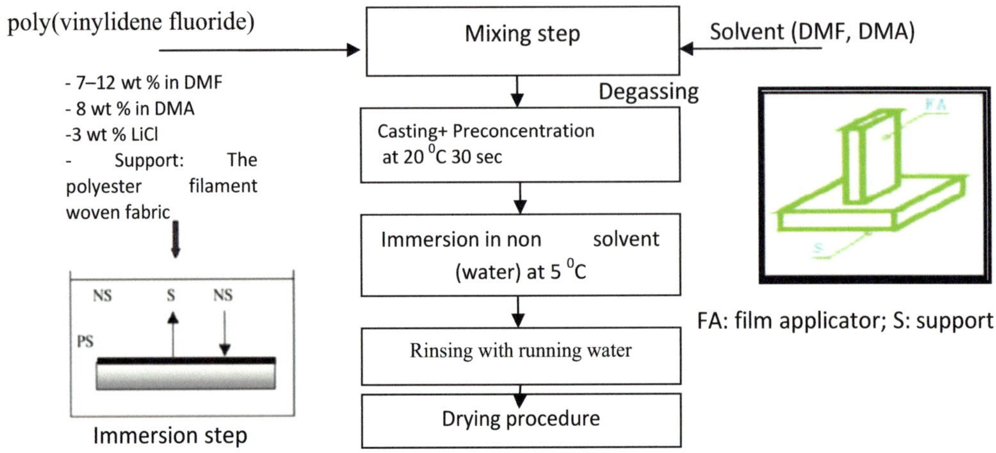

Figure 13. Process flow sheet for the preparation of PVDF membranes for desalination by direct contact membrane distillation [53].

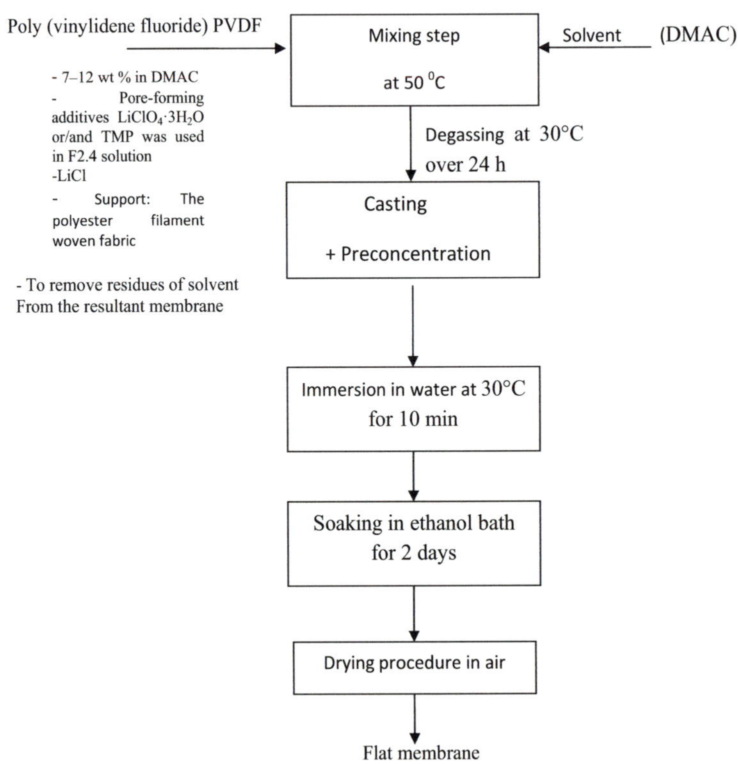

Figure 14. Process flow sheet for the preparation of PVDF membranes for desalination by direct contact membrane distillation [54].

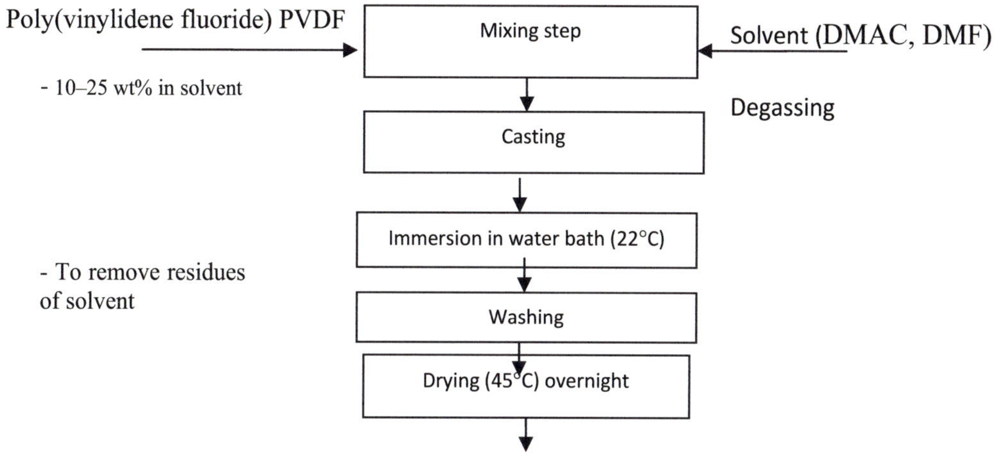

Figure 15. Process block flow diagram for preparing PVDF membrane [55].

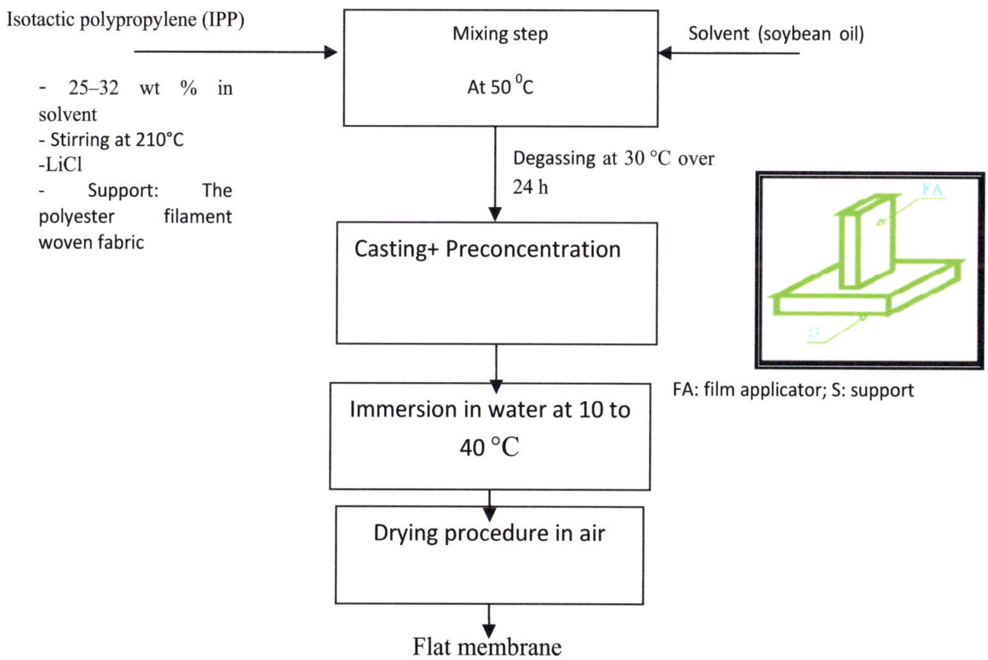

Figure 16. Process flow sheet for the preparation of isotactic polypropylene (IPP) membranes for desalination by direct contact membrane distillation [56].

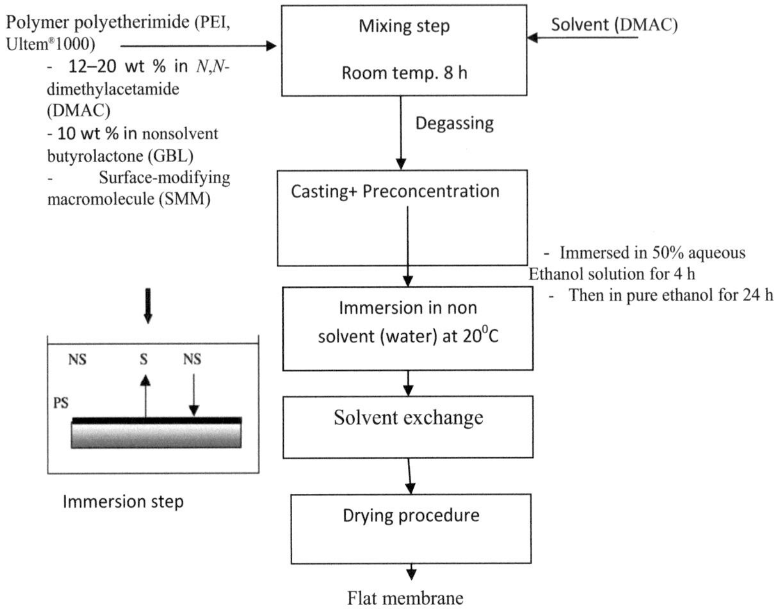

Figure 17. Process flow sheet for the preparation of polyetherimide membranes by direct contact membrane distillation [57].

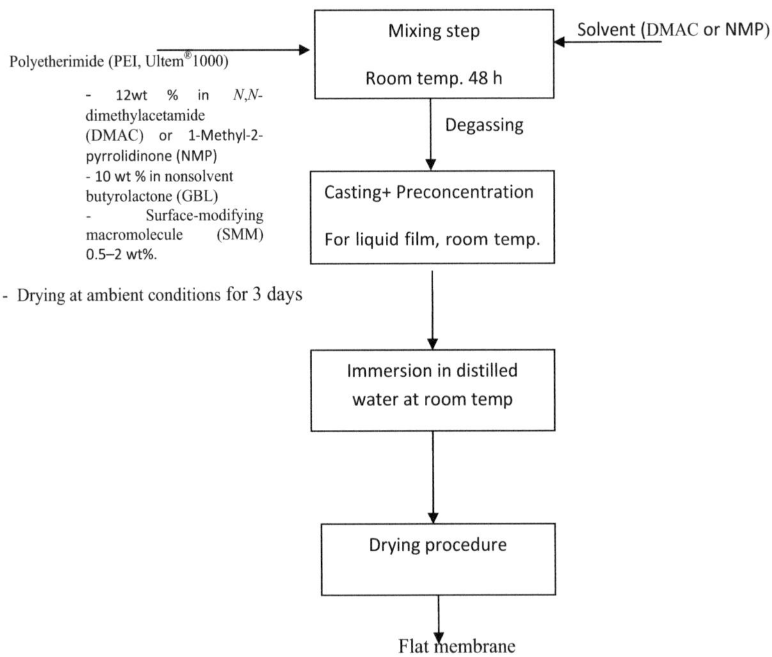

Figure 18. Process flow sheet for the preparation of PEI membranes for desalination by direct contact membrane distillation [58].

http://dx.doi.org/10.5772/60419

5.2. Process flow diagram for RO membrane preparation

The more useable membrane materials for RO are cellulose acetate, polyamide, polyimide, polybenzimidazole, polybenzimidazolone, sulfonated-PSF, and chitosan. The process block flow diagrams for the preparation RO membrane from some of these materials are presented in Figs. (19)–(30).

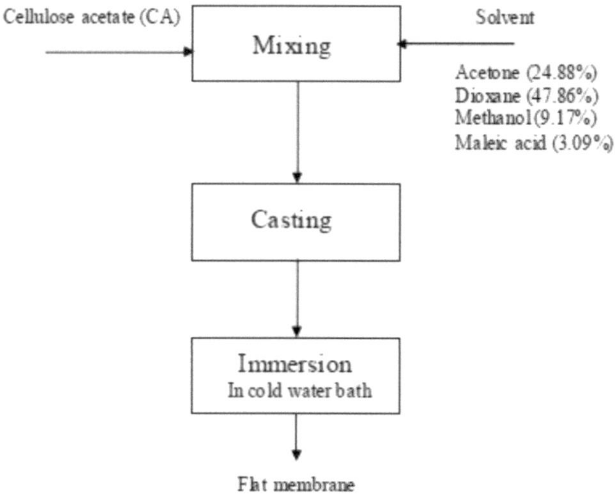

Figure 19. Process block flow diagram for preparing CA membrane [59].

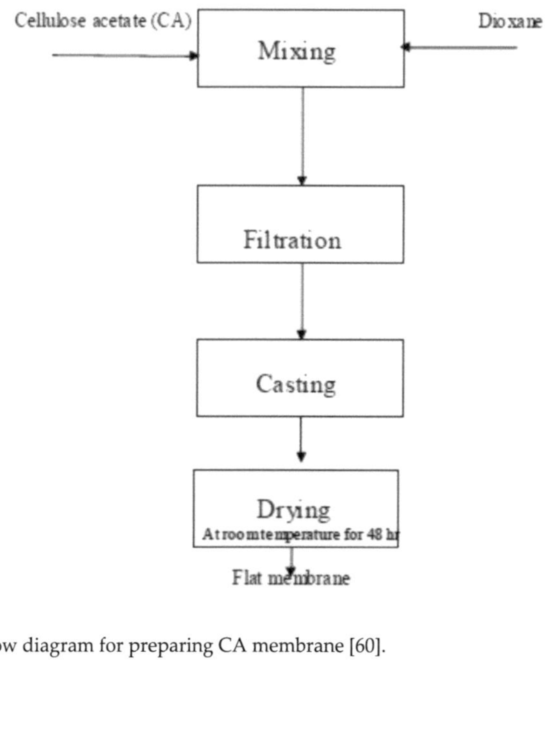

Figure 20. Process block flow diagram for preparing CA membrane [60].

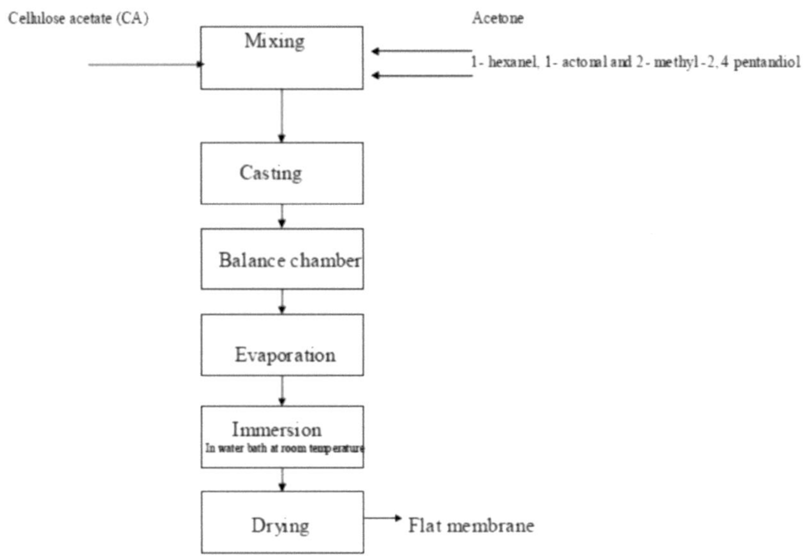

Figure 21. Process block flow diagram for preparing CA membrane [61].

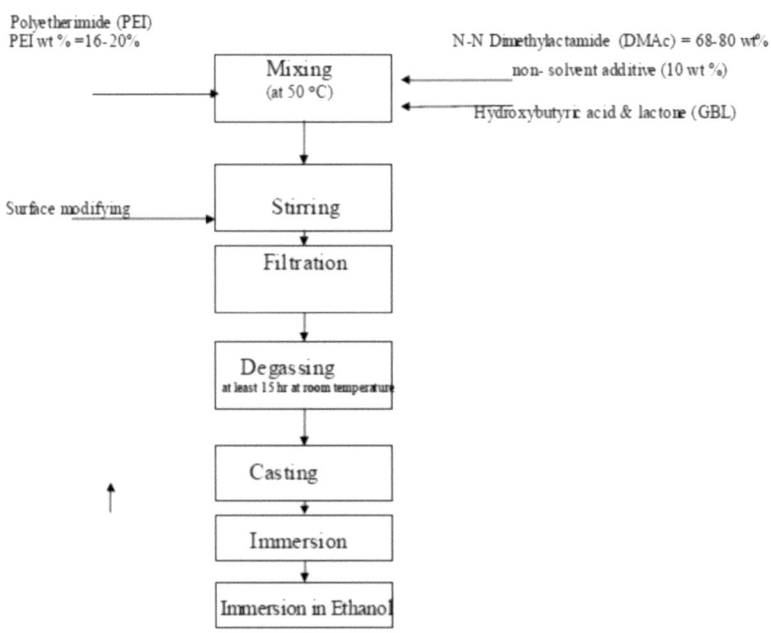

Figure 22. Process block flow diagram for preparing PEI membrane [62].

http://dx.doi.org/10.5772/60419

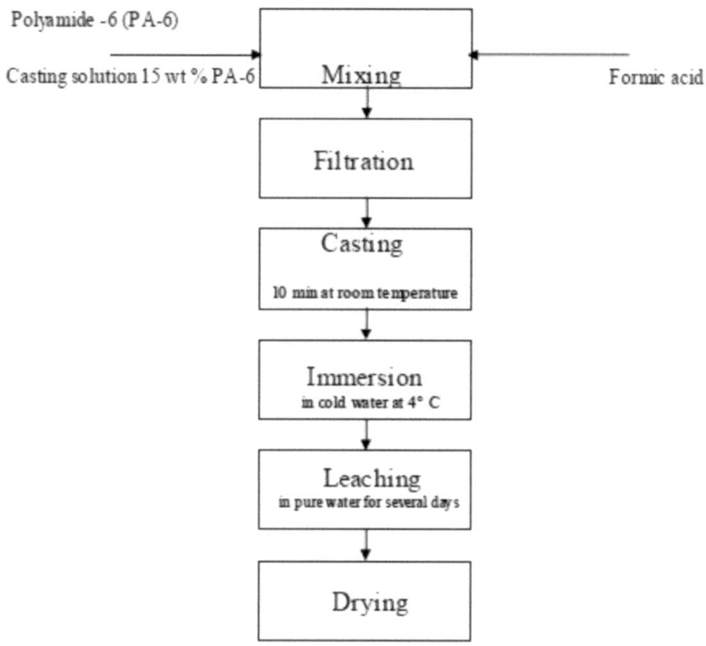

Figure 23. Process block flow diagram for preparing PA-6 membrane [63].

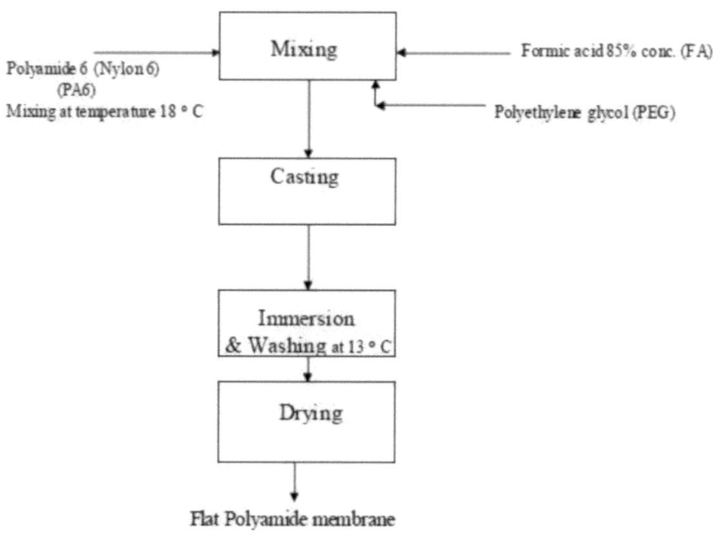

Figure 24. Process block flow diagram for preparing polyamide membrane [64].

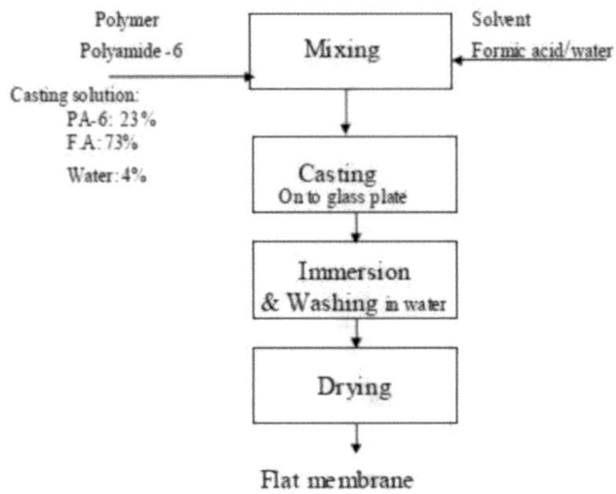

Figure 25. Process block flow diagram for preparing PA-6 membrane [65].

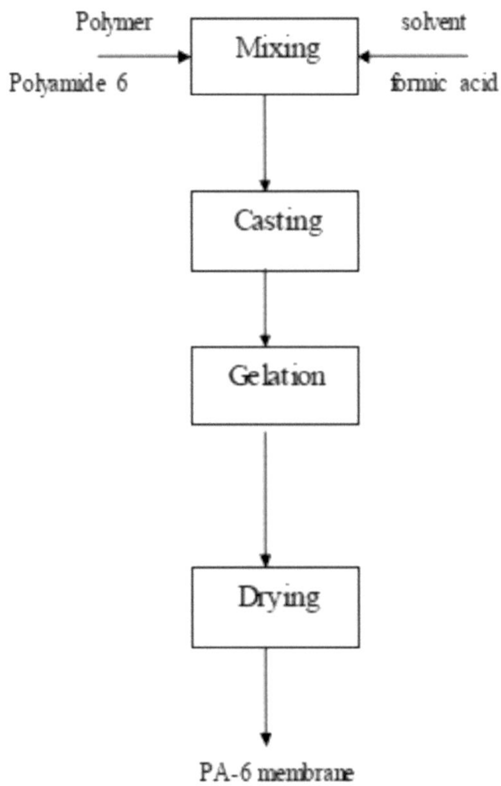

Figure 26. Process block flow diagram for preparing PA-6 membrane [66].

http://dx.doi.org/10.5772/60419

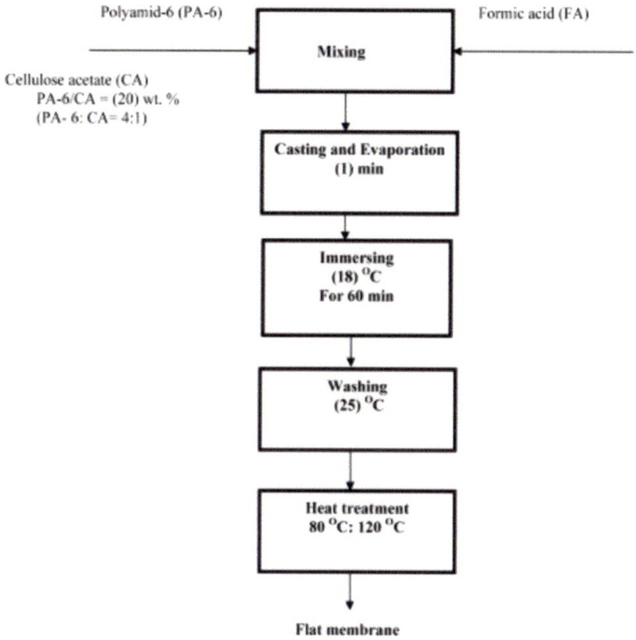

Figure 27. Process block flow diagram for preparing PA- 6/CA membrane [67].

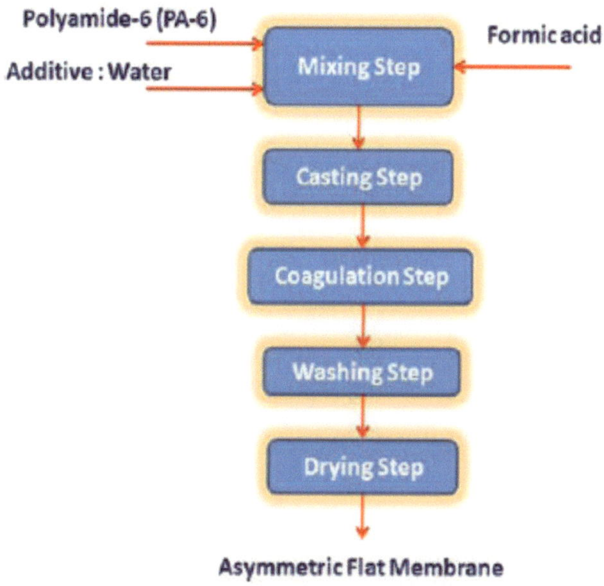

Figure 28. Process block flow diagram for preparing PA-6 membrane [68].

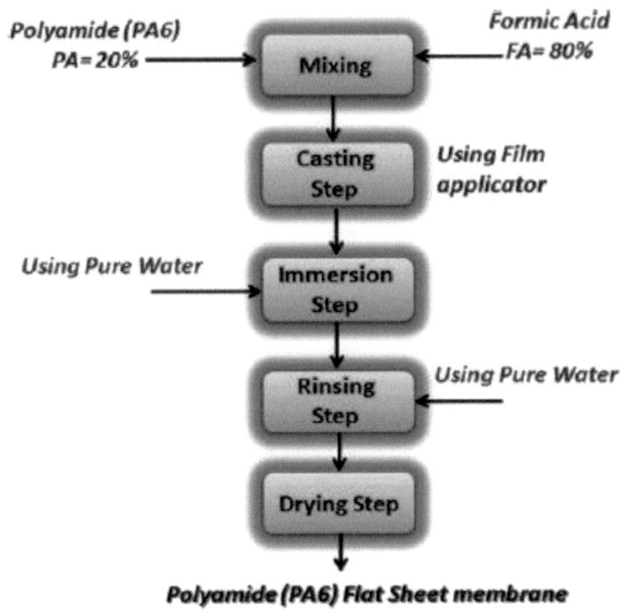

Figure 29. Process block flow diagram for preparing PA-6 membrane [69].

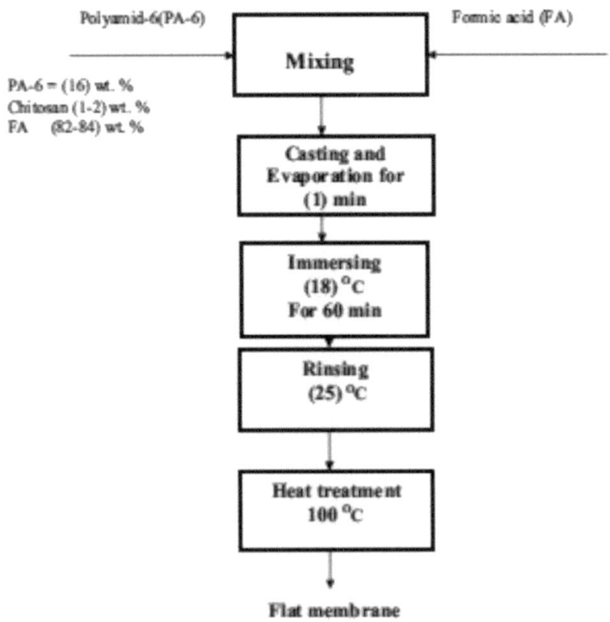

Figure 30. Process block flow diagram for preparing PA-6/Chitosan membrane [70].

http://dx.doi.org/10.5772/60419

6. Conclusion

The preparations of desalination membranes usually involve the phase inversion process, in which a homogeneous casting solution induces phase separation into a polymer-rich phase and a polymer-poor phase by the exchange of solvent with nonsolvent in a coagulation bath. Phase inversion process is the most important technique used to prepare both asymmetric and symmetric polymeric membranes. From a thermodynamic point of view, study on polymer–solvent–nonsolvent system can be well depicted in a ternary phase diagram. The Flory-Huggins theory was found to be a convenient and useful framework for the thermodynamic analysis of component mixing in a membrane preparation system.

In fact, the equilibrium ternary phase diagram system is still a good tool for controlling the membrane morphology and interpreting the membrane structure to be suitable for the required application such as membrane desalination. Significantly, knowledge of phase equilibria (cloud points, binodals, spinodals, and critical compositions) enables one to change the conditions for the preparation of membranes such as the compositions of the casting solution, temperature, and coagulation bath type to obtain an optimum membrane structure. The phase diagram has an important role to report the agreement between experimental work in order to get the required membranes, and the ternary phase diagram miscibility gaps for the evaluations of membrane-forming system.

Nomenclature

DSI	Desalination Systems, Inc.
MD	Membrane distillation
MF	Microfiltration
UF	Ultrafiltration
NF	Nanofiltration
RO	Reverse osmosis
FO	Forward osmosis
PV	Pervaporation
GS	Gas separation
NMP	N-methyl-2-pyrrolidone
SEM	Scanning electron microscopy
DMF	N, N-Dimethylformamide
TFC	Thin-film composite membranes
TGA	Thermogravimetric analysis

SEIP	Solvent exchange immersion precipitation
CA	Cellulose acetate
PMDA	Pyromellic dianhydride
PASA	Poly (amidesulfonamide)
PEAH	Polyetheramide hydrazide
PEG	Polyethylene glycol
PI	Aromatic polyimides such as Kapton™
PEI	Polyetherimide
PPSS	Poly (phenylene sulfide sulfone)
PES	Polyether sulfone
PS	Polysulfone
PTMSP	Poly (1-trimethylsilyl-1-propyne)
PVA	Polyvinyl alcohol
PVDF	Polyvinylidenedifluoride
PA	Propionic acid
PA-6	Polyamide-6
PABH	p-Amino benzhydrazide
PAH	Polyamide hydrazides
PAN	Polyacrylonitrile
PP	Polypropylene
PPO	Polyphenyleneoxide
PTFE	Teflon (polytetrafluoro ethylene)
PVAC	Poly(vinyl acetate)
PVC	Poly(vinyl chloride)
PVDC	Poly(vinylidene chloride)
ZrO_2	Zirconium oxide

Author details

Ayman Taha Abd El-aziem El-gendi

Address all correspondence to: aymantaha2010@yahoo.com

Chemical Engineering & Pilot Plant Department Engineering Research Division, National Research Centre El-Bohouth St., Dokki, Cairo, Egypt

References

[1] Zydney A L., Zeman L J. Microfiltration and ultrafiltration: principles and applications. New York: CRC. ISBN 0-8247-9735-3; 1996.

[2] Mark C P. Handbook of Industrial Membrane Technology. Noyes Publications, Park Ridge, New Jersey, U.S.A; 1990, pp. 61–70.

[3] http://en.wikipedia.org/wiki/Membrane_(selective_barrier).

[4] Baker R W. Membrane Technology and Application. John Wiley & Sons, Ltd; 2004. ISBN: 0-470-85445-6.

[5] Scott K., Hughes R. Industrial Membrane Separation Technology. London: Blackie Academic & Professional, Springer; 1996. ISBN-13: 978-0751403381.

[6] Roudman R., DiGiano F. Surface energy of experimental and commercial nanofiltration membranes: Effects of wetting and natural organic matter fouling. J. Membr. Sci. 2000; 175 61–73.

[7] Inger L A., May B H. Pressure retarded osmosis and forward osmosis membranes: materials and methods, Polymers 2013; 5 303–327.

[8] Mohammad A., Ali N. Understanding the steric and charge contributions in NF membranes using increasing MWCO polyamide membranes. Desalination 2002; 147 205–212.

[9] Drewes J., Reinhard M., and Fox P. Comparing microfiltration-reverse osmosis and soil-aquifer treatment for indirect potable reuse. Water Res 2003;37 3612–3621.

[10] Wagner J. Membrane Filtration Handbook Practical Tips and Hints. Second Edition, Revision 2, Printed by Osmonics, November; 2001.

[11] Semiat R. Water International 2000; 25 54.

[12] Bednarski J., Minamide M., Morin O J. Proc., IDA World Congress on Desalination and Water Science, Madrid 1997; 1 227.

[13] Kronenberg G. Proc., IDA World Congress on Desalination and Water Science, Abu Dhabi 1995; 3 459.

[14] Spiegler K S., El-Sayed Y M. A Desalination Primer, Balaban Desalination 1994; 13.

[15] Matsuura T, Sourirajan S Physicochemical criteria for reverse osmosis separation of monohydric and polyhydric alcohols in aqueous solutions using porous cellulose acetate membranes, J. Appl. Polym. Sci 1973; 17 1043.

[16] Jordi R.-M., Joan B., Veronica G., Maria E M., Fernando V., Ricard D., Jose L. C. Integrating membrane technologies and blending options in water production and distri-

bution systems to improve organoleptic properties. The case of the Barcelona Metropolitan Area, Journal of Cleaner Production 2014; 69 250–259.

[17] Frank W, Smolders C, Calculation of liquid-liquid phase separation in a ternary system of a polymer in a mixture of a solvent and a nonsolvent, Macromolecules 1982; 15 1493–1497.

[18] Frommer M., Lancet M A. Reverse Osmosis Membrane Research, In: Lonsdale H K., Podall H E. (Eds.)., Plenum Press: New York; 1972, p. 85.

[19] Baker RW., Cussler EL., Eykamp W., Koros WJ., Riley RL., Strathman RH. Membrane separation systems: recent developments and future directions, ndc, Membrane Technology and Research, Noyes Publications, Mill Road at Grand Ave., Park Ridge, NJ 07656 (USA) ;1991.

[20] Witte P., Dijkstra P., Berg J., Feijen J. Review phase separation processes in polymer solutions in relation to membrane formation, J. Membr. Sci. 1996; 117 1–31.

[21] Strathmann H. Production of microporous media by phase inversion processes, ACS Symp. Ser 1985; 269 165.

[22] Strathmann H. and Koch K. The formation mechanism of phase inversion membranes, Desalination 1977; 21 241.

[23] Wijmans J. and Smolders C. Preparation of asymmetric membranes by the phase inversion process, In: Londsdale H K. and Pinho M H. (Eds.), Synthetic Membranes: Science, Engineering and Applications, Reidel, Dordrecht, The Netherlands; 1986, pp. 39–56.

[24] Zeman L. and Fraser T. Formation of air-cast cellulose acetate membranes. Part I. Study of macrovoid formation, J. Membr. Sci 1993; 84 93.

[25] Zeman L. and Fraser T. Formation of air-cast cellulose acetate membranes. Part II. Kinetics of demixing and microvoid growth, J. Membr. Sci 1994; 87 267.

[26] Scott K., Hughes R. Industrial membrane separation technology, Chapman & Hall: London; 1996.

[27] Loeb M., Sourirajan S. Sea water demineralization by means of an osmotic membrane, Adv. Chem. Ser 1962; 38 117.

[28] Chun K., Jang S., Kim H., Kim Y., Han H., Joe Y, Effects of solvent on the pore formation in asymmetric 6FDA–4, 40ODA polyimide membrane: terms of thermodynamics, precipitation kinetics, and physical factors, J. Membr. Sci 2000; 169 197–214.

[29] Dyukevick A., Yatskova T. Compaction on ultrafiltration membrane under the effect of working pressure, Khim-Tekhnol-Vody, 1991; 13 544–547.

http://dx.doi.org/10.5772/60419

[30] Bulte A., Naafs E., Eeten F., Mulder M., Smolders C., Strathmann H. Equilibrium thermodynamics of the ternary membrane- forming system nylon, formic acid and water, Polymer 1996; 37 1647–1655.

[31] El-Gendi A., Ahmed S., Talaat H., Preparation and evaluation of flat membranes for phenols separation, Desalination 2007; 206 226–237.

[32] Mulder M. Basic principles of membrane technology (2 ed.). Kluwer Academic: Springer. ISBN 0-7923-4248-8; 1996.

[33] Ismail A., Yean l. Review on the development of defect-free and ultrathin-skinned asymmetric membranes for gas separation through manipulation of phase inversion and rheological factors, J. Appl. Polym. Sci 2003; 88 442–451.

[34] Flory P J., Principles of Polymer Chemistry, Cornell University Press, Ithaca, NY; 1969.

[35] Wijmans J., Baaij J., Smolders C., The mechanism of formation of microporous or skinned membranes produced by immersion precipitation, J. Membrane Sci 1983;14 263–274.

[36] Stern S, Polymers for gas separation; the next decade, J. Membr. Sci 1994; 94 1–65.

[37] Frommer M, Feiner I, Kedem O, Bloch R. The mechanism for formation of skinned membranes, 11, Equilibrium properties and osmotic flows determining membrane Structure, Desalination 1970; 7 393.

[38] Smolders C., Broens L., Altena F., Koenhen D., Asymmetric membrane structures as a result of phase separation phenomena, Desalination 1980; 33–45.

[39] Koenhen D., Mulder M H V., Smolders C A. Phase separation phenomena during the formation of asymmetric membranes, J. Appl. Polym. Sci 1977; 21 199.

[40] Bokhorst H., Altena F., Smolders C. Formation of asymmetric cellulose acetate membranes, Desalination 1981; 38 349.

[41] Chaoyi B., Langer J., Economy J. Chemical modification of P84 copolyimide membranes by polyethylenimine for nanofiltration, J. Membr. Sci 2009; 327 49–58.

[42] Kim C., Kim J., Lee K., Tak T. Preparation of soluble polyimides and ultrafiltration membrane performances. J. Appl. Polym. Sci 1999; 75 1–9.

[43] EL-Gendi A., Ternary Phase Diagram Constraction and Membrane Morphology Evaluation, LAB LAMBERT academic publishing, Germany; 2014. ISBN : 978-3-659-57611-9.

[44] Bottino A., Roda G., Capannelli G., Munari S. The formation of microporous polyvinylidene difluoride membranes by phase separation, J. Membr. Sci 1991;57 1–20.

[45] Kim I., Yoon H., Lee K. Formation of integrally skinned asymmetric polyetherimide nanofiltration membranes by phase inversion process, J. Appl. Polym. Sci 2002; 84 1300–1307.

[46] Shieh J., Chung T S. Phase-inversion poly(ether imide) membranes prepared from water-miscible/immiscible mixture solvents, Ind. Eng. Chem. Res 1999;38 2650–2658.

[47] Smolders C., Reuvers A., Boom R., Wienk I. Microstructures in phase inversion membranes. Part 1. Formation of macrovoids, J. Membr. Sci 1992; 73 259–275.

[48] Amirilargani M., Sadrzadeh M., Mohammadi T. Synthesis and characterization of polyethersulfone membranes, J Polym Res 2010; 17 363–377.

[49] Saljoughi E., Sadrzadeh M., Mohammadi T. Effect of preparation variables on morphology and pure water permeation flux through asymmetric cellulose acetate membranes, J. Membr.Sci 2009; 326 627–634.

[50] Frommer M., Messalam R., Mechanism of membrane formation. VI. Convective flows and large void formation during membrane precipitation, Ind. Eng. Chem., Prod. Res. Dev 1973; 12 328–333.

[51] Strathmann H., Kock K., Amar P., Baker R. The formation mechanism of asymmetric membranes, Desalination 1975; 16 179–203.

[52] Tomaszewska M. Preparation and properties of flat-sheet membranes from poly (vinylidene fluoride) for membrane distillation, Desalination 1996; 104(1-2) 1–11.

[53] Ruiting H., Zhenya G., Kaijie Z. and Guangming Z., Preparation and properties of PVDF-fabric composite membrane for membrane distillation, Desalination 2009; 249(3) 910–913.

[54] Chunsheng F., Baoli S., Guomin L. and Yonglie W., Preparation and properties of microporous membrane from poly (vinylidene fluoride-co-tetrafluoroethylene) (F2.4) for membrane distillation, J. Membr. Sci 2004; 237(1-2) 15–24.

[55] Ortiz J M., Peria L., Mengual L J. Characterization of membrane distillation prepared by phase inversion, Desalination 1995; 100 139–148.

[56] Na T., Qian J., Huanju Z., Jinjin L., Sha C. Preparation and morphological characterization of narrow pore size distributed polypropylene hydrophobic membranes for vacuum membrane distillation via thermally induced phase separation, Desalination 2010 ; 256(1-3) 27–36.

[57] Khayeta M., Menguala J I., Matsuura T. Porous hydrophobic/hydrophilic composite membranes Application in desalination using direct contact membrane distillation, J. Membr. Sci 2005; 252 101–113.

[58] Qtaishat M., Rana D., Khayet M., Matsuura T. Preparation and characterization of novel hydrophobic/hydrophilic polyetherimide composite membranes for desalination by direct contact membrane distillation, J. Membr. Sci 2009; 327(1-2) 264–273.

[59] Combe C., Molis E., Lucas P., Riley R., Clark M M. The effect of CA membrane properties on absorptive fouling by humic acid, J. Membr. Sci 1999; 154 73–87.

[60] Shuguang C., Yanqiao S., Guanwen C. Influence of acetylation degree of cellulose acetate on pervaporation properties for MeOH/ MTBE mixture, J. Membr. Sci 2000; 165 89–97.

[61] Hideto M., Masaaki T., Tsutomu U. Membrane formation and structure development by dry-cast process, J. Membr. Sci 1997; 135 271–288.

[62] Khayet M., Feng C Y., Matsuura T. Morphological study of fluorinated asymmetric polyetherimide ultrafiltration membranes by surface modifying macromolecules, J. membr. Sci 2002; 5502 1–22.

[63] Meier – Haack J., Lenk W., Berwald S., Rieser T., Lunkwitz K. Influence of thermal treatment on the pervaporation separation properties of polyamide -6 membranes, Separation and purification technology 2000; 19 199–207.

[64] Schindler E., (DE): Maier F. (DE). Process for the production of an ultrafiltration membrane from polyamide, U.S. patent 4482514, 1984.

[65] James L T., Martina O., Christin D., Chien-Hsieh S., Carl C G. Polyamide membrane precipitation studied by confocal backscattering microscopy, Polymer V 2002, 43 4153–4157.

[66] Ceynowa – J., Adamczak-P. Enzyme membrane based upon polyamide -6 for oil hydrolysis, J- Appl- Polym –Sci Oct 1992; 46(5) 749–756.

[67] EL-Gendi A., Ali S S., Ahmed S A., Talaat H A. Development of membrane blend using casting technique for water desalination, Membrane Water Treatment 2012; 3 201–209.

[68] EL-Gendi A., Abdallah H. Selectivity performance for polyamide-6 membranes using pervaporation of water/methanol mixtures, Desalination and Water Treatment 2013; 51 3263–3272.

[69] EL-Gendi A., Abdallah H., El-Zanati E., Matsuura T. Pervaporation of methanol from methylacetate mixture using Polyamide-6 membrane, Desalination and Water Treatment 2013; 51 7807–7814.

[70] EL-Gendi A., Deratani A., Ahmed S A., Ali S S. Development of polyamide-6/chitosan membranes for desalination, Egyptian Journal of Petroleum 2014; 23 169–173.

Section 2

Alternative Power for Desalination

Chapter 4

Search for Environmentally Friendly Technology for Processing Molybdenum Concentrates

E.I. Kharin, N.A. Vatolin, B.D. Khalezov and E.A. Zelenin

Additional information is available at the end of the chapter

http://dx.doi.org/10.5772/60671

Abstract

At the Institute of Metallurgy, Ural Branch, Russian Academy of Sciences, a search has been carried out for the oxidative annealing of the molybdenum sulfide concentrate of the new Yuzhno-Shameiskoe deposit with calcium-containing additives. As a result, sulfurous gas transforms into calcium sulfate and does not evolve into the gas phase. In calcine, molybdenum and rhenium remain complete as calcium molybdate and perrhenate. The principles of the selective desalination of molybdenum and rhenium from calcine have been studied. Processes of their recovery from solutions have been studied.

Keywords: oxidation, desalination, molybdenum, rhenium

1. Introduction

Molybdenum rocks, which have found industrial applications, are represented by molybdenite. Molybdenum can be simultaneously recovered from polymetal rocks, which relate to copper–molybdenum, tungsten–molybdenum, lead–molybdenum, and vanadium–molybdenum. After the corresponding steps of enrichment, these rocks are molybdenum concentrates. The enrichment of molybdenum rocks is mainly performed according to cooperative or selective flotation. Oxidized molybdenum rocks, which have poor enrichment ability according to flotation and, finally, all possible waste, tails, and cakes from processing rocks and concentrates that have a large fraction in metallurgy of molybdenum, exist in the largest amounts [1].

The main methods of processing molybdenum sulfide concentrates correspond to the annealing of concentrate with the subsequent preparation of ferromolybdenum according to silicothermy, reduction with carbon, and aluminothermy; annealing with subsequent desalination by various solvents, e.g., ammonia, solutions of hydroxides, and alkali metal carbonates; annealing with the additional charging of the lime or iron scale with the production of calcium molybdate or iron molybdate for subsequent preparation of ferromolybdenum; annealing with the deposition of molybdenum trioxide; sintering with soda, as well as sodium sulfide or sulfate with coal and subsequent desalination, the precipitation of molybdenum sulfide (applied to oxidized molybdenum rocks, e.g., those that contain wulfenites); chloride deposition with sodium chloride or direct chlorination (applied to polymetal rocks containing molybdenum); and hydrometallurgical methods, more specifically, autoclave, using nitric acid at elevated pressure and temperature or use of oxygen, as well as using alkali metals at normal temperatures and pressure.

The first and most important step in the processing of molybdenum concentrate is annealing. When it enters processing, molybdenum concentrate contains 75–95% molybdenum and rhenium sulfide, as well as accompanying copper, iron, zinc, and lead sulfides, and impurities, namely, silicon and aluminum oxides and calcium and magnesium carbonates. Molybdenum concentrate contains 45–55% molybdenum and 30–35% sulfur. Annealing is required in order to eliminate all sulfur. Sulfate and sulfide sulfurs, which exist in the annealed concentrate, are also unacceptable because they easily dissolve and contaminate the solutions obtained in the subsequent hydrometallurgical treatment of calcines. In the formation of oxidative atmosphere, all free carbon, oils, and fluororeagents should be burned from concentrate. Among fluororeagents, the removal of collectors that coat sulfides and oxides by the film, making them hydrophobic, is particularly necessary. A decrease in the wettability of the concentrate results in a decrease in the recovery of molybdenum to solution in hydrometallurgical processing [2, 3].

Rhenium is found in most molybdenum concentrates. In industry, sources of rhenium (by 80%) are molybdenum and copper sulfide concentrates; the remaining amount is obtained by recycling Re-Pt catalysts. Upon the enrichment of copper–molybdenum rocks, Re follows Mo, which is often accompanied by Cu. During flotation, up to 80% of Re enters the concentrate. For example, molybdenum concentrates obtained by enriching porphyry copper rocks contain 0.005–0.075% Re. The oxidative annealing of molybdenum sulfide concentrates is performed at 550–650°C.

In order to anneal molybdenum concentrates, the following setups are conventionally used: muffle and batch-type furnaces with manual raking of calcine, rotary tubular furnaces, multiple-bed furnaces, and fluidized-bed furnaces.

In concentrate, rhenium forms Re_2O_7, which is taken away by the gas flow. The trapping of metal is performed using special wet systems (scrubbers and bubblers) in combination with dry setups (cyclones and sleeve filters), which can trap 60–70% Re. Rhenium is recovered according to various methods. Rhenium is recovered from solutions as Re_2S_7 by sodium sulfide, ammonium sulfide, and polysulfides, as well as in the form of potassium perrhenate $KReO_4$ by KCl solution. Re is selectively sorbed from sulfuric acid solutions with subsequent elution by an aqueous solution of ammonia and the vaporization of elutes, which results in ammonium perrhenate. The combined extraction of rhenium and molybdenum by tertiary amines from sulfuric acid solutions, ammonia reextraction of Re and Mo, the lime purification of reextract from Mo, and the vaporization and formation of ammonium perrhenate are suggested.

Existing national and foreign technologies of processing of molybdenum concentrates and industrial products involve oxidative annealing with the evolution of sulfurous gas to gas phase, more than half of rhenium as Re_2O_7, and part of molybdenum as MoO_3. This creates ecological problems and gives rise to noncollectable losses of metals. In addition, the subsequent hydrometallurgical processing of annealed product results in the formation of waste water and is extremely technically complicated when preparing pure products, e.g., ammonium paramolybdate ($(NH_4)_6Mo_7O_{24} \cdot 4H_2O$) and ammonium perrhenate (NH_4ReO_4). The fact is that, at the desalination of annealed concentrate, in addition to Mo and Re, there are a large number of harmful impurities in solutions, the purification from which leads to the additional noncollectable loss of Mo and Re. As a result, no more than 50–60% Re and 90–97% Mo is extracted to trade products [4–6]. Existing technology for processing molybdenum concentrates is given in Fig. 1.

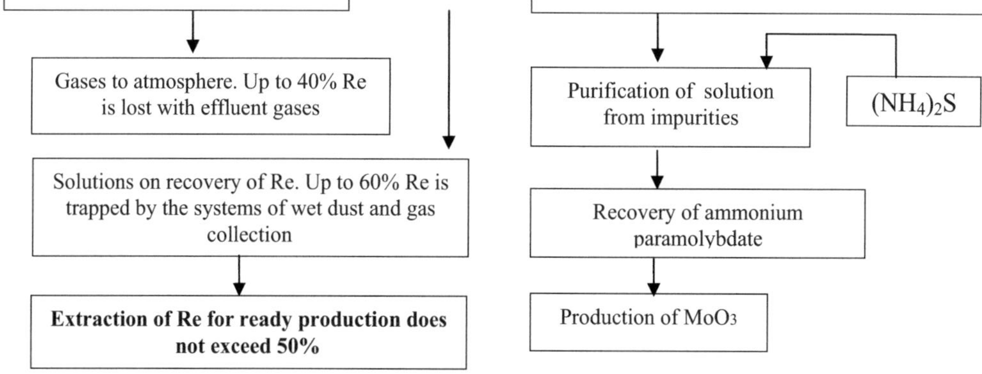

Figure 1. Flow diagram of processing molybdenum sulfide concentrates

Substantial losses of molybdenum and rhenium when processing molybdenum raw material determine the demand for developing improved technology.

2. Results and discussion

At the Institute of Metallurgy, the search for the oxidative annealing of molybdenum sulfide concentrate of the new Yuzhno-Shameiskii deposit was performed (table 1). Concentrate was obtained by enriching rock recovered from an exploratory career [7].

Element	Mo	SiO_2	As	Sn	P	Cu	V	Al	-
wt %	**42,6**	14,3	0,0035	0,0016	0,02	**2,07**	0,2	0,9	-
Element	Ca	Mg	Fe	Pb	W	Zn	S	Cr	**Re**
wt %	0,36	0,459	4,86	0,118	0,044	0,046	31,7	0,011	**0,004**

Table 1. Composition of concentrate

Concentrate was annealed with calcium-containing additives. As a result, sulfurous gas is transformed into calcium sulfate and does not evolve into the gas phase as follows:

$$SO_2 + CaO + 1/2O_2 = CaSO_4$$

In calcine, molybdenum and rhenium remain completely as calcium molybdate and perrhenate:

$$MoO_3 + CaO = CaMoO_4;$$
$$Re_2O_7 + CaO = Ca\left(ReO_4\right)_2.$$

The conditions of annealing are as follows: the temperature must be 550–650°C, the time of annealing must be 60–90 min, and a content of calcium-containing additive in excess of what is stoichiometrically required to bind sulfur, molybdenum, and rhenium. The SO_2 content in effluent gases was determined by an MGL_19.3A gas analyzer. The maximum allowable concentration of SO_2 in effluent gases is 0–2.5 mg/m3.

Effluent gases were passed through alkali solution for trapping rhenium. The rhenium content in the solution was determined qualitatively according to calorimetry based on the formation of a brown–yellow rhenium compound with rhodanide, ReO (CNS). This compound is formed when treating the solution with a mixture of $SnCl_2$ and KCNS [8]. According to the analysis, rhenium was not detected in the solution.

$$ReO_4^- + Sn^{2+} 4CNS^- + 4H^+ \rightarrow \left[ReO_2\left(CNS\right)_4\right]^{3-} + Sn^{4+} + 2H_2O$$

The methods of selective desalination of rhenium and molibdenum from the calcine obtained were worked out. Firstly, rhenium is bleached selectively. Desalination solutions contain 4.74

mg/dm^3 Re, 4.5 mg/dm^3 Mo, 520.5 mg/dm^3 S, and 460.5 mg/dm^3 Ca. Degree of extraction of rhenium to solution of 76% was achieved.

In order to recover rhenium from solutions, sorption method was used. Five sorbents were tested, namely, AH31, ABI7, AMP, Purolite A170 (further A170), and Purolite A172 (further A172). Sorption was performed under dynamic conditions (table 2).

sorbent	Dynamic volume capacity, g/dm^3	Full dynamic volume capacity, g/dm^3
AH-31	57,8	63,8
AMП	68,2	74,2
AB-17	72,4	79,1
A-170	75,0	82,5
A-172	88,1	95,8

Table 2. Capacity characteristics of anion exchangers

As follows from the results of study of sorption on five brands of anionites, A170 and A172 have the highest capacity on rhenium. However, A170 anionite with rhenium sorbs a substantial amount of molybdenum; this does not satisfy to the requirements on rhenium selectivity in the presence of molybdenum. Regarding A172 anionite, it almost does not sorb molybdenum. Desorption of rhenium was performed by 10% aqueous solution of ammonia. Degree of desorption of rhenium from A170 and A172 anionites was less than 98% (table 3) after passing three to four specific volumes of eluting solution (concentration of rhenium in desorbates is 6.3 g/dm^3), and degree of desorption from AH31, ABI7, and AMP anionites was less than 20% (rhenium concentration in desorbates is 0.5–1 g/dm^3).

The results of sorption of rhenium from the leaching solution cakes M.C. on ion exchanger Purolite A-172 shown the high efficiency of the sorption process for the recovery of rhenium from the leaching solution.

Element Name	Re	Mo	S	Ca
Leach solution Re, mg/dm^3	130,0	60,36	6290,0	717,0
Desorbate, mg/dm^3	6260,0	2,35	571,4	226,8

Table 3. Composition of solutions before sorption and desorption on A-172

Of the stripping solution was obtained by crystallization rough ammonium perrhenate with a rhenium content of 69.2 wt. %, which corresponds to the brand AR-0. After purification of crude perrhenate was obtained pure ammonium perrhenate with a rhenium content of 69.3 wt. %, molybdenum, calcium and sulfur is less than 0.001 wt. %, which corresponds to the brand AR-00.

After the desalination of rhenium, cakes were treated with an aqueous solution of soda at a concentration of 250 g/dm^3 for 30–90 min (Fig. 2). Conditions of desalination of molybdenum are as follows: sodium carbonate solution with the concentration from 150 to 200 g/dm^3, temperature is 90°C, time is 60 min, and solid-to-liquid phase ratio is 1:4. Desalination solutions contain 0.75 mg/dm^3 Re, 32.48 g/dm^3 Mo, 53.5 g/dm^3 S, and 40.9 mg/dm^3 Ca. Complete reextraction of rhenium and recovery of molybdenum to solution to the degree of 99.8% was achieved [10]. The technology for processing molybdenum concentrates developed is given in Fig. 3.

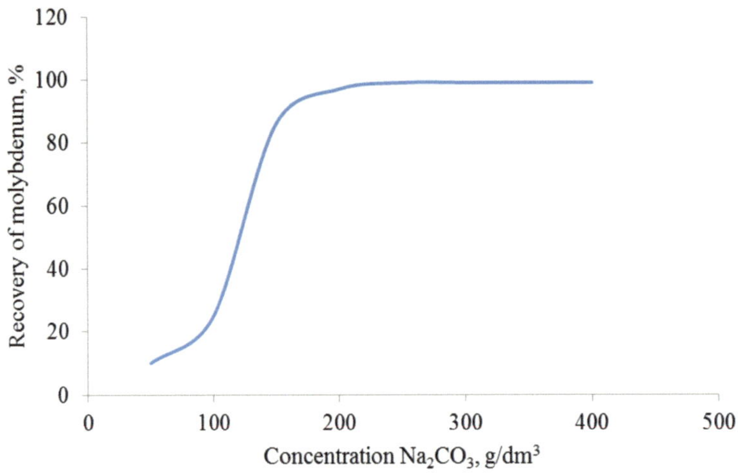

Figure 2. Recovery of molybdenum vs. concentration of soda

3. Conclusions

Existing national and foreign technologies of processing of molybdenum concentrates and industrial products do not enable one to recover rhenium completely and are not ecologically effective. No more than 50 to 60% Re and 90 to 97% Mo are recovered to trade products.

On the laboratory scale, the conditions of the oxidative annealing of molybdenum concentrate with the addition of CaO without the evolution of sulfurous gas in gas phase and the loss of volatile compounds of molybdenum and rhenium were determined. The methods of the selective desalination of rhenium and molybdenum were devised based on the calcine obtained. First, rhenium is bleached selectively; then, molybdenum is bleached by alkali metal carbonate with a concentration of 150–250 g/dm^3. Sodium or potassium carbonate is used as alkali metal carbonate. The degree of recovery of molybdenum takes the value of 99.8%. Degree of selective recovery of rhenium to solution at the first step is 76% and remaining 23.8% is bleached at the second step together with molybdenum. The selective desalination of these metals is a prerequisite for the preparation of high-quality molybdenum and rhenium products.

http://dx.doi.org/10.5772/60671

A principally new direction in the development of environmentally friendly technology of production of molybdenum and rhenium from molybdenum-containing concentrates and industrial products was found. The results of the studies are considered for use at the new Yuzhno-Shameiskii deposit and modernization of existing productions. The demand for molybdenum and rhenium used for doping titanium alloys increases due to the creation of Titanium Valley in the Urals. Molybdenum and rhenium have recently been purchased abroad because national industry does not fabricate this production.

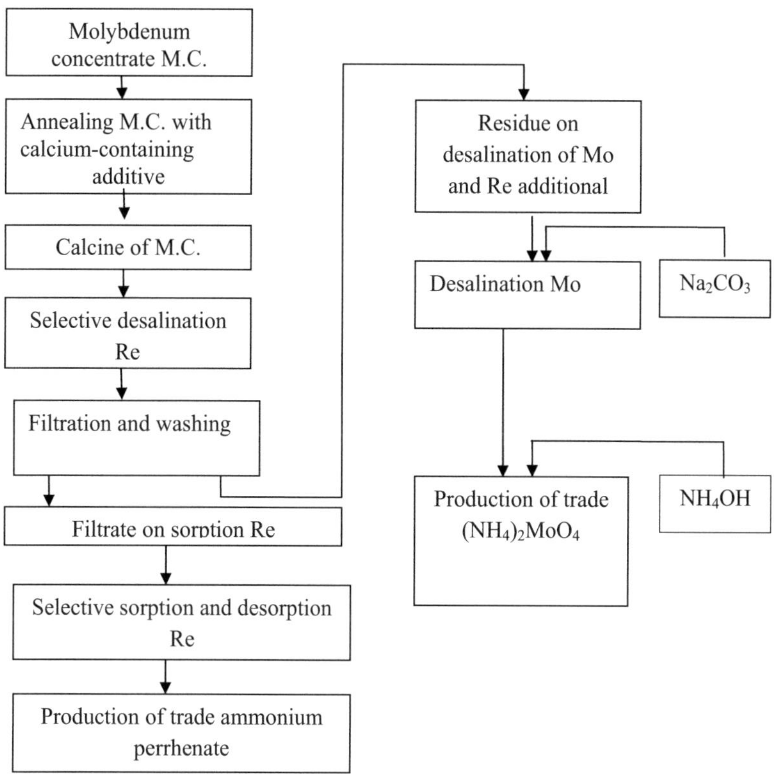

Figure 3. Flow diagram of processing molybdenum concentrates

Author details

E.I. Kharin*, N.A. Vatolin, B.D. Khalezov and E.A. Zelenin

*Address all correspondence to: harin-ei@mail.ru

Institute of Metallurgy, Ural Branch, Russian Academy of Sciences, Yekaterinburg, Russia

References

[1] Korovin, S.S., Bukin, V.I., Federov, P.I., et al., Redkie i rasseyannye elementy (Rare and Dispersed Elements), Moscow: MISiS, 2003, vol. 3.

[2] Zelikman, AN. and Meerson, G.A., Metallurgiya redkikh metallov (Metallurgy of Rare Metals), Moscow: Metallurgiya, 1973.

[3] Krcmenetskii, A.P. and Shaderman, F.I., Two ways to wealth, Met. Evrazii, 2000, no. 3, p. 130.

[4] Palant, A.A., Troshkina, D., and Chekmarev, A.M., Metallurgiya reniya (Metallurgy of Rhenium), Moscow: Nauka, 2007.

[5] Zelikman, AN., Molibden (Molybdenum), Moscow: Metallurgiya, 1970.

[6] Lebedev, K.B., Renii (Rhenium), Moscow: Metallurgizdat, 1963.

[7] Levin, V.Ya., Antonova, L.G., Samsonov, A.V., et al., Geology and specific features of ore genesis in the Yuzhno-Shamciskoe. molybdenum deposit in the Middle Urals, Geol. Rudn. Mestoroz)\d., 1995, vol. 37, p. 530.

[8] Gillerbrand, V.F., Prakticheskoe rukovodstvo po neorganicheskomu analizu (Practical Guide to Inorganic Analysis), Moscow: Goskhimizdat, 1960.

[9] Vatolin, N.A., Khalezov, B.D., Kharin, E.I., and Zelenin, E.A., Rhenium recovery from molybdenum concentrates and industrial products: state of the art and outlooks, "Fundamental'nye i prikladnye problemy nauki, " Trudy I Mezhdunarodnogo simpoziuma, (Proc. 1st Int. Symp. on Fundamental and Applied Problems of Science), Moscow, 2010, vol. 2, p. 132.

[10] Vitolin, N.A., Khalezov, B.D., Lobanov, V.G., and Zelenin, E.A., RF Patent 2393253. Byull. Izobr., 2010, no. 18.

Chapter 5

Solar-Powered Desalination

Emrah Deniz

Additional information is available at the end of the chapter

http://dx.doi.org/10.5772/60436

Abstract

Drinkable fresh water, being a fundamental need of human beings, has become a serious concern for people especially living in crowded cities and countries with limited amount of water resources. Supplying fresh water is an energy intensive task especially when there is need for heating cold water to evaporate for distillation. Most of the big desalination plants around the world use fossil-based fuels as energy source to heat and vaporize the sea water or brackish water in order to produce fresh water.

However, environmental concerns along with decreasing oil reserves and increasing fuel costs call for reconsideration on the fuel types used in many areas as well as in desalination or distillation processes. There are a number of well-known alternative and renewable energy sources in the world like wind, geothermal, solar, biomass, etc. but achievability, availability when needed and energy storage are also very critical issues to operate the distillation plant effectively.

Since most of the desalination processes run on heat, solar energy can be used directly as heat source. Complex desalination systems which are all designed to improve the thermal efficiency and freshwater yield/productivity use many approaches such as multiple stage operation, pre-heating the feed water with the condensation energy, using additional heat sources (waste water from another process), concentrating solar energy to improve the operation temperature of the process, etc.

Solar desalination systems can be a good solution to the water scarcity, especially the Sunbelt region. However, dominant use of clean and renewable energy sources for desalination process and replacing the fossil fuel–based operations require more R&D studies to find more efficient and/or less costly power plants with continuous operation by the means of energy storage solutions.

Keywords: Desalination, solar energy, distillation, water scarcity, desalination methods

1. Introduction

Freshwater and energy are two inseparable and essential commodities for sustaining human life on earth. Rapid population growth and industrialization, especially in developing countries in the recent past, have placed pressing demands for both freshwater and energy [1]. Both are to be conserved and preserved for the sustainable development of the world. On the other hand, there is an acute shortage of both energy and water, especially in the third world countries [2]. Only less than 1% of the water is available for the society for direct use, out of which the maximum fraction has been polluted due to non-manageable industrial developments [2].

Due to the fast increase in the world population, the need for the energy increases rapidly. Various studies have been made to meet this extra energy demand. Most of these studies were focused on alternative energy sources. Solar energy is one of the most popular ones of these sources due to relatively less installation cost and long operation time without any need for maintenance [3]. Basically, all forms of the energy in the world, as we know it, are solar in origin. Oil, coal and natural gas is originally produced by photosynthetic processes, followed by complex chemical reactions, in which decaying vegetation was subjected to temperature and pressure over a long period of time [4].

Solar energy is so powerful and abundant that 30 min of solar radiation falling on earth is equal to the world annual energy demand. It is also cheap, environment friendly and nearly inexhaustible. Most of the world receives powerful solar radiation and has a good chance and opportunity to benefit from solar energy [3].

All ecosystems and every field of human activity depend on clean water and it is one of the most precious resources in today's world. Water is a primary need for life, health and sanitation, which brings it into the international agenda as a very important issue [2]. The lack of potable water poses a big problem in arid regions of the world where freshwater is becoming very scarce and expensive. Clean drinking water is one of the most important international health issues today [5]. Remote and arid to semi-arid regions depend on underground water for drinking. Unfortunately, underground water is not always considered to be fresh drinking water. In some instances, the salinity is too high for being drinkable, and it is called brackish water. In such cases, fresh water has to be either transported for long distances or connected with an expensive distribution water network at extremely high cost for, usually, a small population. Solar desalination can be a good solution for such a problem [6].

Desalination is one of mankind's earliest forms of water treatment, and it is still a popular treatment solution throughout the world today [5]. Distillation is one of the oldest and most rustic desalination techniques. In fact, it reproduces the natural process of the industrial desalination. Solar distillation is a thermal process that represents a sustainable solution to water shortages in the world. The desalination technique by solar energy, based on the principle of greenhouse is not new. Solar energy coupled to desalination offers a promising prospect for covering the fundamental needs of power and water in remote regions, where connection to the public electric grid is either not cost effective or not feasible, and where the water scarcity is severe [5].

http://dx.doi.org/10.5772/60436

2. Hydrologic cycle and water scarcity

The earth has a global water amount of about 1.38×10^9 km^3, 97.5% of which is salty and only 2.5% is fresh water. Most of the fresh water is either frozen in the poles or remains in the soil as moisture or lies in aquifers beyond the reach of human being. A tiny fraction of global water resource (~1%) is available to be used by human beings, animals and plants. Being one of the most critical and vital needs of all plants, animals and human beings, fresh water is created mainly by a continuous evaporation and condensation process called "hydrological (water) cycle" [7].

The water cycle is caused by solar energy which heats the water in oceans, seas, lakes, rivers and lands containing moisture. Water evaporates and becomes water vapour while ice, hail and snow can transit from solid to vapour phase directly which is called sublimation. An average of 16 million m^3 of water is evaporated every second and then the vapour is transported in the atmosphere to where it comes across cooler temperatures which will eventually cause the vapour to condense and precipitate as rain, sleet, snow or hail. Nearly, four-fifth of the evaporated vapour falls back into the oceans and some other flows through the ground surface to the sea. Some water also goes deep into the ground layers and fills aquifers, which are the long-term water storages. Some groundwater comes out as springs through openings in the ground. Eventually, the water returns to the ocean and joins to the water cycle again. Figure 1 shows a schematic illustration of the water cycle (This Figure adapted from http://www.srh.noaa.gov/jetstream/atmos/hydro.htm).

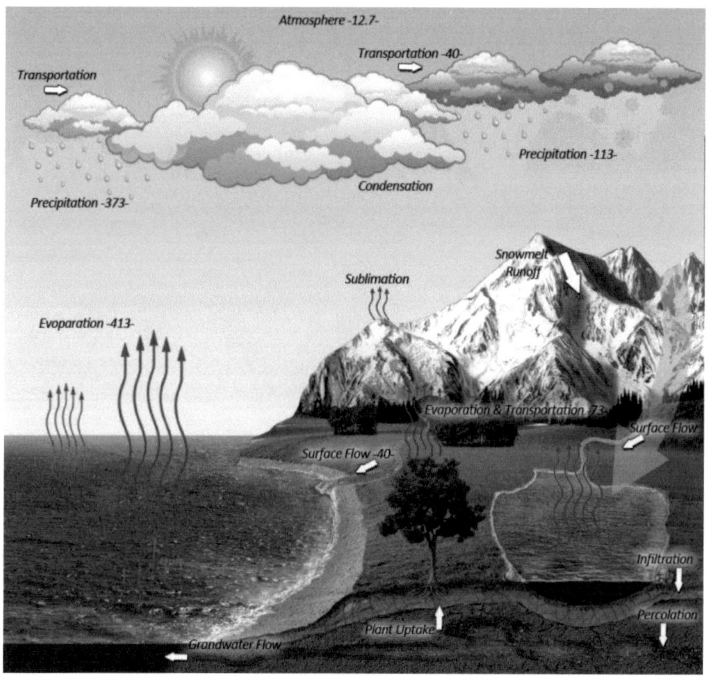

Figure 1. The global water cycle (the units are in 10^3 km^3. year^{-1}) [9]

Due to wind regimes and atmospheric conditions, water vapour and rain are not evenly distributed throughout all the earth land. Therefore some parts of the earth take much more precipitation than the others. This situation causes a failure to provide enough usable water for everyone living in a certain region which can be caused by mainly two reasons. If the water in that region is physically less than the amount of water required in order to meet the demands of people in that region for drinking, home use and agriculture, this situation is called "physical water scarcity". On the other hand, if the water resources are present but they are not effectively used due to lack of investment or absence of qualified personnel or just administrations to process the water to provide people with the water they need, it is called "economic water scarcity" [8].

Although all parts of the earth is exposed to solar radiation throughout a year, the equator zone between 35°N and 35°S latitudes receives much more solar energy compared to the other parts of the earth and is called "Sunbelt region" [10]. About four-fifth of the world population lives in this region. Having abundant solar energy on one side, the Sunbelt region countries experience a serious water scarcity problem. This scarcity is not only physical water scarcity, indeed it is an economic water scarcity especially in Africa, which means lack of investment in water infrastructure or the lack of personnel capacity to adequately meet the drinking and irrigation water demand of the population. Figure 2 shows a schematic illustration of the Sunbelt region and global water scarcity in the world.

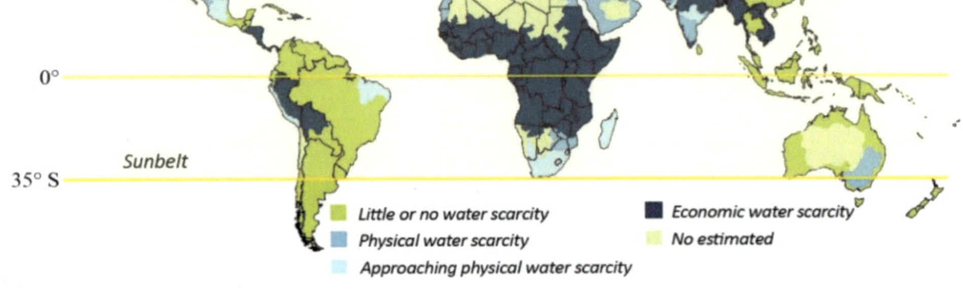

Figure 2. Global physical and economic water scarcity and Sunbelt region [10, 11]

One of the vital factors in combatting the economic water scarcity is a cheap, clean, affordable and environment friendly energy source which will be used to produce and distribute usable and drinkable water. Renewable energy sources such as; solar, wind, geothermal, biomass, etc. is suitable for this purpose.

Sunbelt region is very suitable to use solar energy, which is a very popular and abundant energy source, to produce fresh water from saline water through a number of desalination methods, which will be discussed in detail later. 6 kWh/(m^2.d) of solar energy, which is easily reached in Sunbelt region, is equal to 0.6 l/m^2.d and 220 l/m^2.y of oil. Therefore it is practically possible to use solar energy instead of especially fossil fuels in this region [12].

3. Desalination and solar energy's place among the other energy sources for desalination

Total fresh water consumption in the world can be classified into three categories; about 70% is used for irrigation, 20% is used for industrial purposes, and only 10% of the fresh water is consumed for domestic uses as drinking and cleaning water. In case of a shortage of fresh water, desalination is a way to produce usable and drinkable fresh water from any source of saline water to meet the demand.

Desalination is the process of separating salt from saline water, which is a mixture of pure water and salt, in order to obtain fresh water. Water salinity due to dissolved salts can be expressed in four classes as; fresh water (<0.05%salinity), brackish water (0.05–3% salinity), saline water (3–5% salinity) and brine (>5% salinity) [13]. The most important property of the desalinized water and thus the parameter observed through the process is salinity. Salinity can be expressed in particles per million (ppm) or salt mass fraction (mf_s). 1000 ppm salinity equals to a salinity of 0.1%, or a salt mass fraction of $mf_s = 0.001$.

Although about two-third of the feed water for desalination process is the sea water, waste water (about 6%), river water (about 8%) and brackish water (about 19%) are also used as desalination water especially at places distant from the sea [14]. Actually, feed water with low salt concentration is preferable for the stills, where available, since it causes less contamination and scale formation in the system. The installed capacity of desalinated water system in year 2000 was about 22 million m^3/day and has drastically increased to 71.7 million m^3/day by the year 2010. It can be estimated that 71.7 million m^3/day desalination requires about 650 million tons of oil/year as energy source [15].

This means that; using renewable energy sources in desalination processes instead of fossil fuels, a significant amount of pollution, greenhouse gas and global warming contribution can be avoided. Desalination process is mainly of two types: phase-changing processes and single-phase processes (Table 1). In addition to these two types of desalination, there are hybrid processes that employ both phase change and separation at the same time. Hybrid systems may be comprised of one unit in which both phase change and separation steps take place or they may have two units for two steps. Reverse osmosis combined with MSF or MED are two examples for hybrid process.

Phase-change processes	Single phase processes
1. Multi-stage flash (MSF)	1. Reverse osmosis (RO) – RO without energy recovery – RO with energy recovery (ER-RO)
2. Multiple effect distillation (MED)	2. Electrodialysis (ED)
3. Vapour compression (VC)	
4. Freezing	
5. Humidification/dehumidification	
6. Solar stills – Conventional stills – Special stills – Cascaded type solar stills – Wick-type stills – Multiple-wick-type stills	

Table 1. Desalination processes [16]

Since changing the phase of saline water requires considerable amount of heat, solar energy is a very practical and readily available energy source along with geothermal and wind energy where available. Solar energy, wind energy and geothermal energy are used in renewable energy powered desalination plants with the percentages of 42%, 37% and 21% respectively. Solar power is used as both thermal energy and electricity through photo-voltaic panels.

4. Solar technologies used in desalination

Energy is the most vital need for living. Obtaining the usable forms of energy may cause both economic and environmental problems. Especially the fossil fuels have relatively high costs and environmental impacts which inevitably lead to seek for developing alternative methods. Using renewable energy sources is a good way to cope with energy and environmental problems. Renewable energy sources can easily replace fossil-fuels in the near future especially in stationary plants. Depleting reserves of fossil-fuels and environmental problems make it necessary to use the reserves more carefully. Solar energy has great potential in space heating for buildings owing to its low-grade energy characteristics and is the most important alternative to fossil fuels. Solar systems with feasible design and installation have very short payback periods and meet the energy demand very effectively.

Solar energy is widely used for drying, cooking, distillation, hot water and electricity production which are the very daily life needs of the people. Several methods using solar energy can be used easily to produce potable water from salt water to save people and agriculture from water scarcity. Main requirement of desalination process is thermal energy, and it can be provided through thermal and PV applications of solar energy systems. This energy can be integrated with various types of structure and capacity distillation systems to produce fresh water.

http://dx.doi.org/10.5772/60436

4.1. Salt Gradient Solar Pond (SGSP)

Salt gradient solar pond (SGSP) is a low cost method of capturing and storing solar energy at relatively low temperatures. A SGSP has mainly three layers of water filled in a pond in an order of salinity and relative mass from the bottom to the top.

1. The first and deepest layer of the pond is filled with a very salty water (about 20%), which is called Lower Convective Zone (LCZ). This layer is salty as much as possible to increase the relative mass of the water compared to the next upper layer which is called Non-Convective Layer (NCZ). This is accomplished by gravitational force that traps the water which takes the heat from the solar radiation and from the bottom of the pond, which absorbs the insolation passing through the transparent all UCZ, NCZ and LCZ layers, and tries to go upwards into the NCZ following normal convection process. Since NCZ layer has a relatively light water, convection process is suppressed and inhibited by the gravity of the LCZ water. Thus a transparent convective insulation is formed by the NCZ layer.

2. NCZ layer is structured as multilayer with a salinity gradient from bottom to top in order to slow down convective heat transfer to the upward direction. Slices of water layers with decreasing salinity provide insulation for the heat trapped in LCZ by eliminating convection between the NCZ layers.

3. UCZ is the top water layer of the pond with lowest salinity. UCZ layer is at near ambient temperature. It should be kept thin (about 30 cm) and should be protected from fluctuations caused by wind and wave which leads to mixing of NCZ and UCZ. Wind barriers or mash type covers can be used for big applications.

Solar ponds can provide thermal energy for domestic heating for space and water or power generation and desalination processes. The heat from a solar pond can be used in a Rankine cycle to produce mechanical energy and electricity. Figure 3 illustrates the structure and average heat losses to the ground and atmosphere.

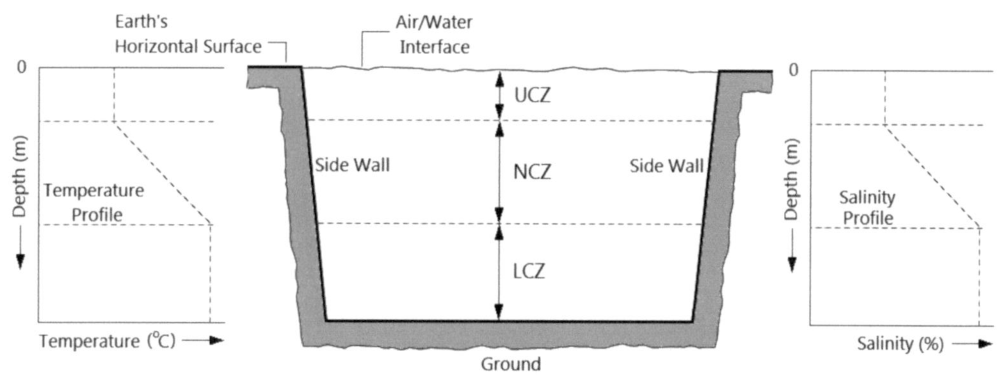

Figure 3. Cross section of the salt gradient solar pond [17]

4.2. Heat pipe systems

A heat pipe is essentially a passive heat transfer device with an extremely high effective thermal conductivity which allows a two-phase heat transfer mechanism resulting in enormous heat transfer capabilities nearly one thousand times that of an equivalent copper piece. The heat pipe in its simplest configuration is a closed, evacuated cylindrical vessel with the internal walls lined with a capillary structure called wick that is saturated with a working fluid.

A heat pipe has three regions namely; evaporator, condenser and adiabatic region. Heat is absorbed through the evaporator part of the heat pipe and transferred into the working fluid to vaporize some fluid. Vaporizing fluid pushes the vapour above towards the condenser part which is always above the evaporator region. The latent heat of evaporation contained in the vapour is transferred to the relatively cold surface of condenser causing the vapour to condensate on the surface of the condenser covered by a porous liner called wick, which serves as a passive pump to draw the fluid back to the evaporator by capillary effect. Then the heat is provided by the condenser to a fluid or gas through its surface. The middle section of heat pipe is called adiabatic region in which vapour travel from the evaporator to the condenser without any significant heat transfer to the pipe wall. Heat pipes can be designed to operate in evacuated tube collector, flat plate collector or directly at different working conditions and temperatures. Heat pipes needs to be installed at a minimum tilt angle of 25° to provide the backflow of the working fluid inside the heat pipe from condenser to evaporator.

4.3. Solar collectors

Solar water heating (SWH) collectors are heat traps that absorb solar energy and transfer the heat into another medium. Changing the shape, design and materials, they have three major parts in common integrated to each other. The first and the most important part is collector which is exposed to solar radiation at an optimum inclination angle allowing to take maximum radiation throughout the solar season. The second part is a transfer medium that transfers the heat collected by the collector to the third part, storage tank. This medium should be fluid and generally made of water-glycol mixture. Storage tank, the third main part, is a simple heat exchanger (liquid-to-liquid) like collector (gas-to-liquid), which transfers the heat from the transfer fluid to the water to be heated and used. The transfer fluids have to be circulated in order to carry the heat continuously coming into the collector as solar radiation. This circulation is accomplished by mainly two ways:

- *Natural circulation* which is driven by the difference between the relative gravities of hot fluid in the collector and relatively cold fluid which gave its heat to the water in the storage tank. This type of design requires the storage tank to be located higher than the collector.
- *Forced circulation* method employs a water pump to circulate the fluid in a closed cycle instead of letting it to circulate slowly by means of the small difference between the gravities of hot water in the collector and warm water in the storage tank.

There are many types of solar water heating (SWH) systems. Stationary type solar water heaters include flat plate collectors (FPC), evacuated tube collectors (ETC) and compound parabolic collectors (CPC). FPC and ETC are widely used for heating domestic use water. These collectors convert solar radiation directly and indirectly into thermal energy. ETCs have a higher efficiency than FPCs but they cost much more than FPCs.

4.3.1. Flat plate collector

Flat plate collectors are the most widely used solar systems today. They are made of three main parts. An insulated collector case holds the tubes, metal plate and glass cover. A sheet metal plate of aluminium, copper or steel can be painted or coated black. Metal tubes (usually aluminium or copper) that bonded onto the metal plate so as to provide good heat conduction, and are all connected to a common tube at both ends called header tubes. Bottom side of the collector case is well insulated to minimize thermal loses and the top side which is exposed to the sunlight is covered with a glass tightly to ensure a high level of greenhouse heating inside the collector. The pipes and copper are enclosed in an insulated metal frame, and topped with a sheet of glass (glazing) to protect the absorber plate and create an insulating air space.Figure 4. shows a cross section of a flat plate collector and solar water heaters with natural circulation and heat pipe.

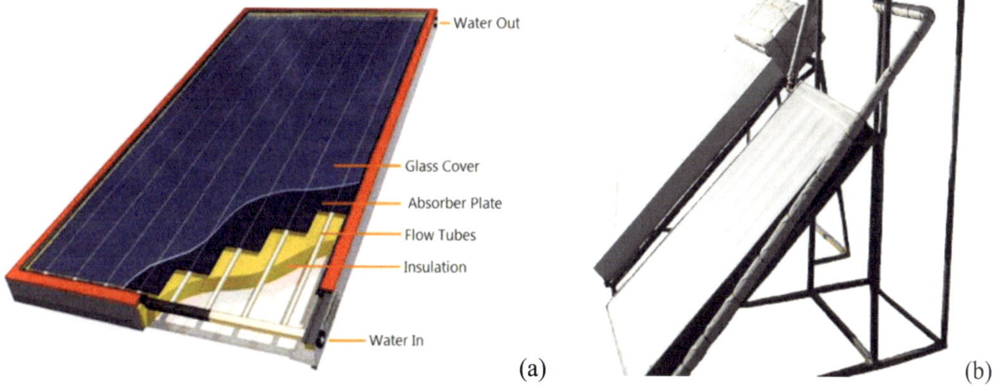

Figure 4. (a) Cross section of the flat plate solar collector [18], (b) solar water heaters with natural circulation and heat pipe [19]

Sunlight falling onto the collector surface passes through the glass cover and hits to the black plate and tubes inside the collector casing. The heat absorbed by the tubes and the plate is transferred to the fluid circulating inside the tubes. The fluid which can either be a working fluid which circulates in a closed loop between the collector and storage tank where it transfers its energy to the water that will be used, or running water can be directly routed through the collector tubes. The circulation of the water inside the collectors can be driven by the difference between the specific gravity of heated water inside the collector and cooled water inside the storage tank, which is called natural circulation. This type of collector has some installation requirements such as installing the storage tank higher than the collector to provide cold water to flow downward replacing the hot water. Forced circulation on the other hand makes it possible to install the storage tank to almost any place lower than the collector which allows more aesthetic and convenient designs on the roof.

4.3.2. Evacuated tube collector

Evacuated-tube collector is a later generation of flat plate collectors which was first seen in 1970s. Main difference of evacuated tube from the flat plate collector is that it employs a vacuumed glass tube with an absorber inner surface. Vacuum layer serves as insulation much more superior than the air trapped between the glass cover and absorber plate in the flat plate collectors. Evacuated tube collectors are mainly two types as direct flow and heat piped system. Using a reflector improves the heat absorption performance of the collector.

Heat pipe evacuated-tube collectors use a copper heat pipe attached to an absorber plate and a vacuum tube. Tubes can be changed one by one without dismantling and emptying the whole system which makes it easy to perform installation, maintenance and repair tasks more easily. Some heat pipe collectors have overheat protection system that blocks the way fluid flows from the condenser to the evaporator region by a temperature triggered spring.

4.4. Concentrating solar systems

Concentrating solar systems are mainly used for power generation. They concentrate the solar energy to a point or a line at which heat energy is collected at medium or high temperatures depending on the type of the system and used for power generation in a conventional heat-driven power plant. Installed global capacity of concentrating solar thermal power plants according to years are 0.4 GW in 2004, 2.5 GW in 2012 and 3.4 GW in 2013, clearly showing an increasing trend [20]. Most popular types of concentrating solar power technologies include Linear Fresnel, dish, parabolic trough and solar towers Figure5.

Parabolic trough collector technology (PTC) is just like a parabolic semi-pipe which is cut off longitudinally and oriented from north to south. The parabolic reflective surface made of a polished metal or mirror concentrates the solar radiation onto a single focal line at which a tube is located to contain a thermal fluid for energy absorption and transfer to an associated plant that employs a heat engine to generate electricity. Parabolic trough is equipped with a solar tracking mechanism which allows the focal line hold on the pipe throughout the daytime. The temperature of the thermal fluid inside the tube (thermal oil, pressurized water or molten salt) can rise up to 400, 500 and 550°C, respectively. The hot working fluid can be used in Rankine cycle to produce mechanical energy or electricity. Pressurized water is useful for producing steam.

Linear fresnel collector technology (LFC) is another example of line focus technique like parabolic trough technology. It costs less than parabolic trough which is because of requiring

http://dx.doi.org/10.5772/60436

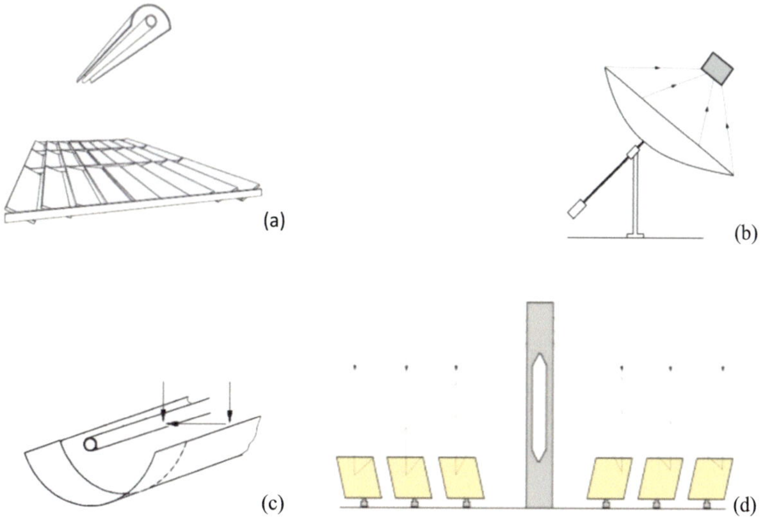

Figure 5. Concentrating solar power technologies (a) Linear fresnel collector, (b) dish collector, (c) parabolic trough collector and (d) solar towers[21]

a lighter structure for flat reflectors. The receiver is also fixed and does not rotate with the reflector assembly which is tracking the sun from morning to evening. Only moving part of this system is the reflector units which lay longitudinally at north south direction. These modular units are rotated throughout day at the exact angle to keep the focal line on the receiver pipe located above the central line. Cosine loss is a drawback of this technology, caused by the intervention of module sides as to shade the solar light on the next mirror especially in the morning and afternoon. Flat mirror modules are not capable of focusing light as good as a parabolic mirror. Receiver located above the mirrors' plane shades onto the mirrors except for the noon time. There are several options of heat transfer fluid in this system as in parabolic trough, but water is widely used because linear Fresnel collectors are well suited to produce steam. Collectors are able to produce steam at 250°C temperature and 50 atm. pressure directly, without using exchanger.

Tower solar power technology (TSP) is mainly a combination of a central receiver mounted on top of a tower and many mirrors distributed around the tower on the ground as to form arrays of sun-tracking mirrors which reflect the solar irradiation to the receiver unit on the tower. These mirrors called heliostat can track the sun at two axis. Solar towers can reach high temperature concentrations since it is a point focus technology instead of linear focusing. The heat is absorbed by transfer fluid similar to above-mentioned solar systems. Water can be directly converted to superheated steam and used in a Rankine heat engine and to power an electric generator. Another advantage of the tower technology compared to parabolic trough collector is its ability to operate with various heat transfer fluids such as molten salt, open air, superheated steam, and pressurized air. A brayton cycle can also be driven by hot pressurized air.

Solar dish collector technology (SDC) is a unified version solar towers which hold the parabolic mirrors and a Rankine or Sterling engine attached to the receiver that is located at the focal point of the mirrors to utilize the concentrated heat. The mirror and engine assemble is mounted on a single body equipped with two axis solar tracking mechanism. Focal concentration ability of solar dish makes it possible to achieve as high as 1000°C of receiver temperature. Attaching the engine directly to the receiver plate eliminates loses during the transfer of heat from the receiver to the generator which makes solar dishes more efficient than the other systems. On the other hand, it is not so easy to integrate solar dishes with energy storage systems and other energy sources.

Among these concentrating solar technologies, parabolic trough is the most widely used technique in the world today. PTC and TSP technologies are able to store heat more than 10 hours through direct or indirect energy storage systems. There are 76 concentrating solar power projects in the world with 2.88 GWe of total capacity. Although PTC plants are very dominant among the currently operational solar power plants (95.7%), under construction projects will increase the ratio of LFC from 2.07% to 5.74%, TSP from 2.24% to 20.82% and SDC from 0% to 0.052%, decreasing the ration of PTC from 95.7% to 71.43% [22].

4.5. Photo-voltaics

Photovoltaic (PV), as its name implies, is an extraordinary phenomenon that converts light to instantly ready direct current. Semiconductor materials inherently have this physical property and are easily used in production of PV cells. PV cells have two or more layers of semiconducting material, commonly silicon. When the photons in sunlight hits onto this semiconductor layers electrical charge is generated and this charge can be harvested by using metal contacts resulting in DC current. The smallest unit of this arrangement is called solar cell. Solar cells have a very small output capacity but they can easily be connected to each other to form a bunched structure called PV panel. PV panels can also be connected to each other in any size and number to produce a desired power output. PV panels have no emission, no noise and no moving parts. Also, their installation and maintenance tasks are very easy [23]. The main drawback of PV panels is their cost which is decreasing rapidly. Affordable prices boosted the use of PV panels in recent years. Global installed capacity of PV plants was 2.6 GW in 2004, increased to 100 GW in 2012 and become 139 GW by the year 2013 [20].

5. Solar desalination

Desalination of sea water or other salty ground waters is a practical and proven method of producing fresh water where it is needed. The main issue for this desalination process is a low cost, environment friendly, readily available energy to drive the process. Solar energy is one of the best sources of this type and it is abundant throughout the year especially in solar belt region at which most of the water scarcity is suffered by about 5 billion people. Solar water desalination is a well-known and proven technique which has been used for a long time at remote areas and places suffering from shortage of potable quality water. There are many variations of solar driven desalination systems. Figure 6 shows a pull classification and integrated big picture view of desalination processes and the place of renewable energy among

http://dx.doi.org/10.5772/60436

the other methods. These systems can be classified mainly into two groups as direct and indirect desalination systems which will be described below.

5.1. Indirect systems

Most of the large solar desalination plants are driven by indirect solar energy. Indirect solar desalination systems can be classified into thermal, mechanical or electric driven technologies. Solar energy is collected through concentrating (PTC, LFC, TSC, SDC) or non-concentrating (FPCs, HPC, SP) collectors to run thermal desalination processes such as MSF, MED, thermal vapour compression (TVC) and membrane desalination (MD). Another indirect use of solar energy in desalination system is producing electricity from solar irradiation via PV panels and use to run ED which is the only desalination technology using electricity directly to produce fresh water. RO and freezing desalination techniques require mechanical energy which can be obtained from solar energy through heat engines (Rankine, sterling and brayton) or PV panels. Figure7 shows the shares of desalination technologies in indirect solar desalination plants installed worldwide.

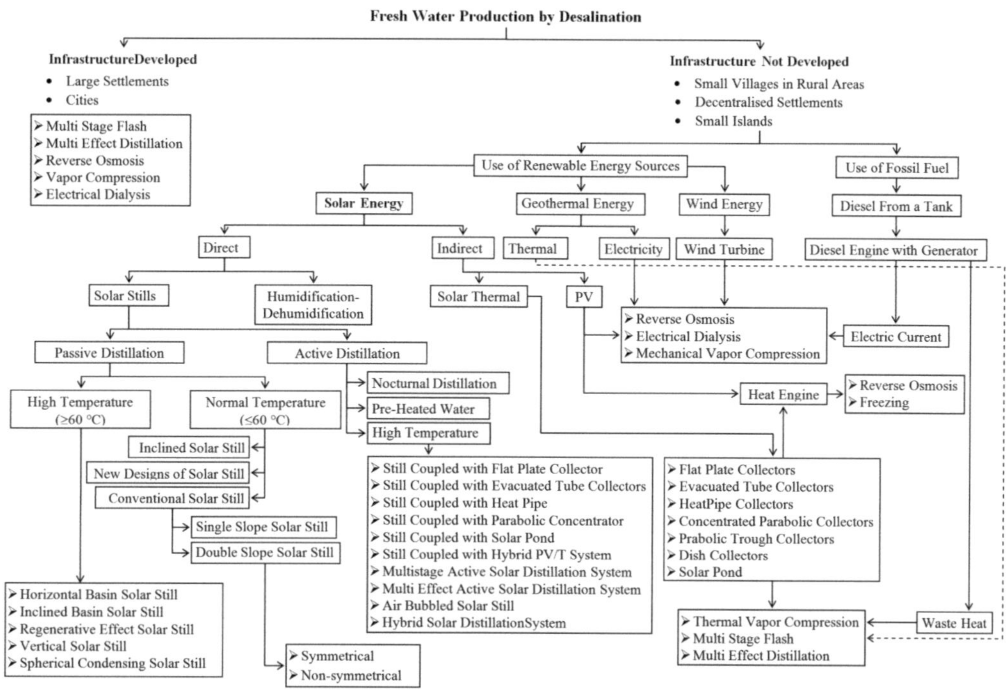

Figure 6. Desalination techniques used for fresh water production [1, 2, 12, 24]

Reverse osmosis (RO), which is the dominant indirect solar desalination techniques (about 52%) has the potential to improve the sustainability of desalination process by replacing solar energy with fossil fuels and reducing operational cost significantly [24]. It is a pressure driven process which forces the salt water through a semi permeable membrane where concentrated brine is separated from the feed water producing fresh water as the output of the membrane. RO membranes can separate more than 98% of the salt contained in the sea water. Required feed water pressure for brackish and seawater is 10–15 bar and 55–65 bar, respectively. Typical RO desalination systems can recover 45–50% of seawater and 90% of brackish. Membrane which is the core element of RO process losses its performance due to fouling and scaling [25].

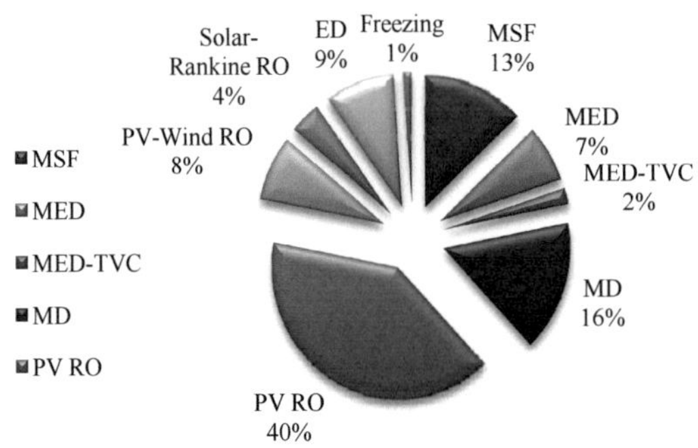

Figure 7. Shares of desalination technologies in indirect solar desalination plants installed worldwide [24]

Pressurized feed water required for RO process can be provided by using either sterling or Rankine engine using solar heat, or using electric motor powered by PV panels. Since the usability of solar energy depends on season or weather conditions, sustainable and continuous production of fresh water requires taking some additional measures like energy storage such as thermal energy storage and battery, or hybridization with other energy sources like wind, geothermal and diesel, or with another desalination method combined with RO. Although the pre-treatment of water before the RO membrane reduces the energy efficiency, it is still more efficient than phase change thermal processes. Separated brine water with high pressure can be rejected after passing through a pressure exchanger to recover some of the wasted energy. Figure8illustrates a basic RO system powered by a solar-heated Rankine turbine. The waste heat of the cycle may be reused to preheat the feed water.

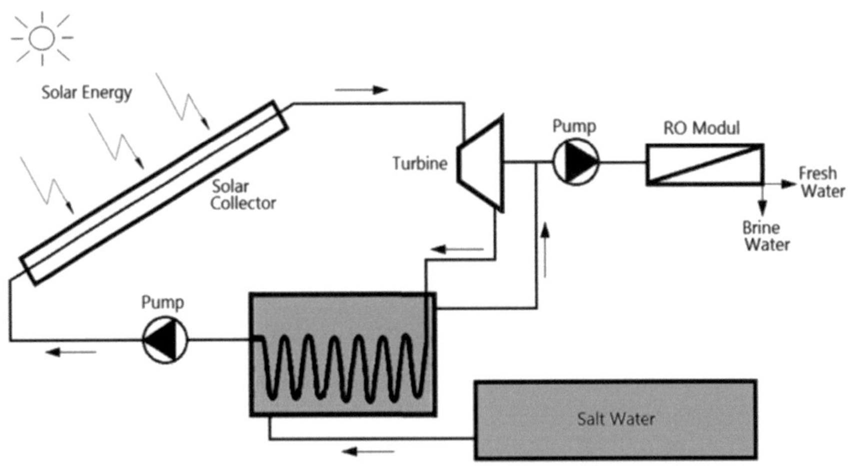

Figure 8. Basic diagram of a reverse osmosis system powered by a solar-heated Rankine cycle [26]

http://dx.doi.org/10.5772/60436

Multiple stage flash distillation (MSF) system has a number of adjacent vessels with an internal heat exchanger and collector for condensed water. Each of these vessels is called stage. Stages have their own inside pressure from high to low in order. Different pressure in each vessel means different boiling points with a decrement of 2–5°C from high to low. Figure 9 shows a schematic view of one-stage and two-stage flash distillation systems. Operation of an MSF plant at brine temperatures as high as possible theoretically increases the efficiency of the plant. However, avoiding scale formation and accelerated corrosion of metal surfaces in contact with seawater require limiting the top brine temperature at about 120°C.

As shown in Figure 9, cold salt water travels through the stages from cold side to hot side absorbing heat inside the vessel to cause condensation of steam and formation of distilled water. The vessels here serve as an evaporator as the brine evaporates inside. Although higher number of stages increases efficiency, it also increases the installation costs. Therefore, there are about 19–28 stages in modern large MSF plants. Preheated salt water exits the hottest (or the first) stage and enters to the collector to absorb additional heat and enters to the first stage vessel. The first stage vessel is adjusted to a certain pressure that the entering hot brine is over the boiling temperature for that pressure. Therefore, a portion of the incoming brine water suddenly evaporates which is called the "flash". Steam produced by this flash hits to the condenser above and becomes liquid condensate which drops on a fresh water collector and taken out as fresh water through a controlled valve. Demister is used to trap the water particles that may burst up during the flash and mix with the fresh water [27].

MSF plants can be integrated to any heat sources including solar concentrating (PTC, LFC, TSC, SDC) collectors, solar pond and flat plate, evacuated and heat piped collectors and any type of waste heat at moderate temperatures (from a steam or gas turbine power plant etc.).

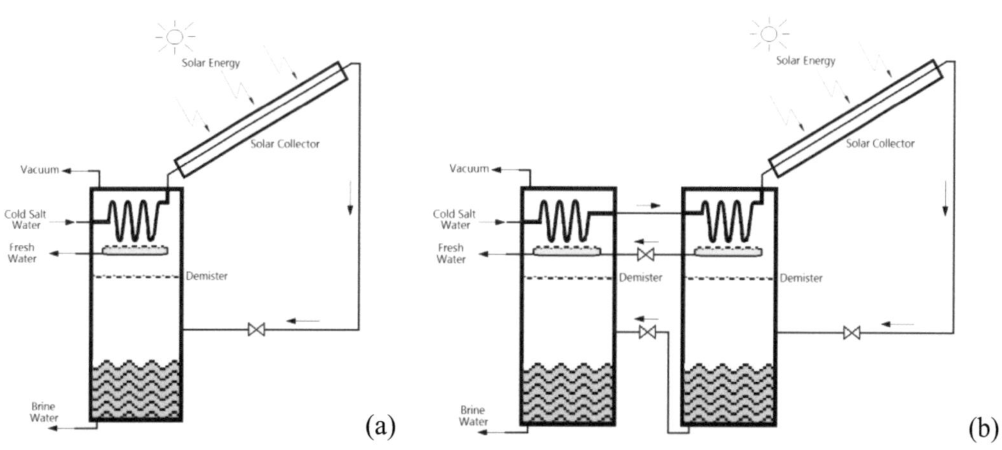

Figure 9. Schematic view of (a) one stage flash distillation, (b) two stage flash distillation [27]

Multi effect distillation (MED) units are a practical and promising way of water distillation because of its ability to use renewable energy (solar, wind, geothermal, etc.) and reuse low-grade waste heat from any source (from a steam or gas turbine power plant, etc.). Solar-assisted

MED process consumes both thermal energy (thermal vapour compression) and mechanical energy (mechanic vapour compression) to produce distilled water [28]. Figure 10 shows the typical arrangement of an MED with thermo compression (MED-TVC). Steam is produced using a thermal energy source and ejected through a thermo-compressor into a distillation cell drawing some vapour from the last cell of MED-TVC system. In solar-assisted MED desalination systems heat exchangers are widely used, and circular tubes are the most commonly adopted heat transfer elements. The heat and mass transfer processes play important roles, which usually lead to bulky horizontal or vertical tube arrays heat exchangers. Salt water is sprayed onto these exchangers in each cell. Hot steam passes through the heat exchanger in the first cell and condenses to become distilled water. Latent heat of condensing water is transferred to the sprayed salt water and some of this water evaporates while the rest accumulates at the bottom of the cell. Vapour produced in the first cell is transferred into the heat exchanger of the next cell and transfers its energy to the salt water sprayed on the exchanger of second cell just like in the first cell. Same process repeated until the last cell and the vapour produced in the last cell is sucked by thermo-compressor to recycle the vapour.

MED system is also able to operate by mechanical vapour compression (MED-MVC) when there is no usable heat. A mechanical compressor sucks the vapour from the last cell producing vacuum which promotes evaporation and compresses the vapour before sending to the heat exchanger of first cell with elevated temperature caused by the compression which also increases the evaporation rate in the first cell. MED-MVC system operates similar to MED-TVC cycle except for the mechanical compressor.

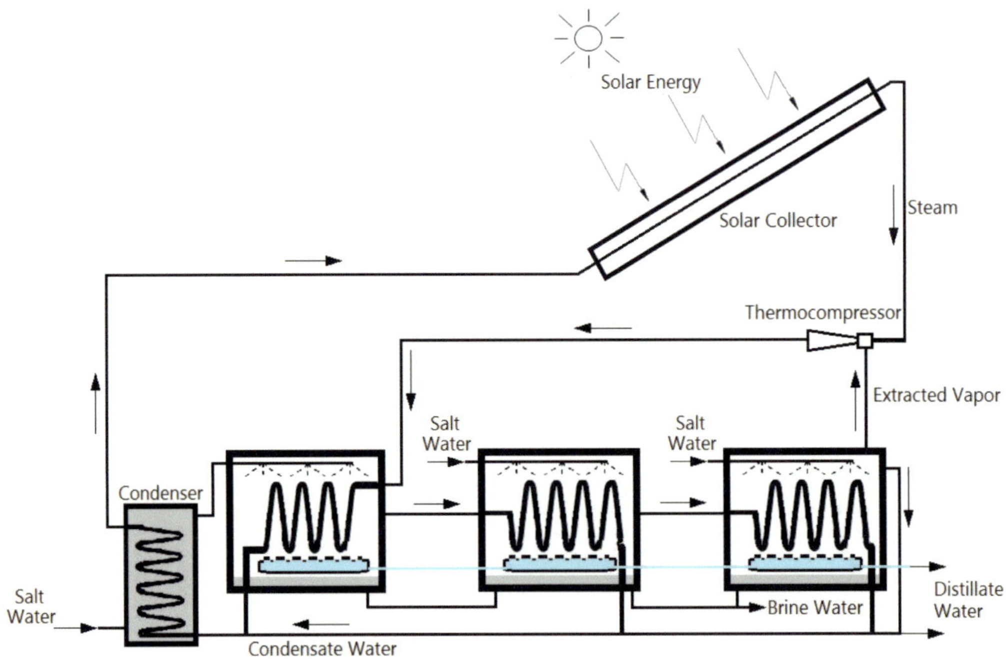

Figure 10. Typical arrangement of a multiple effect solar distillation with thermo-compression (MED-TVC) [28]

5.2. Direct system applications

Direct solar desalination methods make use of the heat energy contained in the solar irradiation directly to produce fresh water without association with any other mechanical or electrical devices. Direct systems are low cost and suitable for small applications. Since their operation temperature and steam pressure is low, they have smaller production rates than indirect desalination systems. There are mainly two types of direct desalination technique such as humidification-dehumidification method and solar stills. Solar stills have also two variants as active and passive distillation methods.

5.2.1. Solar Humidification-Dehumidification (HD-DHD)

A gigantic scale HD-DHD method is used in the nature for millions of years to produce fresh water from the seas and oceans using the sun as the heat source [29]. As shown in Figure 1, water evaporates and humidifies the above air. Air flows in the atmosphere carries the vapour to where it will condense and dehumidify to form fresh precipitation called rain, snow or hail. HD-DHD distillation method is the small scale replication of this process.

The productivity of HD-DHD system is about five times the productivity of an equivalent basin type solar still at the same climatic conditions. HD-DHD process is also named as the multiple-effect humidification-dehumidification process; multiple-effect humidification (MEH) or solar multistage condensation evaporation cycle (SMCEC).

In HD-DHD system atmospheric air is heated through a solar air heater. Because the water vapour holding capacity of air increases by the temperature (about 100 gr vapour/ kg dry air at 60°C and 500 gr vapour/kg dry air at 80°C) [30]. Warm dry air enters in to the humidification chamber and absorbs vapour from sprayed salt water. It passes through a pipe into dehumidification in which cold salt water passes through another pipeline which acts as a condenser for incoming warm humid air. Thus the salt water is preheated by the heat, recovered from condensing vapour, to utilize evaporation in the humidification chamber and the warm air is dehumidified leaving fresh water at the bottom of dehumidification chamber. Figure 11 shows the schematic illustration of HD-DHD system. Building and operating an HD-DHD system is simple safe and low-cost making it a very suitable desalination process among the small capacity plants [31].

5.2.2. Solar stills

Solar stills can be used to produce fresh water from salt water in a very cheap, simple and easy way [32]. They are preferred for small-scale fresh water needs of people in remote places instead of transporting fresh water. A very fortunate aspect of the solar water distillation technique is that when fresh water demand is at its peak, solar insolation is also high (especially hot seasons) [6].

Solar irradiation passes through a cover and falls onto the black bottom surface of the still causing the surface and contained salt or brackish water to warm up. Heated water evaporates and rises up until it touches to the inner side of the cover where it condenses and forms fresh

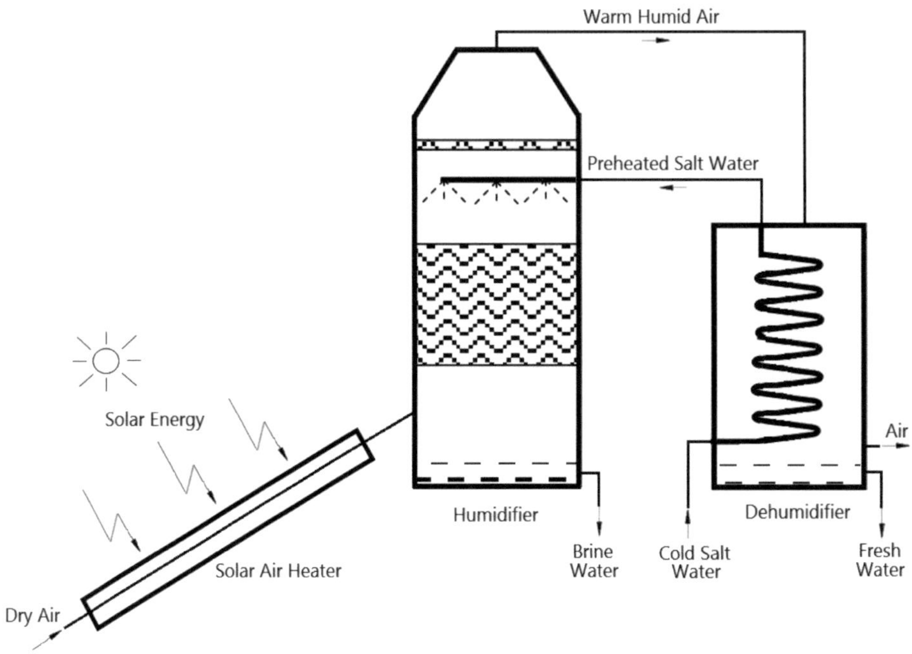

Figure 11. Schematic diagram of humidification-dehumidification system [29]

water drops. Since the cover of the still is designed with a tilt angle, these drops are moved by gravitational forces towards the distilled water collecting channels. Figure 12 shows the schematic view of a double-slope symmetrical basin still (also known as, roof type or green-house type). The weak point of solar stills is the significant amount of heat loses because its large surfaces are in contact with the ground and air.

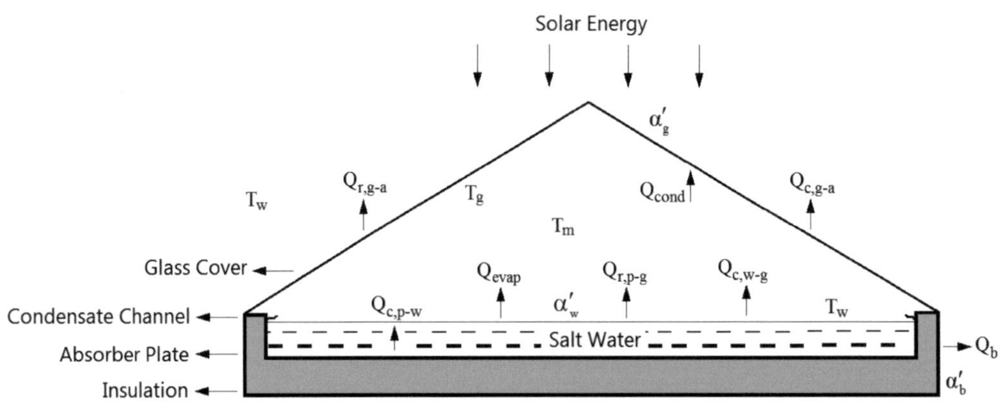

Figure 12. Distribution of the solar energy falling on a double slope symmetrical basin still [33, 34]

Ground side can be insulated to some extent. However, the upper side which has to be exposed to solar irradiation cannot be insulated and there is serious amount of heat lose through radiation, convection and condensation of vapour on the cover surface. The performance of a solar still is closely related to the thermo physical properties of the material to be used in the still, tilt angle of cover, spacing between cover and water surface, insulation, vapour tightness and absorbance-transmittance properties of still, etc. as well as operating parameters such as water depth in the basin, initial water temperature, water salinity, etc.

Figure 12 shows the distribution of the solar energy falling on a basin still system. T_a, T_b, T_g and T_w in the figure are ambient temperature, basin temperature, glass temperature and water temperature, and α'_b, α'_g and α'_g are; the solar fluxes absorbed by the basin liner, glass cover and the water mass respectively.

h_{wg}, h_{ga} and h_{wb} are the heat transfer coefficients from the water surface to glass, from the glass to the environment and from the water to basin liner respectively, given by Tiwari [35]. h_{ew} is the coefficient of heat loss by evaporation from water surface, P_g is the glass saturated partial pressure, and P_w is the water saturated partial pressure [35, 36].

$$h_{wg}=8.71+h_{ew} \tag{1}$$

$$h_{ew}=4.0\frac{p_w-p_g}{T_w-T_g} \tag{2}$$

$$\begin{aligned} P_g &= e^{\left(25.317-\frac{5144}{T_g}\right)} \quad \text{a} \\ P_w &= e^{\left(25.317-\frac{5144}{T_w}\right)} \quad \text{b} \end{aligned} \tag{3}$$

$$h_{ga}=5.7+3.8V \tag{4}$$

$$h_{wb}\approx 130\ \text{W/m}^2{}^\circ\text{C} \tag{5}$$

Energy balance

Following Kumar et al. [37], the energy balance equations for different components of an active solar still are given as follows:

Glass cover

Sum of the radiation absorbed by the glass surface and the heat transferred from the water to glass surface is equal to the heat transferred from glass surface to the ambient.

$$\alpha'_g I_{eff} + h_{wg}\left(T_w - T_g\right) = h_{ga}\left(T_g - T_a\right) \tag{6}$$

Water mass

Sum of the usable energy coming from the collector, the radiation absorbed by the water mass and the heat transferred from the basin liner (glass cover) to the water equals to the sum of heat stored in the water and the heat transferred from the water surface to the glass surface.

$$\dot{Q}_u + \alpha'_w\left(1-\alpha'_g\right)I_{eff} + h_{wb}\left(T_b - T_w\right) = \left(MC\right)_w \frac{dT_w}{dt} + h_{wg}\left(T_w - T_g\right) \tag{7}$$

Basin liner

Solar radiation absorbed by the basin liner is equal to the sum of the heat transferred to the water by convection and the heat transferred from the surface to the ambient.

$$\alpha'_b\left(1-\alpha'_g\right)\left(1-\alpha'_w\right)I_{eff} = h_{wb}\left(T_b - T_w\right) + h_b\left(T_b - T_a\right) \tag{8}$$

If the rate of useful energy coming from the collector (W) is zero ($\dot{Q}_u$=0), Equation 7 becomes energy balance equation for a passive solar still. Evaporative heat transfer correlation is given as follows:

$$Q_{ew} = h_{ew}\left(T_w - T_g\right) \tag{9}$$

Hourly output of still is [35, 38]

$$\dot{m}_{ew} = \frac{h_{ew}\left(T_w - T_g\right)}{L}.3600 \tag{10}$$

The efficiencies were calculated by the following equation [35, 39]:

$$\eta(\%) = \frac{Q_{ew}}{I_{eff}.A} \tag{11}$$

Solar stills are mainly of two types according to their operation modes and modifications as active or passive solar stills. Active solar stills typically use a secondary external heat sources such as; collector/concentrator panel, solar pond, hybrid PV/T systems, waste thermal energy from any chemical/industrial plant, etc. If there is no supplementary external heat source, the system is called a passive solar still [40].

5.2.2.1. Passive distillation

Passive distillation systems are divided into two groups such as high temperature (≥60°C) and normal temperature (≤60°C) distillation systems. High temperature passive distillation systems are horizontal basin still, inclined basin solar still, regenerative effect solar still, vertical solar still and spherical condensing solar still. Normal temperature passive distillation systems are inclined solar still, new designs of solar still and conventional solar still. Basin type is the most widely used solar stills today. Basin type solar stills have been modified into several types according to their cover designs such as; single slope, double slope, V type and hemispherical as shown in Figure 13. Average distillate production rate of a standard single-basin still is between 0.005 and 0.011 m^3 m^{-2} day^{-1} depending on the insulation quality [32].

Different designs of basin type solar stills have been developed and tried to find an optimum solar still which; can be transported to the site and assembled easily, does not require rare materials that cannot be found easily, has an acceptable service life, can operate by itself without any need for external power support, has a rainfall collecting facility and does not pollute or contaminate the fresh water and of course with low cost.

The basic components of a solar still are briefly described below:

Glazing should transmit the solar irradiation to coming inside the still and resist to thermal radiation going outside. The glazing also needs to be abrasion resistant and hydrophilic. Readily available, easy to handle and assemble glazing material is preferable. Commonly used materials are glass or treated plastic.

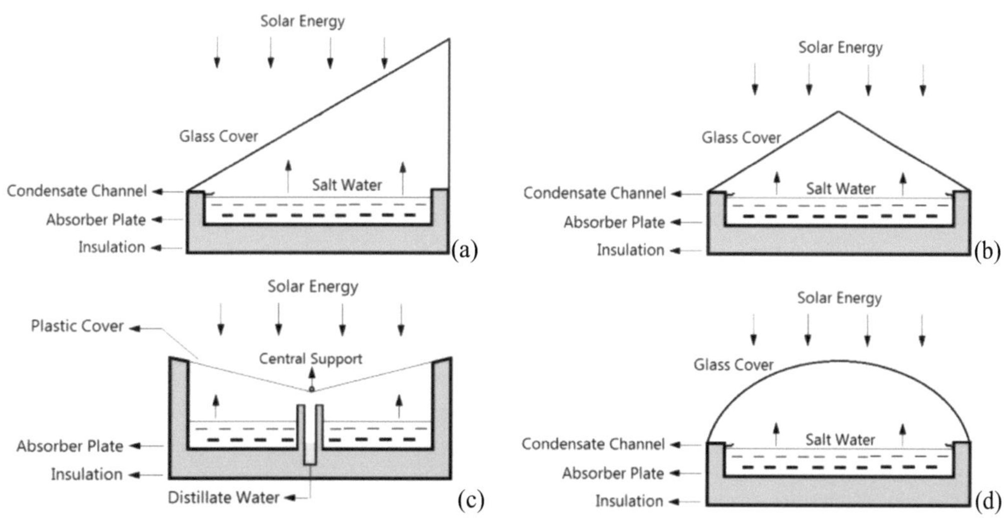

Figure 13. Common design of solar stills: (a). single-slope basin still, (b). double-slope basin still, (c). V-type solar still, (d). Hemispherical type solar still [41]

Liner is used to absorb the solar irradiation and give the heat to the water. Since it is in contact with warm salt water and basin tray, it should be impermeable to warm salt water, durable and easily cleanable. Preferred materials are asphalt matt, black butyl rubber, black polyethylene etc.

Sealant should be easy to apply, durable and low cost. Common materials are putty, tars, tapes silicon and sealant.

Basin tray forms the main base of the system. Therefore it should have long life high level of corrosion resistance and low cost. Preferred materials are wood, galvanized iron, steel, aluminium, asbestos cement, masonry bricks, concrete, etc.

Condensate channel is the channel through which condensed fresh water is collected and directed to distillate water tank. *Preferred materials are* aluminium galvanized iron, concrete, plastic material, etc.

Side walls make the still robust and rigid along with providing thermal resistance against the heat transfer from the heated salt water in the basin to the outside. It should be made of a material that can hold the top cover without any failure for long years. A low thermal conductivity is also a key property of side walls. Mostly preferred side wall materials include wood, concrete, reinforced plastic, etc.

Integrating a separate external condenser with the still as in Figure 14 decreases the convective heat loss through the still cover and provides an effective heat sink for the condensing vapour which increases the distillate yield by about 50–70%.Some of the evaporating water condenses on the cover surface and a fraction of the vapour passes to the condenser chamber by the effect of pressure difference causing the pressure in the still chamber to drop. Lowered pressure decreases the formation rate and number of vapour droplets on the inside surface of the still cover which allows more solar radiation to reach the water in the still basin and improves evaporation.

Using nano fluids along with external condenser is another contemporary method to further increase the productivity by about 115% [42]. External condenser accumulates the latent heat of condensing vapour which can be used to preheat salt water before entering the still or to prolong the distillation process during night hours. Using external condenser makes it possible to use a cover with very low inclination Figure 14a [43].

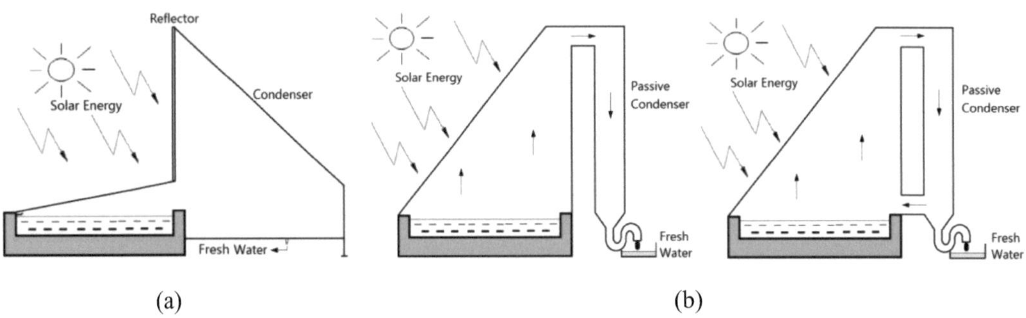

Figure 14. (a) Schematics of the solar still with minimum inclination, coupled to an outside condenser [44], (b) solar still with passive condenser [45]

Conventional basin type solar stills have significant disadvantages; horizontal water surface inevitably causes cosine loses especially at higher latitudes and large thermal capacity of the water in still basin limits fresh water output. Many researchers and new designs have been done to minimize or overcome these drawbacks of solar stills [46].

Stills with inclined absorber surfaces are reported to have significantly higher productivity compared to basin-type stills. In an inclined still, water flows from the top to the bottom of the absorber surface. To maintain uniform thickness of water, a wick is used to draw water by capillary effect. The productivity of a solar still is affected by the temperature difference between the water and condensing surfaces. A higher temperature difference between these surfaces yields higher productivity. To maintain this temperature difference, various methods were proposed [47].

In inclined stills feed water coming from the tank comes to the collector in pipes, passes through the drilled holes and drops onto the black absorber plate to evaporate by taking heat from the plate which is heated by solar irradiation. Vapour goes up and touches to the glass cover which is cool enough to condense on. Fresh water distillate accumulates on the inner surface of the glass cover and flows downwards to the condensate outlet port by gravitational forces.

Water droplet falling onto the absorber plate does not distribute perfectly on the absorber plate surface. Using a wick on the absorber plate helps to distribute water more evenly over the absorber plate using capillary effect which improves evaporation rate. Another way of improving the performance is to cool glass cover to ease condensation. Shaded plate is a simple yet effective solution (Figure 15). One fourth of the glass surface is shaded from the top leaving a gap of several centimetres between the shading plate and the glass. This arrangement provides a chimney effect in this gap and improves convective heat transfer to the atmosphere which cools down this part of the glass and increases the condensate production rate [47].

A good way to solve the cosine loss problem of basin type solar still is to design an inclined structure with cascaded weirs (Figure 16) [48]. Salt water is fed from the top and condensate is collected from the bottom end. Feed water flows through the weirs and fills all the weirs evenly. There is a small distance between the cover and absorber plate which quickens the saturation and condensation processes making the cascade system more efficient more than other solar stills.

Weir-type cascade solar stills do not suffer from dry spot or channelization problems since the brine is forced to flow each step one by one without leaving any dry surface on the absorber plate. Water flow way is longer than a normal or wicked inclined type stills and accordingly the solar exposition time is longer which increases the efficiency. It has the advantages of both basin type and inclined type solar stills. Further development of these weir-type cascade stills include using wick on each cascaded steps and phase change material (typically paraffin wax) beneath the absorber surface to store energy when it is abundant and give it back to the salt water when it is needed in cloudy days or evening times [48].

Another well-designed still is a combination of a glass cylinder and a tray or trough inside the cylinder (Figure 17a). Salt water is fed into the tray and the water travels through the tray. Incoming solar energy heats the tray and water to cause evaporation of salt water and consequently condensation of the vapour on the cylinder surface. Water droplets slip down

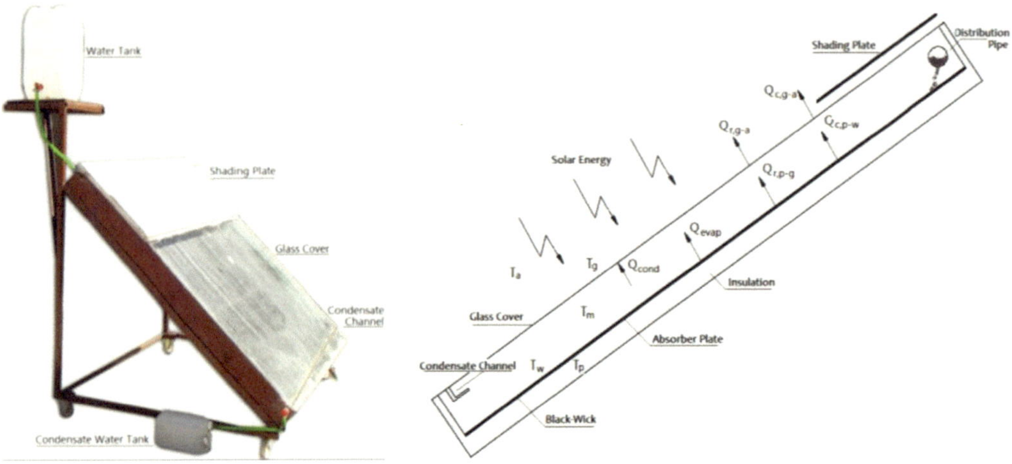

Figure 15. Inclined solar water distillation system and schematic diagram of the system [47]

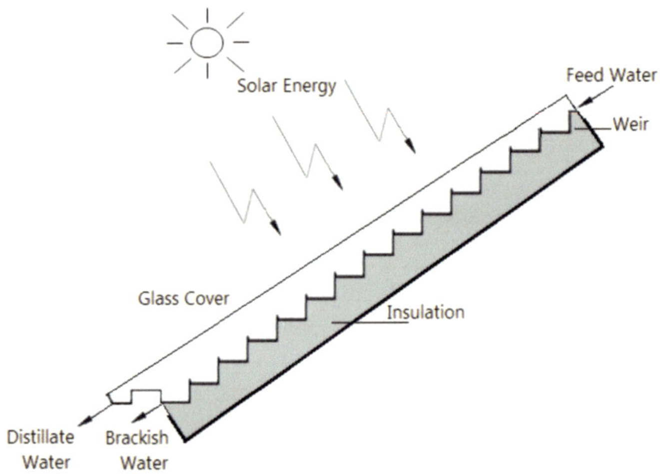

Figure 16. Schematic view of a weir type cascade solar still [49]

and accumulate at the bottom of glass cylinder and collected through the fresh water outlet. Cylindrical tube type stills are compact, robust and have high yield per unit are compared to the conventional basin still design [50]. There is a similar design in which wick is used in the tray to absorb salt water and diffuse throughout the tray with capillary effect. Wick lies along an incline, with the upper edge dipped in a saline water reservoir and there are two outlets, one for the excess water and the other is for condensate. Capillary suction of the cloth fibers used as wick produces a thin water film which can easily evaporate by the incoming solar radiation. The condensing surface area of the cylindrical glass over the evaporation tray is much more than that of a flat surface and this results in a relatively colder glass cover faster condensation rate [46].

http://dx.doi.org/10.5772/60436

A similar approach uses a sphere instead of cylinder as the still housing. A black painted of covered metallic plate is located at the centre plane of the spherical glass (Figure 17b). Spherical solar stills works like cylindrical ones and they are about 30% more efficient than an equivalent conventional solar still. Spherical stills have even more condensation area per evaporation surface compared to cylindrical solar stills but it is not scalable as easy as cylindrical ones [46].

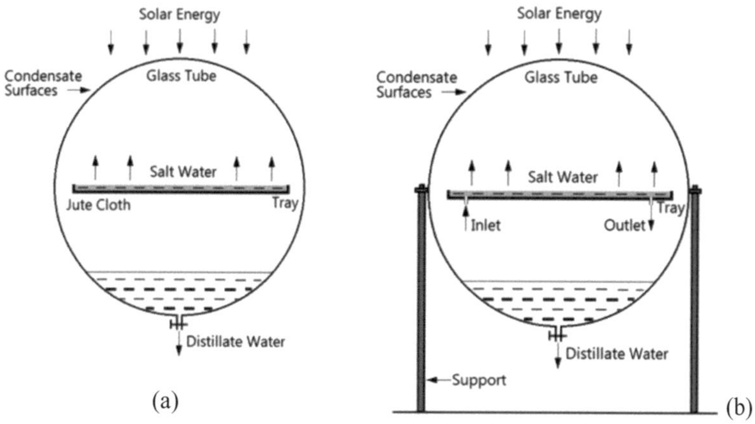

Figure 17. Schematic representation of (a) tubular solar still (front view) [51], (b) a spherical condensing solar still [46]

5.2.2.2. Active distillation

Many investigations have been conducted in attempt to improve the efficiency and productivity of solar stills. Some of these techniques are decreasing the depth of water in the basin, mixing black dye with the salt water, using better insulation to minimize the heat losses, improving the vapour tightness, proper orientation of the still as to receive more solar irradiation, etc.

Apart from the above-mentioned passive methods, there are a number of active methods of improving thermal efficiency such as integrating a still with a solar heater of concentrator. Active solar stills receive additional thermal energy from an outer source to the water in the basin which improves the rate of evaporation. A detailed classification of active and passive solar stills is given in Figure 6. Active solar stills are classified according to the integration type and operation principles of the solar stills. The main classification categories are: nocturnal distillation, pre-heated water distillation and high-temperature distillation solar stills.

Nocturnal production solar stills are able to operate when there is no sunlight. This can be achieved by mainly two ways: storing extra energy during day-time and using the stored energy at night, and making use of waste heat from another source. In order to store energy, still basin is filled to a depth that is more than required for full evaporation in an average day. At the end of the day some warm water would still remain in the basin of the still and continue to evaporate during no-sunshine hours of the day, which is called nocturnal distillation. This evaporation can be provided by feeding hot water from another heat source during night.

Pre-heated water application solar stills make use of waste heat from an external plant such as paper industries, chemical industries, thermal power plants and food processing industries to heat the water in the basin through a heat exchanger or use the warm water directly in the basin to improve evaporation rate.

High temperature distillation stills increase the basin water temperature from about 20–50°C to 70–80°C by coupling an external solar system such as FPC, ETC, HPC, PSC, SP and PV/T hybrid system. In addition to these methods, some other methods with different operation properties are used with high temperature distillation method like multistage active, multi-effect air-bubbled solar still and hybrid solar distillation [52]. Figure 18 shows how a passive distillation system (a) can be converted to an active distillation system by addition of an external energy supply plant which can be integrated by natural circulation (b) or forced circulation (c).

Latent heat of condensation is one of the most significant heat losses of solar distillation systems. Finding a way for the re-utilization of this heat would greatly increase the thermal efficiency of the solar distillation system, which is defined as the daily production per square meter. One of those ideas is re-using the latent heat of condensation at the cover of a basin to heat the water in another basin [46]. Such a design is called multi-basin solar stills (Figure18). In multi-basin design, two or more basins are constructed like the floors in an apartment building. The bottom-most basin is covered by an absorber plate while the upper basins are transparent to allow solar radiation to reach the bottom plate. Condensing vapour at each basin cover heats up the cover by the latent heat of condensation. Heated cover of a basin forms basin of the upper still section and heats the water on it by that latent heat by re-utilizing the waste heat. Each section has its own condensate collection and salt water feeding channels.

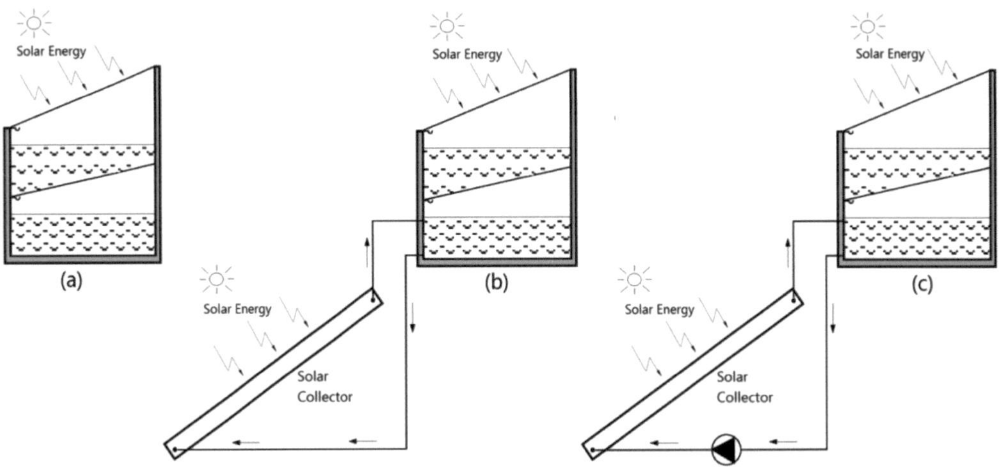

Figure 18. Schematic view of a double basin solar still. (b) Double basin still coupled to a collector in the natural circulation mode. (c) Double basin coupled to a collector in the forced circulation mode [52]

As shown in Figures 19 and 20, flat plate collectors and evacuated collectors can be used in active solar distillation systems. Solar collectors have high efficiency and improve the amount of distillation. However, the collector should be used in closed cycle to avoid precipitation of salt and other contaminants in the tubes and demolish the performance of the collector. Instead, a heat exchanger should be used (Figure 20) to transfer the heat to the basin water of the still [34].

http://dx.doi.org/10.5772/60436

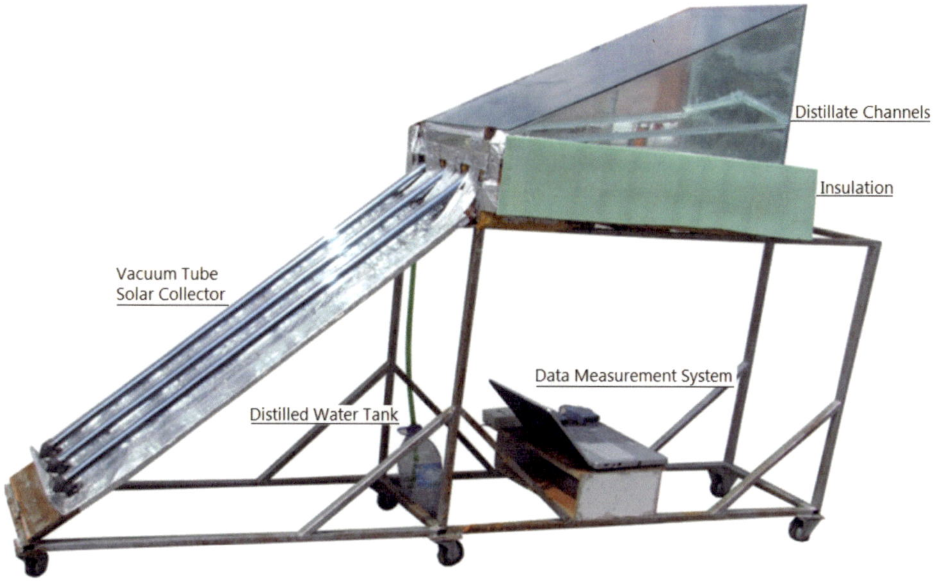

Figure 19. Vacuum tube collector assisted solar distillation system [34]

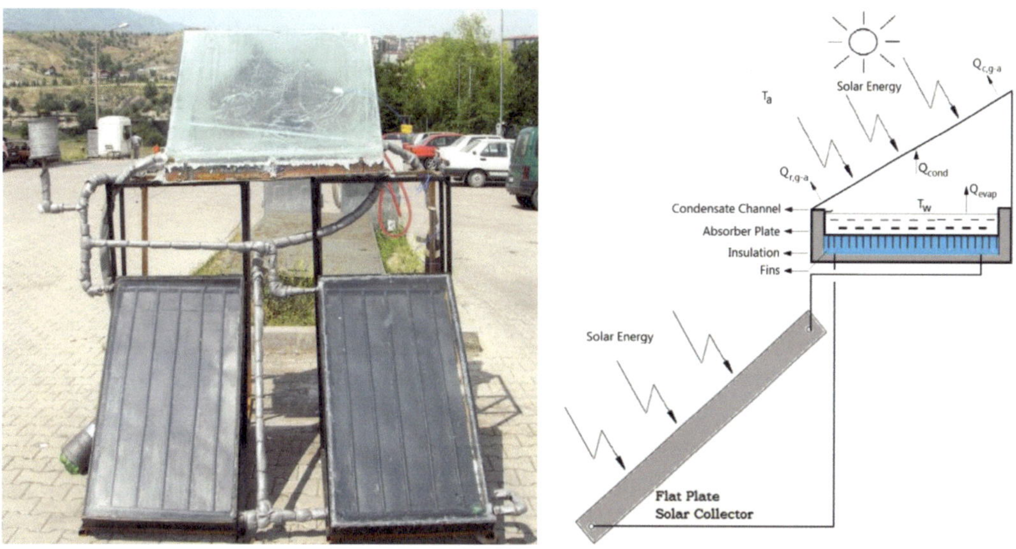

Figure 20. Flat plate solar collector assisted active distillation system and its schematic view

Solar stills can be successfully integrated with parabolic solar concentrators (Figure 21). Solar tracking parabolic concentrators concentrate the solar irradiation falling on a large area onto a small receiver area at high temperatures. High temperature and low heat loss area of the still basin which is located on the focal point of parabolic concentrator greatly improves the efficiency of the still. Salt water can be supplied to the still by natural or forced circulation [52].

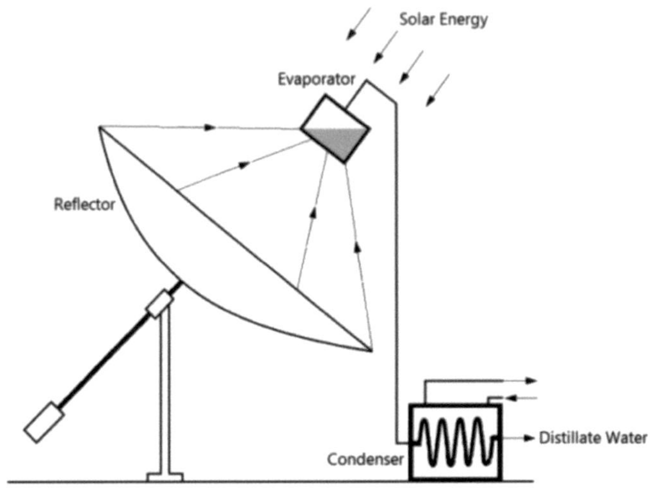

Figure 21. Schematic of a concentrating collector still [52]

Since the most critical stages of distillation process are evaporation and condensation, any measures that helps these two stages increases the efficiency significantly. A clever idea for promoting the evaporation at a certain temperature is forced air bubbling which causes an instantaneous atomization of water towards the air and a rapid evaporation. If it is possible to pre-heat the air that will be used for bubbling evaporation would be much better since the air that will carry the vapour also has the extra heat that evaporation process requires instantly during the bubbling effect (Figure 22). Another effective way of improving the still efficiency is cooling the cover surface [46, 52].

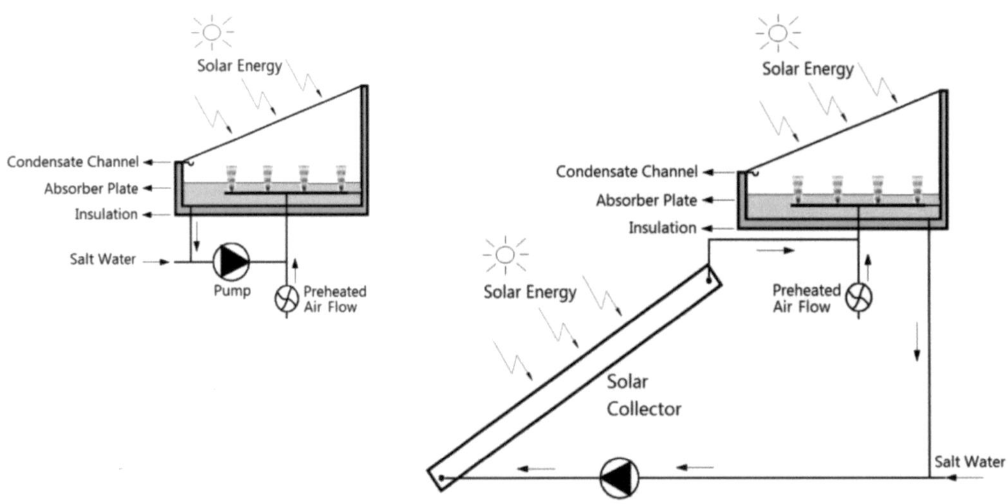

Figure 22. Air-bubbled solar still [46]

http://dx.doi.org/10.5772/60436

In a recent design, evacuated solar collector is hybridized with wicks/solar still to improve the productivity of still (Figure 23). Using single layer or double layer wick on absorber plate and integrating a feed water tank to feed hot salt water which is heated by solar collector during the daytime made up of a great combination of wick, inclined solar heating and energy storage which boosted the thermal performance and operation time of the still [53].

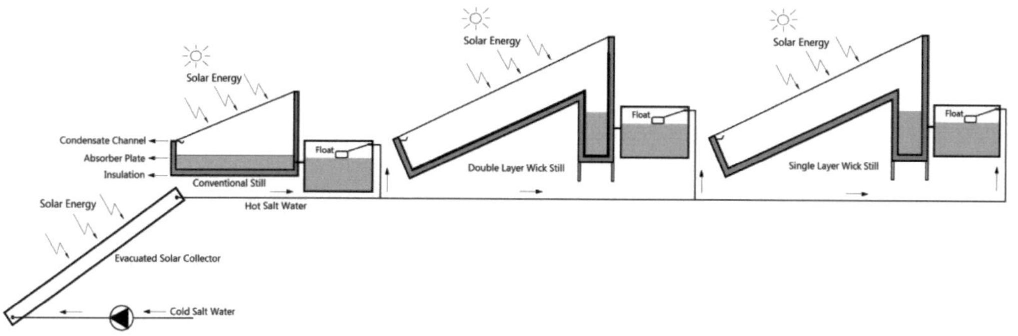

Figure 23. Schematic diagram of hybrid desalination system using wicks/solar still and evacuated solar water heater [53]

5.2.2.3. General considerations on solar stills

Making a general consideration of the solar stills, some common results can be concluded:

- Fresh water yield of a solar still depends on several parameters like solar insolation, ambient air temperature, wind speed, atmospheric humidity, sky conditions etc.
- The inclination angle of cover should be equal to the latitude of the installation site to minimize the cosine loses.
- Increasing water depth in the still basin decreases the distillate production rate.
- Each type of solar still has its own advantages and disadvantages. None of them is perfect.
- Increasing salinity of the salt water decreases the distillate production rate especially at low concentration range.
- Due to inclination angle single slope passive solar stills are better than double slope passive solar stills.
- FPC with forced circulation mode increase the performance solar still more than thermo syphon mode.
- Wind can slightly increase the total production.
- Although from morning to noon the highest distillate output are obtained at the highest water temperature times, more outputs can be observed at less water temperatures which can be attributed to the cooling glass cover before the system itself.

- Double slope passive solar still has a higher thermal efficiency than double slope active solar still.
- In active double effect solar stills lower basin gives the maximum yield because of the high noon temperature.
- The amount of fresh water produce in the still is closely related with thermal conductivity of condensing cover material.

6. Conclusion

Drinkable fresh water, being a fundamental need of human being, has become a serious concern for people especially living in crowded cities and countries with limited amount of water resources. Supplying fresh water is an energy intensive task especially when there is need for heating cold water to evaporate for distillation. Most of the big desalination plants around the world use fossil-based fuels as energy source to heat and vaporize the sea water or brackish water in order to produce fresh water. However, environmental concerns along with decreasing oil reserves and increasing fuel costs call for reconsideration on the fuel types used in many areas as well as in desalination or distillation processes. There are a number of well-known alternative and renewable energy sources in the world like wind, geothermal, solar, biomass etc. but achievability, availability when needed and energy storage are also very critical issues to operate the distillation plant effectively.

A very fortunate aspect of the issue is the coincidence of fresh water need and abundant solar energy. Most of the places with severe water scarcity are about the equator region where people needs more water due to highest degrees of air temperature in the world. Therefore, especially for these countries located in solar belt region, solar energy is a readily available, cost-free and abundant energy source which does not require transportation or very sophisticated technology to benefit from. Since most of the desalination processes run on heat, solar energy can be used directly as heat source which decreases the installation and operation costs significantly. Apart from simple solar powered stills, of course there are more complicated, yet much more efficient solar powered stills. Complex desalination systems which are all designed to improve the thermal efficiency and freshwater yield/productivity use many approaches such as: multiple stage operation, pre-heating the feed water with the condensation energy, using additional heat sources (waste water from another process), concentrating solar energy to improve the operation temperature of the process etc.

Solar desalination systems can be a good solution to the water scarcity of especially the Sunbelt region. However, dominant use of clean and renewable energy sources for desalination process and replacement of the fossil fuel-based operations require more R&D studies to find more efficient and/or less costly power plants with continuous operation by the means of energy storage solutions.

http://dx.doi.org/10.5772/60436

Nomenclature

A_c	Area of collector (m^2)
A	Absorptivity
CPC	Compound parabolic collectors
c_w	Specific heat of water in solar still (J/kg K)
ED	Electro dialysis
ETC	Evacuated tube collector
FPC	Flat plate collector
F_R	Heat removal factor
h_b	Overall heat transfer coefficient from basin liner to ambient air through bottom and side insulation (W/m^2 °C)
HD-DHD	Humidification-dehumidification
h_{ew}	Heat loss coefficient by evaporation from water surface (W/m^2K)
h_{wg}	Heat transfer coefficient from the water surface to the glass (W/m^2K)
h_{ga}	Heat transfer coefficient from the glass to the ambient (W/m^2K)
h_{wb}	Heat transfer coefficient from the water to the basin liner (W/m^2K)
HP	Heat pipe
I_c	Solar intensity on the glass cover of the solar collector panel (W/m^2)
I_{eff}	Effective solar radiation intensity (W/m^2)
L	Latent heat of vaporization (J/kg)
LCZ	Lower Convective Zone
LFC	Linear Fresnel collector technology
MED	Multiple effect distillation
MED-MVC	Mechanic vapour compression MED
MED-TVC	Thermal vapour compression MED
MEH	Multiple-effect humidification
M_w	Mass of water in basin (kg)
$\dot{m}_{ew}$	Hourly output of still (kg/m^2h)
mf_s	Salt mass fraction
MSF	Multi-stage flash
NCZ	Non convective zone
P_g	Glass saturated partial pressure (N/m^2)
P_w	Water saturated partial pressure (N/m^2)
PTC	Parabolic trough collector
Q_{ew}	Evaporative heat transfer (W)

$\dot{Q}_u$	Rate of useful energy from collector (W)
RO	Reverse osmosis
SDC	Solar dish collector technology
SGSP	Salt gradient solar pond
SMCEC	Solar multi-stage condensation evaporation cycle
SWH	Solar water heating
T	Transmittance
T_a	Ambient air temperature (K)
T_b	Basin temperature (K)
T_g	Still glass cover temperature (K)
T_w	Still water temperature (K)
TSP	Tower solar power technology
UCZ	Upper convective zone
U_L	Overall heat transfer coefficient (W/m^2 °C)
V	Wind speed (m/s)
VC	Vapour compression
α'_b	Solar flux absorbed by the basin liner
α'_g	Solar flux absorbed by glass cover
α'_w	Solar flux absorbed by water mass
Subscripts	
c	Collector

Author details

Emrah Deniz*

Address all correspondence to: edeniz@karabuk.edu.tr

Department of Mechanical Engineering, Karabuk University, Karabuk, Turkey

References

[1] Gude VG, Nirmalakhandan N, Deng S. Renewable and sustainable approaches for desalination. Renewable and Sustainable Energy Reviews. 2010;14(9):2641–54.

[2] Shankar P, Kumar S. Solar distillation – A parametric review. VSRD International Journal of Mechanical, Automobile & Production Engineering. 2012;2(1):17–33.

[3] Yılmaz S, Deniz E. The comparison of efficiencies between two phased protected zone and indirect solar water heaters. Technology. 2009;12(4):259–66.

[4] Kalogirou SA. Solar thermal collectors and applications. Progress in Energy and Combustion Science. 2004;30(3):231–95.

[5] Qiblawey HM, Banat F. Solar thermal desalination technologies. Desalination. 2008;220(1):633–44.

[6] Akash BA, Mohsen MS, Nayfeh W. Experimental study of the basin type solar still under local climate conditions. Energy Conversion and Management. 2000;41(9):883–90.

[7] Chahine MT. The hydrological cycle and its influence on climate. Nature. 1992;359(6394):373–80.

[8] Managing Water Under Uncertainty and Risk [Internet]. UNESCO; 2012. Report No.: 4. (accessed 10 February 2015): http://unesdoc.unesco.org/images/0021/002156/215644e.pdf

[9] Trenberth KE, Smith L, Qian T, Dai A, Fasullo J. Estimates of the global water budget and its annual cycle using observational and model data. Journal of Hydrometeorology. 2007;8(4):758–69.

[10] Ondraczek J, Komendantova N, Patt A. WACC the dog: The effect of financing costs on the levelized cost of solar PV power. Renewable Energy. 2015;75:888–98.

[11] Bilton AM, Wiesman R, Arif AFM, Zubair SM, Dubowsky S. On the feasibility of community-scale photovoltaic-powered reverse osmosis desalination systems for remote locations. Renewable Energy. 2011;36(12):3246–56.

[12] Koschikowski J. Water desalination: When and where will it make sense? [Internet]. Fraunhofer Institute for solar energy systems ISE; 2011. (accessed 10 February 2015): https://ec.europa.eu/jrc/sites/default/files/jrc_aaas2011_energy_water_koschikowski.pdf

[13] Gupta B, Mandraha TK, j Edla P, Pandya M. Thermal modeling and efficiency of solar water distillation: A review. American Journal of Engineering Research (AJER). 2013;2(12):203–13.

[14] Mezher T, Fath H, Abbas Z, Khaled A. Techno-economic assessment and environmental impacts of desalination technologies. Desalination. 2011;266(1):263–73.

[15] Li C. Innovative desalination systems using low-grade heat [Internet] [PhD Thesis]. University of South Florida; 2012. (accessed 10 February 2015): http://scholarcommons.usf.edu/cgi/viewcontent.cgi?article=5322&context=etd

[16] Kalogirou SA. Seawater desalination using renewable energy sources. Progress in Energy and Combustion Science. 2005;31(3):242–81.

[17] Kurt H, Özkaymak M, Deniz E. Experimental investigation of heat storage characteristic of urea and borax salt gradient solar ponds. Pamukkale University Engineering College, Journal of Engineering Sciences. 12:449–55.

[18] Acar B. Düz ısı borulu güneşli su ısıtma sistemleri ile birleşik ısı borulu güneşli su ısıtma sistemlerinin performans değerlerinin kıyaslamalı olarak belirlenmesi [Msc Thesis]. [Fen Bilimleri Enstitüsü]: Zonguldak Karaelmas Üniversitesi; 2007.

[19] Deniz E, Buğutekin A, Kaşifoğlu S, Öz ES, Binark AK. Güneş kollektörlerinde soğutucu akışkan R-404a kullanımı ve sistem verimine etkileri. Isparta, Turkey; 2006. sayfa 158–66.

[20] Renewables 2014, Global Status Report [Internet]. (accessed 10 February 2015): http://www.ren21.net/Portals/0/documents/Resources/GSR/2014/GSR2014_KeyFindings_low%20res.pdf

[21] Ghaffour N, Bundschuh J, Mahmoudi H, Goosen MF. Renewable energy-driven desalination technologies: A comprehensive review on challenges and potential applications of integrated systems. Desalination. 2015;356:94–114.

[22] Baharoon DA, Rahman HA, Omar WZW, Fadhl SO. Historical development of concentrating solar power technologies to generate clean electricity efficiently–A review. Renewable and Sustainable Energy Reviews. 2015;41:996–1027.

[23] Tzen E, Perrakis K, Baltas P. Design of a stand alone PV-desalination system for rural areas. Desalination. 1998;119(1):327–33.

[24] Ali MT, Fath HE, Armstrong PR. A comprehensive techno-economical review of indirect solar desalination. Renewable and Sustainable Energy Reviews. 2011;15(8): 4187–99.

[25] Ghermandi A, Messalem R. Solar-driven desalination with reverse osmosis: the state of the art. Desalination and Water Treatment. 2009;7(1-3):285–96.

[26] García-Rodríguez L, Delgado-Torres AM. Solar-powered rankine cycles for fresh water production. Desalination. 2007;212(1):319–27.

[27] Saidur R, Elcevvadi ET, Mekhilef S, Safari A, Mohammed HA. An overview of different distillation methods for small scale applications. Renewable and Sustainable Energy Reviews. 2011;15(9):4756–64.

[28] Hamid MOA, Zhang B, Yang L. Application of field synergy principle for optimization fluid flow and convective heat transfer in a tube bundle of a pre-heater. Energy. 2014;76:241–53.

[29] Narayan GP, Sharqawy MH, Summers EK, Lienhard JH, Zubair SM, Antar MA. The potential of solar-driven humidification–dehumidification desalination for small-scale decentralized water production. Renewable and Sustainable Energy Reviews. 2010;14(4):1187–201.

[30] Kabeel AE, Hamed MH, Omara ZM, Sharshir SW. Water desalination using a humidification-dehumidification technique—A detailed review. 2013 (accessed 10 February 2015): http://file.scirp.org/Html/8-2000200_33877.htm

[31] Parekh S, Farid MM, Selman JR, Al-Hallaj S. Solar desalination with a humidification-dehumidification technique—a comprehensive technical review. Desalination. 2004;160(2):167–86.

[32] Hanson A, Zachritz W, Stevens K, Mimbela L, Polka R, Cisneros L. Distillate water quality of a single-basin solar still: laboratory and field studies. Solar Energy. 2004;76(5):635–45.

[33] Haruna IU, Yerima M, Pukuma AD, Sambo II. Experimental investigation of the performance of basin type single-slope solar still. (accessed 10 February 2015): http://www.ijstr.org/final-print/mar2014/Experimental-Investigation-Of-The-Performance-Of-Basin-Type-Single-slope-Solar-Still.pdf

[34] Deniz E. An experimental and theoretical analysis of a vacuum tube solar collector-assisted solar distillation system. Energy Sources, Part A: Recovery, Utilization, and Environmental Effects. 2012;34(17):1637–45.

[35] Mowla D, Karimi G. Mathematical modelling of solar stills in Iran. Solar Energy. 1995;55(5):389–93.

[36] Badran OO, Abu-Khader MM. Evaluating thermal performance of a single slope solar still. Heat and Mass Transfer. 2007;43(10):985–95.

[37] Kumar S, Tiwari GN. Estimation of convective mass transfer in solar distillation systems. Solar Energy. 1996;57(6):459–64.

[38] Tripathi R, Tiwari GN. Thermal modeling of passive and active solar stills for different depths of water by using the concept of solar fraction. Solar Energy. 2006;80(8): 956–67.

[39] Rubio E, Fernández JL, Porta-Gándara MA. Modeling thermal asymmetries in double slope solar stills. Renewable Energy. 2004;29(6):895–906.

[40] Tiwari GN, Singh HN, Tripathi R. Present status of solar distillation. Solar Energy. 2003;75(5):367–73.

[41] AnfasMukram T, Suneesh PU. Experimental analysis of active solar still with air pump and external boosting mirrors. International Journal of Innovative Research in Science, Engineering and Technology. 2013;2(7):3084–904.

[42] Kabeel AE, Omara ZM, Essa FA. Enhancement of modified solar still integrated with external condenser using nanofluids: an experimental approach. Energy Conversion and Management. 2014;78:493–8.

[43] El-Bahi A, Inan D. Analysis of a parallel double glass solar still with separate condenser. Renewable Energy. 1999;17(4):509–21.

[44] El-Bahi A, Inan D. A solar still with minimum inclination, coupled to an outside condenser. Desalination. 1999;123(1):79–83.

[45] Muftah AF, Alghoul MA, Fudholi A, Abdul-Majeed MM, Sopian K. Factors affecting basin type solar still productivity: A detailed review. Renewable and Sustainable Energy Reviews. 2014;32:430–47.

[46] Arjunan TV, Aybar H, Nedunchezhian N. Status of solar desalination in India. Renewable and Sustainable Energy Reviews. 2009;13(9):2408–18.

[47] Deniz E. An investigation of some of the parameters involved in inclined solar distillation systems. Environmental Progress & Sustainable Energy. 2013;32(2):350–4.

[48] Zoori HA, Tabrizi FF, Sarhaddi F, Heshmatnezhad F. Comparison between energy and exergy efficiencies in a weir type cascade solar still. Desalination. 2013;325:113–21.

[49] Sivakumar V, Sundaram EG. Improvement techniques of solar still efficiency: A review. Renewable and Sustainable Energy Reviews. 2013;28:246–64.

[50] Chen Z, Yao Y, Zheng Z, Zheng H, Yang Y, Hou L, vd. Analysis of the characteristics of heat and mass transfer of a three-effect tubular solar still and experimental research. Desalination. 2013;330:42–8.

[51] Ahsan A, Fukuhara T. Mass and heat transfer model of tubular solar still. Solar Energy. 2010;84(7):1147–56.

[52] Sampathkumar K, Arjunan TV, Pitchandi P, Senthilkumar P. Active solar distillation —A detailed review. Renewable and Sustainable Energy Reviews. 2010;14(6):1503–26.

[53] Omara ZM, Eltawil MA, ElNashar EA. A new hybrid desalination system using wicks/solar still and evacuated solar water heater. Desalination. 2013;325:56–64.

Section 3

Hybrid Desalination Systems

Chapter 6

Cogeneration Power-Desalting Plants Using Gas Turbine Combined Cycle

M.A. Darwish, H.K. Abdulrahim, A.A. Mabrouk and A.S. Hassan

Additional information is available at the end of the chapter

http://dx.doi.org/10.5772/60209

Abstract

The gas-steam turbine combined cycle (GTCC) is the preferred power plant type because of its high efficiency and its use of cheap and clean natural gas as fuel. It is also the preferred type in the Arab Gulf countries where it is used as cogeneration power-desalting plant (CPDP). In this chapter, descriptions and analysis of the GTCC components are presented, namely, the gas turbine cycle (compressor, combustor, gas turbine), heat recovery steam generator, and steam turbine. Combinations of the GTCC with thermally driven desalination units to present CPDP are presented. A parametric study to show the effect of using GTCC on several operating parameters on the CPDP is also presented, as well as cost allocation methods of fuel between the two product utilities (electric power and desalted seawater are also presented).

Keywords: multi stage flash, multi effect with thermal vapor compression, reverse osmosis, steam power plants, gas turbines, combined cycle, cost allocation, energy and exergy analyses

1. Introduction

The efficiency of power plant (PP) using gas turbine (GT) combined cycle (GTCC) is higher than that of steam cycle PP prevailed as baseload-type plant before 2000, and the GT power cycle that was used as peak load and starting units in these steam PPs. The efficiency of the steam turbine (ST) plants is in the range of 35–40 %, the GT is in the range of 30–36 %, and

GTCC is in the range of 45–58 % (Figure 1). This is the main reason for the GTCC to become the preferred-type PP in the Gulf Cooperation Countries (GCC) and worldwide, besides using clean and cheap natural gas (NG) as a fuel. Increasing the PP efficiency reduces the emissions of greenhouse gases (GHG), sulfur dioxide, and nitrogen oxide. The GTCC (Figure 2a) includes an upper GT cycle (i.e., compressor, combustion chamber (cc), gas turbine) and bottom steam cycle (i.e., heat recovery steam generator (HRSG), steam turbine (ST), and condenser). The GT cycle produces almost 2/3 of the GTCC electric power (EP) output, with mainly NG used as fuel supplied to the cc. Hot gases exhausted from the GT (typically at 475–600 °C) are directed to HRSG to produce steam. This steam, when at high-enough temperature and pressure, is directed to ST to produce more work without adding more fuel. The ST cycle produces about 1/3 of the GTCC power output. Processed heat (in the form of steam) produced from the HRSG or extracted (or discharged) from the ST can be used to operate desalting plant (DP), district heating absorption cooling, and/or other processes.

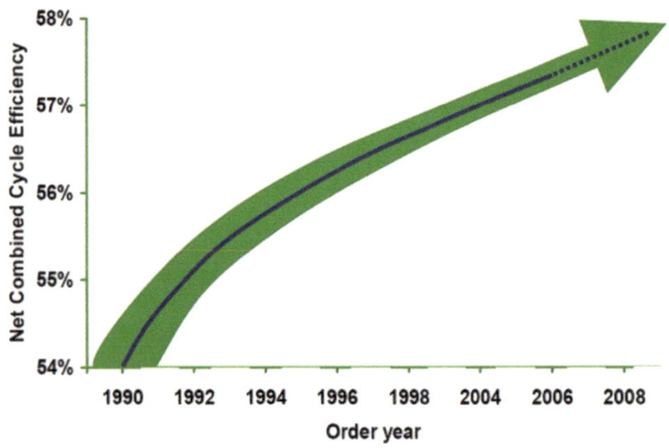

Figure 1. SGT6-5000F CC efficiency evolution [2]

Most PPs in the GCC are cogeneration power-desalting plants producing both desalted seawater (DW) and EP in single plants called CPDP. The desalting units are supplied with its needed low-pressure (LP) steam by extracting (or discharging) steam from the ST of the GTCC or directly from the HRSG when the ST is not operated or does not exist. The ST used in the GTCC can be extraction-condensing steam turbine (ECST) or back-pressure steam turbine (BPST) discharging all of its steam to the DP. Examples of recently installed CPDP using GTCC are Shuaiba North in Kuwait, (Figure 2b), Jebel Ali in the UAE, and Ras Girtas and Mesaieed in Qatar.

In Kuwait, all power plants were of steam type before 2003 and were combined with mainly multistage flash (MSF) desalting plants (DP) up to 2003 to form CPDP; see Table 1. The use of ST for power production in Kuwait followed the 1980s general world trend of using ST in the PP, when the share of GTCC plants was very limited. In the 1990s, the share of GTCC increased very rapidly in the world due to extensive improvements in the GTs. These improvements in GT resulted in reliable GTCC technology and low capital cost of the GTCC plants compared to the ST cycle of the same capacity. NG availability at low cost in many parts of the world and

http://dx.doi.org/10.5772/60209

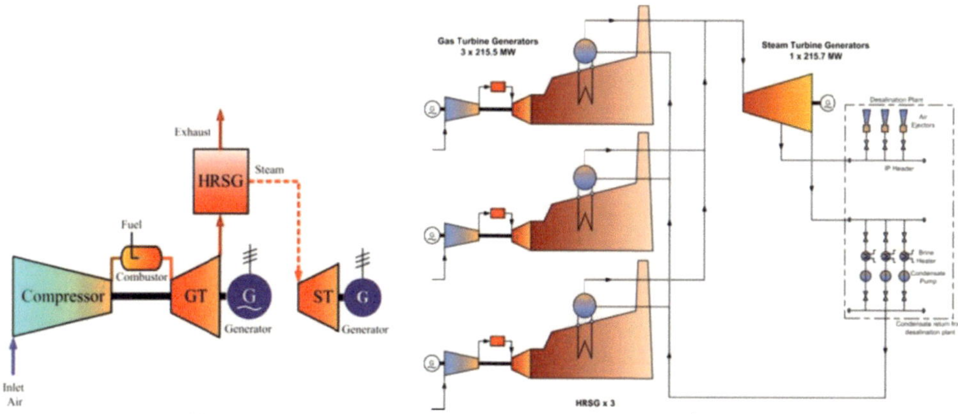

(a) (b)

Figure 2. a: Gas turbine combined with steam turbine forming GTCC [1] b: Schematic diagram of Shuaiba North (GTCC) combined with MSF desalination plant

the high efficiency of the GTCC (and thus the use of less fuel with less impact on the environment) promoted the share of the GTCC all over the world. Today, the GTCC-type PP becomes the preferred choice of PPs in most areas in the world, particularly in the GCC as shown in Table 2. Moreover, the GTCC equipment costs are less than that of the conventional ST plants.

Plant name	Date of commissioning	Capacity of steam power plant	Number and size of MSF units
Shuwaikh	1982		3 units × 6.5 MIGD
Shuaiba	1971–1975	6 turbine × 120 MW	6 units × 4.4 MIGD
Doha East	1978–1979	7 turbine × 150 MW	7 units × 6 MIGD
Doha West	1983–1985	8 turbine × 300 MW	4 units × 6 MIGD
			12 units × 7.2 MIGD
Az Zour South	1988–2001	8 turbine × 300 MW	16 units × 7.2 MIGD
Sabiya	2003	8 turbine × 300 MW	8 units × 12.5 MIGD

Table 1. Cogeneration steam power-desalination plants in Kuwait up to 2003 [1]

Project	Contractor	Year	Desalination	MIGD	Power plant	MW
Al-Fujairah	Doosan	2002	MSF/RO	12.5×5/37.5	GTCC	656
Al-Taweelah A2	Siemens	2001	MSF	12.5×4	GTCC	710
Al-Taweelah B	Siemens	2004	MSF	12.5×6	GTCC	720
Al-Shuweihat S1	Siemens	2002	MSF	17.0×6	GTCC	1500
Umm Al-Nar B	Hanjung	2000	MSF	12.5×5	GTCC	850
Jebel Ali K II	Fisia Itali	2003	MSF	13.5×3	GTCC	880
Shuaiba II	Doosan	2003	MSF	10.0×10	GTCC	500

Table 2. Recent combined cycle plants with MSF units in the United Arab Emirates and Saudi Arabia [1]

This article presents description of GTCC power plant units, detailing the GT, HRSG, ST, and their combining arrangements with DP to form CPDP. Methods of allocating the fuel supplied to the CPDP using GTCC between the EP and DW are presented.

2. The gas turbine cycle

In the past, STs were the prevailing type of PPs used to satisfy the baseload. GTs were included in these PPs to satisfy peak load and daily operation at time of high power demand lasting few hours in summer months and as emergency system. GTs for peak-load applications operate for short periods, few hours, e.g., 2–500 h/y, with no concern to thermal efficiency, but fast loading and start reliability are the concerns. Emergency GT units have to reach full load in a very short time; and aeroderivative GTs were originally designed to be capable of producing full power from cold metal in 120 seconds. The GT is usually classified as heavy frame industrial and aeroderivative types. Aeroderivative gas turbines use advanced aircraft engine to provide flexible, lightweight, and compact GTs. Heavy frame-type GTs are usually slower in speed, narrower in operating speed range, heavier, larger, have higher air flow, and slower in start-up. Traditionally, preference has been to place heavy frame industrial units in easily accessible baseload applications.

Developments of GT in terms of high unit capacity (up to 400 MW; see Figure 3a), reliability, and efficiency (up to 38 %) extend their use to cover base-load. Many simple GT cycle power plants are operating in the GCC, e.g., 497 MW Ras Abu Fontas (RAF) in Qatar including 6 GT×32 MW, 6×48 MW, and 2×9 MW of GT units. It has also 10×7 MIGD capacity multistage flash (MSF) desalting units with steam supplied from HRSG. The plant was commissioned in 1980. The desalting capacity was reported as 55 MIGD in 2010. The GT combined with ST forms GTCC of higher efficiency than either GT or ST cycle. Another diagram of GT is given in Figure 3b.

Figure 3. a: Siemens SGT5-8000H of 400 MW capacity [3] b: Cutaway diagram of a Westinghouse 501D5A gas turbine [4]

http://dx.doi.org/10.5772/60209

Simple GT cycle (Figures 4a–d) consists of a compressor, turbine, and generator usually mounted on single shaft and combustion chamber (cc). When started, the generator is usually operated as a motor to get sufficient rotor speed. Then, the GT is ignited, and power supply to the generator-motor is switched off. The GT accelerates until it reaches its nominal speed, and generator is synchronized and connected to the power grid. The GT is operated at constant speed to keep constant frequency at the generator output. The load changes are compensated by the adjustment of the input fuel flow to the combustor.

Another arrangement (Figures 5a, b) is to put the compressor and portion of the turbine on one shaft while the other part of the turbine and the generator on another shaft. The compressor with the first part of the turbine is called the gas generator (GG); and the GG output is equal to the compressor-consumed power. The other part of the turbine and generator is called the free turbine, which produces the net GT power output and gives more flexibility. Figure 4b shows GE's LM2500 Base aeroderivative gas turbine package that has dual fuel (oil and gas) capability; fast load response; 16-stage axial-flow compressor; annular combustor; two-stage, high-pressure, and single-rotor gas turbine; and highly efficient six-stage power turbine.

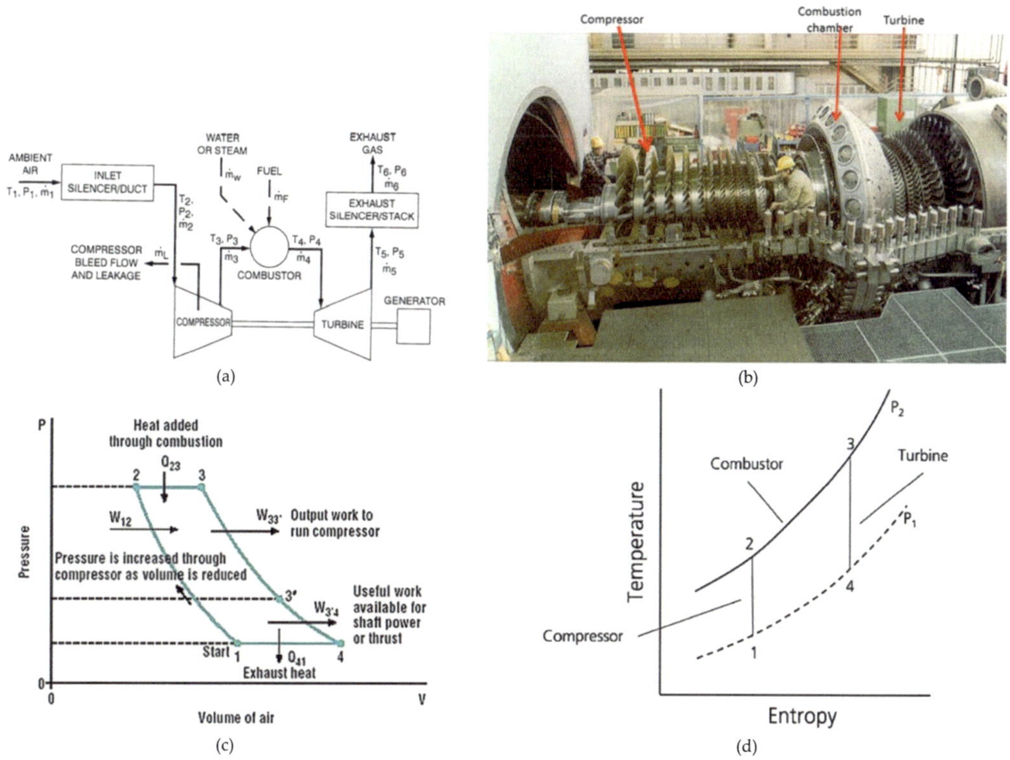

Figure 4. a: Gas turbine open (Brayton) cycle with its operating variables [5] b: Single-shaft gas turbine unit components [5] c: Gas turbine cycle presentation on P-v diagram [5] d: Gas turbine cycle presentation on T-s diagram

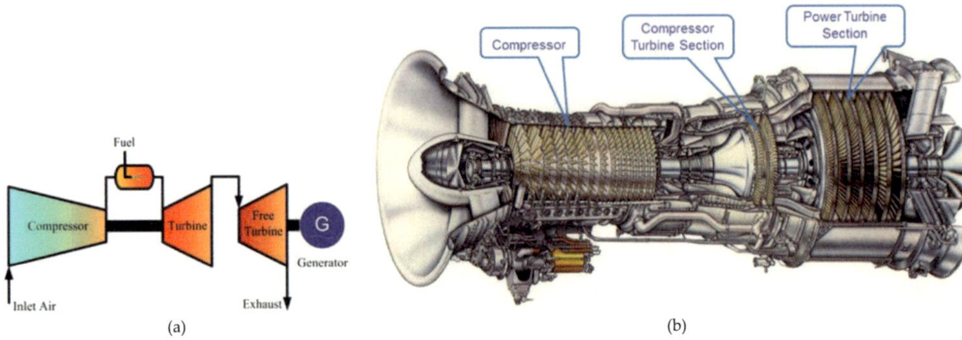

Figure 5. a: Two-shaft GT. b: Two-shaft GE LM2500 aeroderivative GT [6]

2.1. Analysis of ideal gas turbine cycle

The simple ideal gas turbine open cycle (see Figure 4c) known as Brayton cycle consists of four processes:

1. Isentropic compression of ambient air (working fluid) from pressure P_1 to P_2 by a compressor, with P_2/P_1= rp, called pressure ratio.
2. Heat transfer to the working fluid by mixing fuel with the compressed air and combusted in the cc from 2 to 3; usually P_2 is assumed equal to P_3 for ideal cycle, i.e., isobaric process.
3. Isentropic expansion of the working fluid in GT turbine from 3 to 4.

The exhausted hot gas is released from the turbine to the atmosphere, and fresh air is used to start or continue the cycle. In fact it is not a real cycle, but process 4-1 can be considered as an isobaric process of heat rejection to atmosphere. The cycle can be represented on both pressure-specific volume (P-v) and temperature-entropy (T-s) diagrams as shown in Figure 4d.

When the air is considered as ideal gas, the property relations for process 1-2 are

$$PV = RT, \quad P_1V_1 = RT_1, \quad P_2V_2 = RT_2$$
$$PV^k = C, \quad P_1V^k{}_1 = P_2V^k{}_2 = C$$

where

$$k = C_p / C_v,$$

and then

$$\frac{P_2V_2}{P_1V_1} = \frac{P_2}{P_1}\left(\frac{P_2}{P_1}\right)^{-\frac{1}{k}} = \frac{RT_2}{RT_1}$$

The required compressor-specific process (work per kg) is

$$w_{c,is}(1 \rightarrow 2) = -\int_1^2 vdp = \int_1^2 C\frac{C}{P^{1/k}}dP$$
$$= \frac{k}{1-k}(P_2v_2 - P_1v_1) = -\frac{k}{k-1}P_1v_1(\frac{P_2v_2}{p_1v_1} - 1) = -\frac{k}{k-1}RT_1((\frac{P_2}{p_1})^{\frac{k-1}{k}} - 1)$$

This is negative work (work consumed by the compressor) and is equal also to

$$\left|w_{c,is}(1 \rightarrow 2s)\right| = (h_{2s} - h_1) = c_p(T_{2s} - T_1)$$

T_{2s} and h_{2s} are the absolute temperature and enthalpy at point 2 if the expansion is isentropic ($q_{1\text{-}2} = 0$).

Heat addition process from 2 to 3 in the combustion chamber is considered ideal with no pressure loss (isobaric), $P_2 = P_3$. The heat input q_{in} between 2 and 3 is equal to the enthalpy increase: $q_{in} = h_3 - h_{2s} = C_p(T_3 - T_{2s})$.

It is noticed here that T_3 is the highest temperature in the cycle and is called the turbine inlet temperature (TIT). The amount of specific heat input per kg of air is also equal to

$$q_f = \frac{m_f}{m_a}(LHV),$$

where m_f is the mass flow rate of the fuel input and LHV is the fuel low heating value (heat generated per kg of fuel, when the water vapor in the combusted gases is in vapor state.

It is also noticed here that $w_{2\text{-}3} = 0$.

The property relations of isentropic expansion process in the turbine can be expressed as

$$PV = RT, \quad P_3V_3 = RT_3, \quad P_4V_4 = RT_4$$
$$PV^k = C, \quad P_3V_3^k = P_4V^k{}_4 = C$$

The turbine isentropic work is expressed by

$$w_{t,is}(3 \rightarrow 4s) = \int_3^4 vdp = \int_1^2 C\frac{C}{P^{1/k}}dP = \frac{k}{1-k}(P_4v_4 - P_3v_3) = -\frac{k}{k-1}P_3v_3(\frac{P_3v_3}{p_4v_4} - 1)$$
$$= \frac{k}{k-1}RT_3((\frac{P_3}{p_4})^{\frac{k-1}{k}} - 1)$$

This work can also be expressed by enthalpy change as

$$\left|w_{t,is}(3 \to 4s)\right| = (h_3 - h_{4s}) = c_p(T_3 - T_{4s1})$$

Since part of the turbine work is used to drive the compressor, the net work output ($w_{net} = w_t - w_c$) is expressed as

$$w_{net}(ideal) = w_t - w_c = (h_3 - h_{4s}) - (h_{2s} - h_1) = c_p(T_3 - T_{4s}) - c_p(T_{2s} - T_1)$$

$$w_{net} = w_t - w_c = c_p T_3 (1 - \frac{T_{4s}}{T_3}) - c_p T_1 (\frac{T_{2s}}{T_1} - 1)$$

$$= c_p T_3 (1 - \frac{1}{rp^{\frac{k-1}{k}}}) - c_p T_1 (rp^{\frac{k-1}{k}} - 1)$$

It is noticed that the net heat ($q_{in} - q_{out}$) is equal to the net work, or

$$\oint \partial Q = \oint \partial W = q_{in} - q_{out} = w_{out} - w_{in}$$

$$= c_p(T_3 - T_{4s}) - c_p(T_{2s} - T_1) = c_p(T_3 - T_{2s}) - q_{out}$$

$$q_{out} = c_p(T_{4s} - T_1) = q_{4,1} = c_p \cdot (T_1 - T_4)$$

The ideal cycle efficiency (net work/heat in) is expressed by

$$\eta = \frac{w_{net}}{q_{in}} = \frac{q_{in} - q_{out}}{q_{in}} = 1 - \frac{q_{out}}{q_{in}} = 1 - \frac{T_{4s} - T_1}{T_3 - T_{2s}} = 1 - \frac{T_1(\frac{T_{4s}}{T_1} - 1_1)}{T_2(\frac{T_3}{T_{2s}} - 1)}$$

It is noticed that

$$\frac{T_{2s}}{T_1} = (\frac{P_2}{P_1})^{\frac{k-1}{k}}, \quad \frac{T_3}{T_{4s}} = (\frac{P_3}{P_4})^{\frac{k-1}{k}}, \quad P_1 = P_4, P_2 = P_3$$

$$\frac{T_{2s}}{T_1} = \frac{T_3}{T_{4s}}, \quad T_{2s} T_{4s} = T_1 T_3, \quad \frac{T_{2s}}{T_3} = \frac{T_1}{T_{4s}},$$

Then

$$\eta c = 1 - \frac{T_1}{T_2} = 1 - \frac{1}{rp^{\frac{k-1}{k}}}$$

where

$$rp = (\frac{P_2}{P_1}) = (\frac{P_3}{P_4})$$

It is noticed that the efficiency depends also on T_3 (TIT), pressure ratio rp, and k, the ratio of specific heats at constant pressure to that at constant volume and is equal to 1.4 for air, $k = \frac{C_p}{C_v}$. The dimensionless work output can be expressed by

$$\frac{w_{net}}{C_p T_1} = \frac{T_3}{T_1}(1 - \frac{T_{4s}}{T_3}) - (\frac{T_2}{T_1} - 1), \frac{T_{2s}}{T_1} = \frac{T_3}{T_1}(1 - \frac{1}{rp^{\frac{k-1}{k}}}) - (rp^{\frac{k-1}{k}} - 1)v$$

Process (2-3): isobaric heat supply, $q_{2\text{-}3} = h_3 - h_2$, $w_{mech,2,3} = 0$. (4)

Process (3-4): isentropic expansion, $w_t = c_p(T_3 - T_{4s})$. (5)

State (4-1): isobaric heat release, $q_{4,1} = c_p \cdot (T_1 - T_{4s})$. (9)

There are differences between the ideal Brayton cycle and real gas turbine cycle. In the real cycle, the following are included:

1. Difference between the ambient air condition and the compressor inlet condition at point 1.

2. There are heat loss and friction losses in the compression and expansion processes, and thus these are not really isentropic processes.

3. Gas properties vary with temperature and not constants as assumed in the ideal cycle.

4. There is stagnation pressure loss in the combustion chamber and incomplete combustion.

5. Some of the air discharged from the compressor is extracted to the turbine for cooling.

The losses in compressors are usually expressed through the following:

1. By the compressor efficiency defined by

η_c = ideal work of compression/actual work of compression

2. By assuming the compression is adiabatic but with friction

2.2. The GT performance

The simple cycle in an *h-s* diagram including losses is shown in Figure 6a.

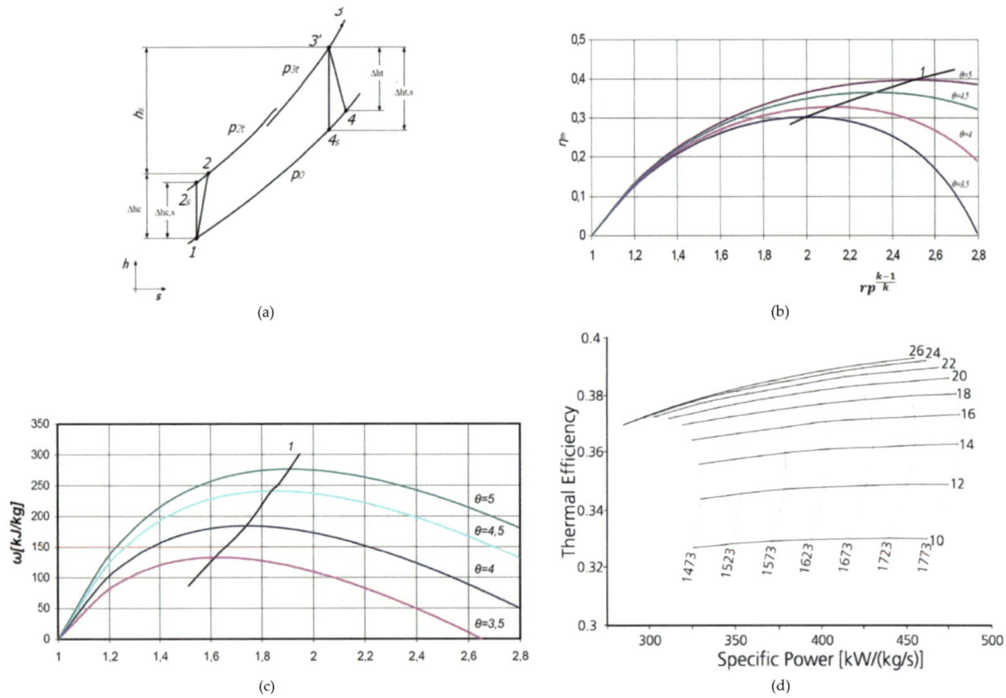

Figure 6. a: Enthalpy-entropy (h-s) diagram for ideal and practical gas turbine cycle [7] b: Dependence of the thermal efficiency ηth of the cycle on the parameters rp, k, and θ for $\eta_{t,is}$= 0.88 and $\eta_{c,is}$ = 0.86. Line 1 joins points of maximum efficiency for each curve [7] c: Dependence of the specific work of the cycle on the parameters π, κ, and θ for η_T =0.88 and η_C =0.86. Line 1 joins points of maximum specific work for each curve [7] d: Thermal efficiency vs specific power for varying pressure ratios (10–26) and combustor outlet temperature (1,473–1,773 K) for a gas turbine [8]

$$\eta_c(\textit{cycle efficiency}) = \left(h_{3'} - h_4\right) / \left(h_3 - h_2\right)$$

$$\eta_{t,is}(\textit{turbine isentropic efficiency}) = \left(h_{3'} - h_4\right) / \left(h_{3'} - h_{4s}\right)$$

$$\eta_c(\textit{compressor isentropic efficiency}) = \frac{\mathrm{w}_{c,is}}{\mathrm{w}_c} = \frac{h_{2s} - h_1}{h_2 - h_1}$$

$$= \frac{\frac{k}{k-1} RT_1 \left[\left(\frac{P_2}{P_1}\right)^{\frac{k-1}{k}} - 1 \right]}{h_2 - h_1} = \frac{k}{k-1} RT_1 \left[\left(\frac{P_2}{P_1}\right)^{\frac{k-1}{k}} - 1 \right] / C_p\left(T_2 - T_1\right)$$

The losses in turbines are usually expressed by the turbine efficiency defined by

η_t = actual work of expansion/ideal work of expansion.

http://dx.doi.org/10.5772/60209

Assuming that θ is the ratio of the turbine inlet temperature and compressor inlet temperature, which in this case is $\theta = T_3/T_1$,

$$\eta_c = \frac{\frac{T_3}{T_1}\eta_{t,is}\left(1-\frac{1}{rp^{\frac{k-1}{k}}}\right)-\frac{1}{\eta_{c,is}}(rp^{\frac{k-1}{k}}-1)}{\left(\frac{T_3}{T_1}\right)-1-\frac{1}{\eta_{c,is}}(rp^{\frac{k-1}{k}}-1)}$$

The efficiency of the thermodynamic cycle depends mainly on the TIT (T_3) or its dimensionless parameter $\theta = T_3/T_1$ as well as the pressure ratio rp as shown in Figures 6b–d. The highest cycle temperature is limited by the material and cooling of the first turbine stages; pressure ratio can be optimized to maximize the efficiency for a specific combustor temperature. Besides optimization of the efficiency, the gas turbine is also optimized for power output (Figure 6d). The optimization sets the conditions for the combustor. For the gas turbine cycle in Figure 6 at a combustor outlet temperature at 1,743 K, the optimal pressure ratio for specific power is 14:3 bar and the optimal pressure ratio for efficiency is 25:1 bar. These values are engine specific but show the tendency for optimization. The efficiency of the thermodynamic cycle depends mainly on the TIT (T_3) or its dimensionless parameter $\theta = T_3/T_1$ as well as the pressure ratio rp.

The thermal efficiency always increases with the increase of θ or the TIT, T_3, which has limitation with the materials. The pressure ratio rp ($P_2/P_1=P_3/P_4$) affects the cycle efficiency, which increases with rp until it reaches a maximum and then starts to fall. The optimal compression ratio changes with alteration of the compressor and turbine efficiencies.

The specific work, defined by the work per unit mass of the air, increases T_3 and reaches a maximum for a certain rp as shown in Figure 6c.

Two distinct losses occur in the combustion chamber: combustion inefficiency and pressure loss.

The first implies an imperfect conversion of the chemical energy in the fuel/air mixture into thermal energy. It is defined as

$$\eta_{cc} = \frac{\overline{c}_p\left[(\dot{m}_{air}+\dot{m}_f)T_{t3}-\dot{m}_{air}T_{t2}\right]}{\dot{m}_f\Delta h_u}$$

The typical combustion efficiency is around 0.99 or better.

The thermal efficiency of a real gas turbine cycle is lower than the one of the ideal cycle. In the T-s diagram or in the P-v diagram, respectively, the differences are obvious since there are no more isentropic changes possible.

2.3. Gas Turbine (GT) components

GTs are operating according to Brayton cycle and using the following components.

2.3.1. Air intake

The air to compressor should pass through an air filter to prevent dust from entering the machine and is accelerated in a duct to the compressor. The inlet duct in front of the compressor is usually designed as a diffuser. This decelerates the air at the inlet and converts part of the air kinetic energy into pressure.

Figure 7a shows an air filter installed at the air inlet to the compressor. The inlet air duct can contain an air cooling system. The compression in the GT is a constant volume process. So, the air temperature decrease would increase the air density and mass flow rate, decrease the specific power consumed by the compressor (per unit mass), and increase the GT power output. Figure 7b shows the effect of compressor inlet temperature on the GT output power and heat rate. The air inlet temperature can be decreased by evaporative cooling, fogging, and chilled water system as shown on the psychometric chart given in Figure 7c.

Figure 8a shows an inlet air to compressor using evaporative cooling which used relative humidity and wet bulb temperature that are rather low. This system has the advantage of low capital and operation cost as it can operate on raw water and uses air washer that cleans the inlet air. Figures 8b–d show an inlet air to compressor using fogging system. It is also an evaporative cooling system that is used when relative humidity and wet bulb temperature are rather low. This system uses demineralized water and increases GT performance better than the previous evaporative cooling system.

Figure 9a shows mechanical refrigeration system (direct type) used hot in areas and can bring the air temperature to any specific requirement irrespective of ambient temperature and humidity ratio. This system has the advantage of increasing the GT performance better than evaporative cooling and fog system. However, this system has high initial capital cost and high operation and capital cost. Figure 9b shows the absorption refrigeration system (direct type), which is similar to that of Figure 9a, but with absorption cooling system operated mainly with steam or hot water substituting the mechanical refrigeration system. This systems has also the advantage of increasing the GT performance better than evaporative cooling and fog system, but at higher initial capital cost and high operation and capital cost.

2.3.2. GT compressor

The main parameters of a compressor are the required pressure ratio (rp), volumetric flow rate, consumed power, and permissible shaft length. The used compressors types in GT application are axial, centrifugal, and combination of both. Axial compressors have more stages to reach the same compression ratio achieved by centrifugal type, and thus, axial compressors have a longer shaft than centrifugal ones. Axial compressors have lower changes of flow direction during compression and thus better efficiency (82–90 %) compared to centrifugal (72–82 %). Axial compressors handle much wider range of volume flows, are used

http://dx.doi.org/10.5772/60209

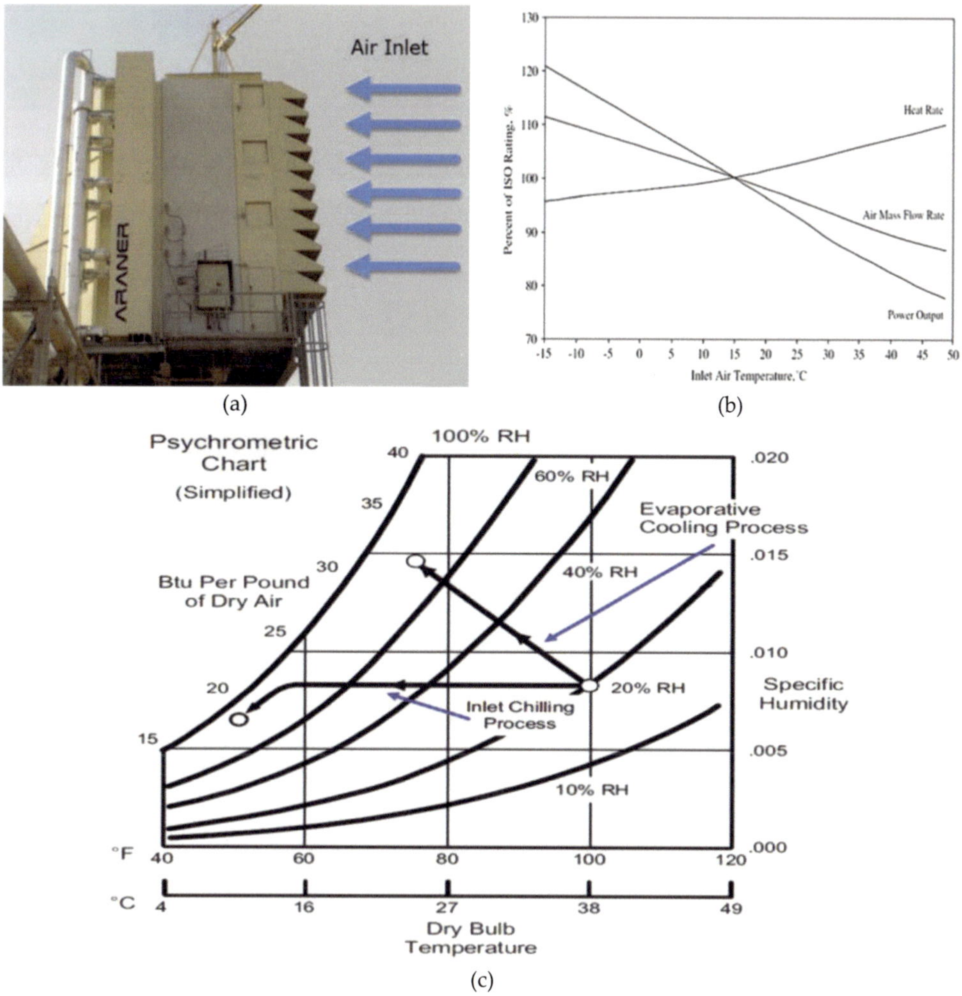

Figure 7. a: Turbine inlet air cooling filter-house modification to place the cooling coil coming from ammonia compression chiller plant [9] b: Typical inlet air cooling impacts on combustion turbine performance [1] c: Psychometric chart showing evaporative cooling process and chilled water cooling process [10]

in all heavy utility gas turbines, have much lower tendency for flow separation at the inlet blades, and are more reliable in the case of fast load changes. Centrifugal compressors have small-size, short shafts, used only in small gas turbines (less than 5 MW) and high rotor speeds. Combination of axial and centrifugal compressors utilizes axial compressor reliability and the centrifugal compressor high-pressure ratio.

In centrifugal compressor (Figure 10a) the air (to be compressed) enters the impeller center and moves outward by centrifugal force to the compressor discharge diffuser. The rotating impellers accelerate the air velocity, and the air kinetic energy is converted to an increase in static pressure by slowing the flow through a diffuser before being discharged.

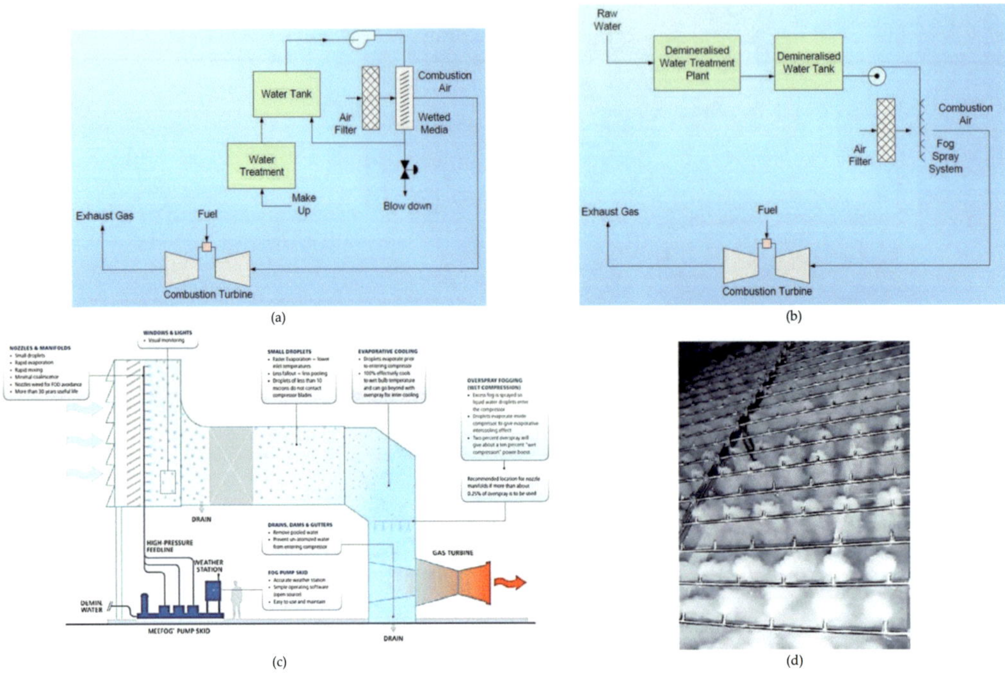

Figure 8. a: An inlet air to compressor using evaporative cooling which used relative humidity and wet bulb temperature that are rather low [11] b: An inlet air to compressor using fogging system which used relative humidity and wet bulb temperature that are rather low and using demineralized water [11] c: Fog system produces billions of microfine (10-micron average) droplets at 2,000 psi that create a much larger overall evaporative surface, which allows the droplets to evaporate and cool the airflow far more quickly than larger, heavier droplets. This results in faster, more effective evaporation and cooling with significantly lower drain water rates [12] d: MeeFog™ array for a frame 7FA gas turbine, Mee Industries – Fogging Systems for Offshore Gas Turbines

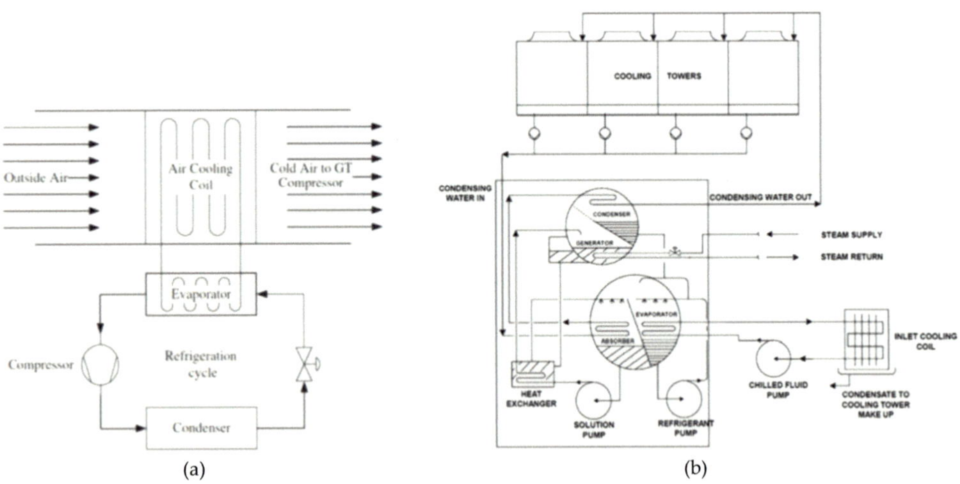

Figure 9. a: Mechanical refrigeration system (direct type) used in areas where relative humidity is rather high [1] b: Absorption refrigeration system (direct type) used in areas where relative humidity is rather high [1]

http://dx.doi.org/10.5772/60209

Axial compressors have moving (rotor) and fixed (stator) blades (Figure 8b). The arrays of blades are set in rows, usually as pairs: one rotating and one stationary. While rotating airfoils (known as blades or rotors) accelerate the fluid, the stationary airfoils (known as stators or vanes) decelerate the air, i.e., slow it down, and its kinetic energy is converted to pressure energy. The stators redirect the flow direction for the rotor blades of the next stage. The discharge velocity is almost equal to the suction velocity. This process is repeated by several stages depending on the desired output pressure.

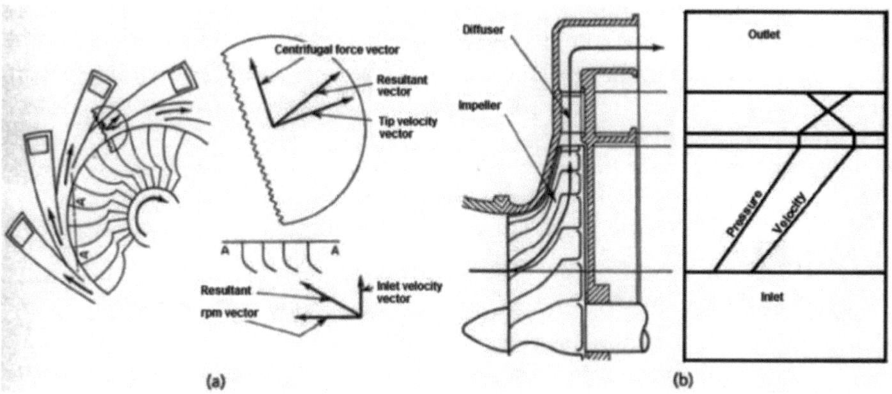

(a)

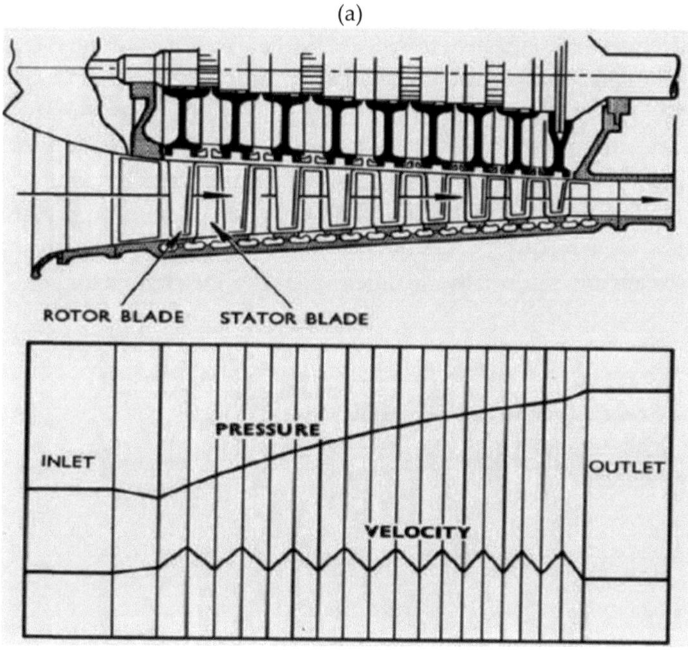

(b)

Figure 10. a: Centrifugal-compressor flow, pressure, and velocity changes; (a) airflow through a typical centrifugal compressor and (b) pressure and velocity changes through a centrifugal compressor [13] b: Schematic diagram of an axial flow compressor and pressure and velocity profile [14]

The direction of flow is parallel to the direction of the rotation. The design of compressor blades is different than those of turbines. The compressor blades have divergent profile and act as diffuser to increase air pressure. The turbine blades have convergent profile which works as a nozzle, reducing air pressure by changing its pressure energy into kinetic energy. More on axial compressor design is given in Ref. [15]. Although an axial stage may not offer as much of pressure ratio as a centrifugal stage of the same diameter, a multistage axial compressor offers far higher pressure ratio (and therefore mass flow rates and resultant power) than a centrifugal design.

Separation of the air flow from the surface of the blades of the first compressor stage is real problem in axial and centrifugal compressors. Flow separation from the surface of single blades generates high turbulence in the grid and can partly block the flow path of the incoming air aerodynamically. This effect, called a rotating stall, stresses the whole gas turbine structure with oscillating pressure waves.

2.3.3. GT combustor

The compressed air leaving the compressor is directed to the combustion chamber (cc), called combustor, where fuel such as natural gas (or petroleum liquids) is injected. In a combustor (Figure 11a) the fuel chemical energy is converted to thermal energy. So, the combustor combines and mixes air and fuel, ignites them, and contains the mixture during combustion. The combustor contains basically four zones – primary zone, secondary zone, dilution zone, and various wall jets – to manage heat transfer at the combustor boundary as shown in Figure 11b. Air entering the combustor is distributed to four major injection points. The first is through swirl vanes positioned at the combustor front face and typically surround the fuel injection port. The swirl vanes impact a circumferential velocity component to the air and thereby thrust the air radially outward as the air enters the combustor (Figure 11c). This creates a pressure void at the center line and induces a backflow to fill the centerline pressure deficit. This effectively creates, as a result, a recirculation flow that extends approximately one duct diameter downstream and defines the "primary zone" of the combustor.

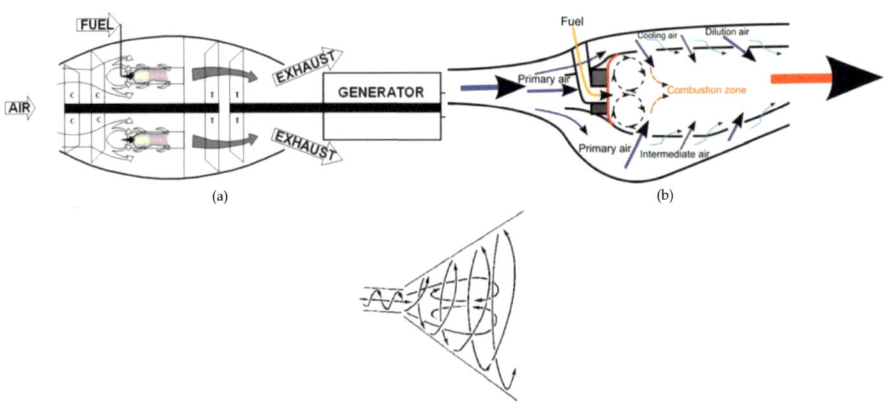

Figure 11. a: Stationary gas turbine electric power generator [16] Figure 11b: Schematic illustration of a general combustor [8] c: Circulation created by air swirler

http://dx.doi.org/10.5772/60209

The combustors are classified as:

1. Annular (continuous chamber that encircles the air in a plane perpendicular to the air flow) (Figure 12a)
2. Can-annular (similar to the annular but incorporates several can-shaped combustion chambers rather than a single continuous chamber) (Figures 12b, c)
3. Silo (silo, frame-type, combustor has one combustion chamber mounted externally to the gas turbine body)

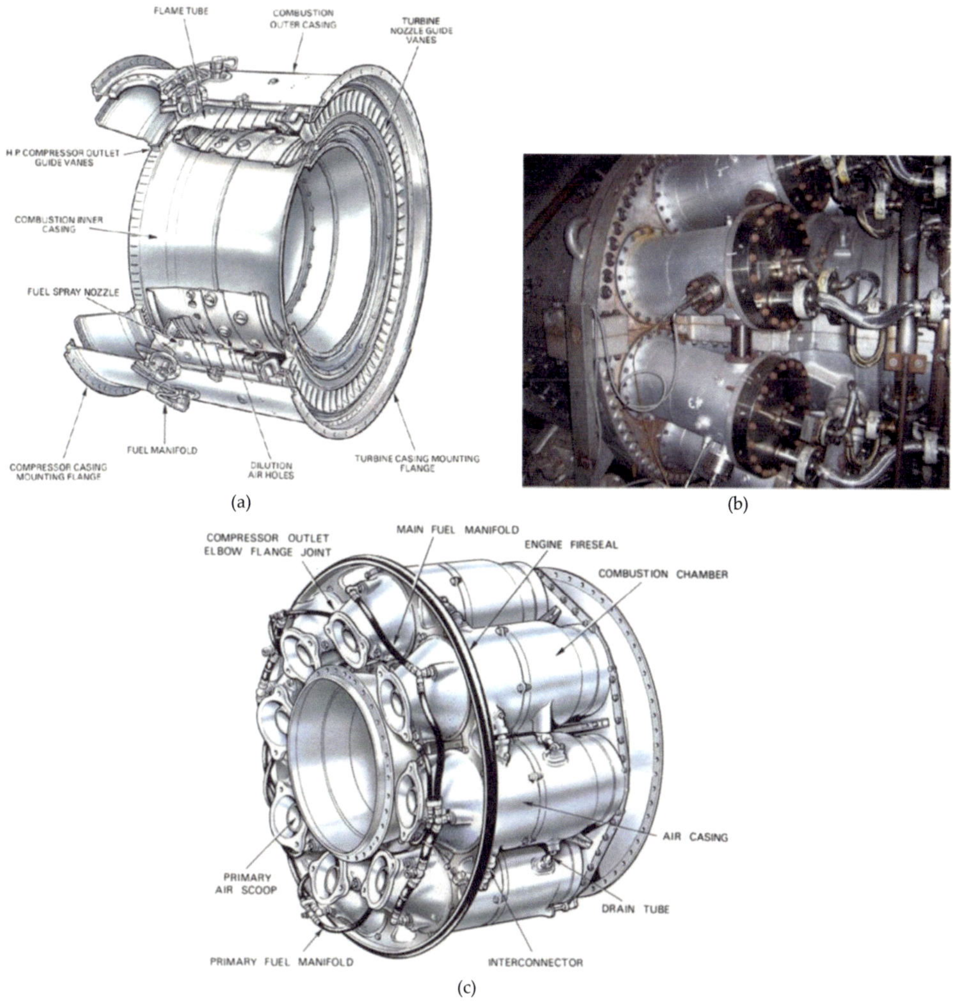

Figure 12. a: Annular combustion chambers [17] b: Gas turbine combustor arrangement [5] c: Several combustors arranged equidistant on the same pitched circle diameter, and each consists of an inner flame tube or liner cylinder mounted on the same axis inside an outer casing cylinder, called tubular combustors [18]

The combustion process in the GT combustor can be classified as diffusion flame combustion or lean-premix staged combustion. In the diffusion flame combustion, the fuel/air mixing and combustion take place simultaneously in the primary combustion zone, and this generates regions of near-stoichiometric fuel/air mixtures where temperatures and NO_x generation are very high. In lean-premix combustion, fuel and air are thoroughly mixed in an initial stage resulting in a uniform, lean, unburned fuel/air mixture which is delivered to a secondary stage where the combustion reaction occurs [19]. The combustion process starts with mixing the fuel with air supported by natural or forced turbulences in the airflow through the combustor. Continuous and stabilized combustion process is affected by the speed of fuel and air particles to the reaction zone, transport of flue gas from there, the speed of the chemical reaction in the reaction zone, and the residence time of any particle in the reaction zone. When the air-fuel mixing is slow compared to the chemical reaction rates, the mixing time controls the burning rate.

In diffusion flames, fuel and oxygen are mixed in the reaction zone through molecular and turbulent diffusion and have wide stability rate of combustion process. It has the advantages of relatively simple design of the fuel nozzles. Since the local conditions at the flame front are rich in fuel, diffusion combustion is insensitive against combustion instabilities and keeps on burning and generates regions of near-stoichiometric fuel-air mixtures with very high temperatures even at very lean conditions. The high temperature by diffusion flames leads to the production of large quantities of thermal NO_x.

To reduce the reaction temperatures and/or the formation of thermal NO_x, premix combustion is developed, where fuel and air are homogeneously mixed in an initial stage to become lean, unburned fuel-air mixture which is delivered to a secondary stage where the combustion reaction takes place. Manufacturers use different types of fuel-air staging, including fuel staging, air staging, or both; however, the same staged, lean-premix principle is applied. Gas turbines using staged combustion are also referred to as Dry Low NOX combustors. The majority of GT currently manufactured are lean-premix staged combustion turbines.

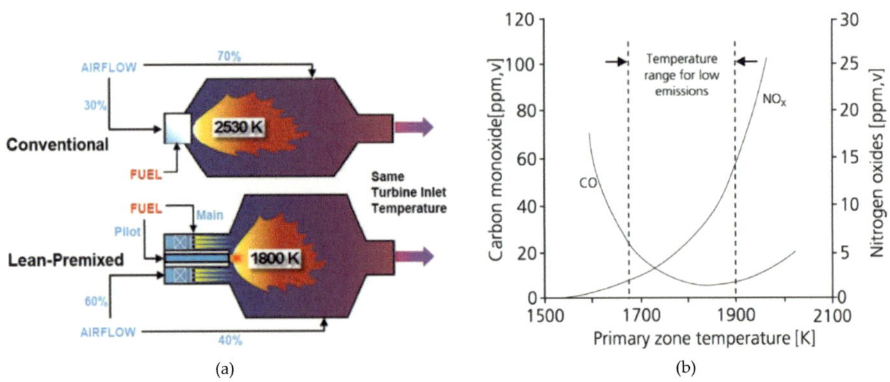

Figure 13. a: Schematic of a conventional and a lean-premix combustor [20] b: Primary zone temperature influence on NOX and CO emissions [8]

In premix, mixing of fuel and air occurs far before the reaction zone. Depending on the burner design and the flow velocity, the time from the fuel injection to the moment of ignition is within several milliseconds. This time is used to create a mostly homogeneous mixture, with a fuel concentration within the ignition range of the specific fuel for the given compressor discharge temperature. The typical adiabatic flame temperature, to which a premix combustion system is adjusted, is at 1,750 K. At this temperature, the formation of NO_x is still on an acceptable level, while the heat transfer from the flame is high enough to ensure the ignition of the fresh mixture (Figures 13a, b).

In general, there is an operation window for low emissions that range from the primary zone temperatures 1,670 K to 1,900 K (Figure 13b). The upper temperature limit is set by the temperature dependence of NOX and the lower limit by carbon monoxide. The increase in CO for lower temperatures is related to poor combustion and the lean blowout limit for the burner.

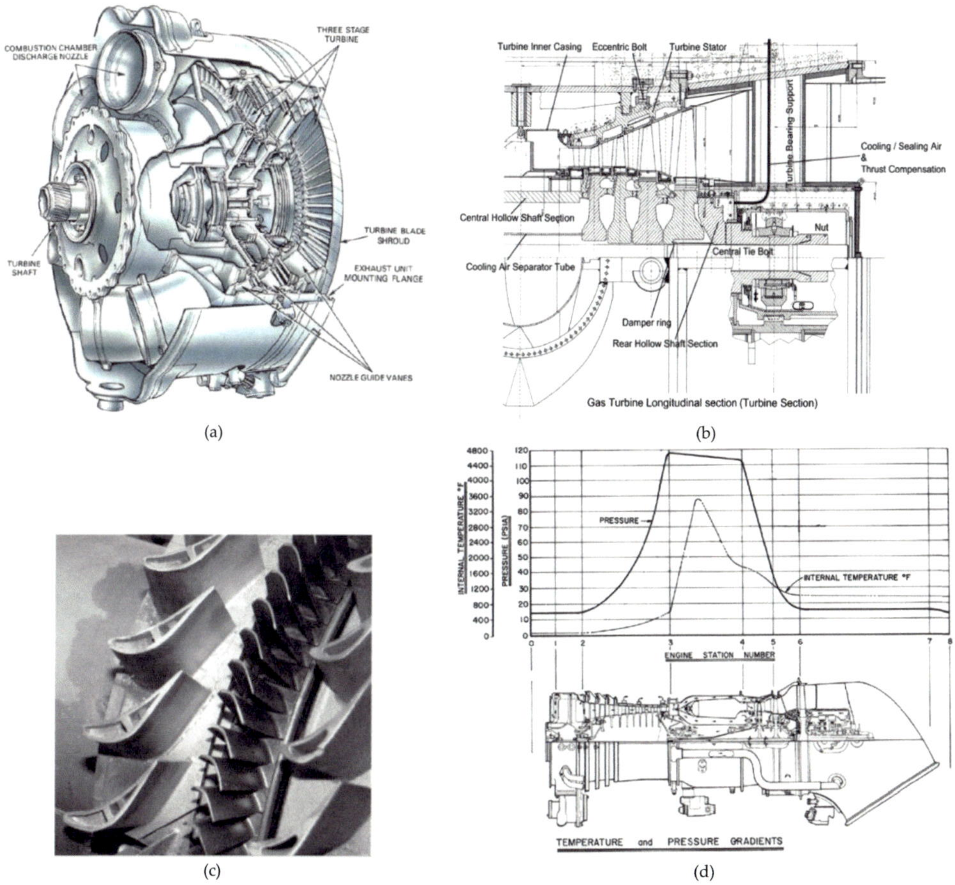

Figure 14. a: A triple-stage turbine with single-shaft system [17] b: The gas turbine section of the Siemens V94.2 gas turbine. c: Turbine stage with stators to the left which have the main function to act as nozzles to increase the velocity of the gas primarily in the tangential direction, by converting pressure energy to kinetic energy. To the right of the stators are the rotors, which have the function to convert the kinetic energy to power by causing a rotation of the shaft [4] d: Temperature and pressure throughout gas turbine [18]

2.3.4. GT turbine

The hot gases produced in the combustor are expanded in the turbine (Figures 14a–d) to give mechanical energy that operates the compressor, and the balance produces the electric power (EP). The turbine, similar to the compressor, can be axial or centrifugal type. The axial type is easier to cool, as the turbine is exposed to high thermal stresses by the hot gases entering the turbine. The turbine cooling is crucial as it provides the potential of raising the TIT and thus the efficiency. Gas turbines can be particularly efficient when heat content of the hot gases from the turbine is recovered in HRSG to power a conventional ST in GTCC. The hot gases from the GT can also be used for space or water heating or drive an absorption chiller for cooling the inlet air and increase the power output. Figure 14d shows that the hot gases leaving the GT are high enough to generate steam.

3. The Combined Gas-Steam Turbine Cycle (GTCC)

3.1. The GTCC overview

The exhaust gases leaving the GT can have high temperature (up to 600 °C) and use a heat recovery steam generator (HRSG) to generate steam. This steam can operate thermally driven desalting units such as multistage flash (MSF) (Figures 15a, b)) or multi-effect thermal vapor compression (ME-TVC) desalting systems or can operate steam turbine (ST). Combination of GT, HRSG, and ST cycle forms GTCC (Figure 15c) of much higher efficiency than single-cycle PP using GT or ST. A schematic diagram of steam turbine (Rankine) cycle components that can be combined with GT is shown in Figure 16a. Large steam turbine is usually divided into high-pressure (HP), intermediate-pressure (IP), and low-pressure (LP) cylinders (Figure 16b). In GTCC, the GT cycle is called the upper cycle, the steam turbine is called the bottom cycle, and both cycles are shown on T-s diagram in Figure 17a. Modern ST power generation, as shown in Figure 16a, is based on the Rankine cycle which includes the ideal basic cycle processes of (a) isentropic expansion in the steam turbine (ST) from 3-4 and from 5-6; (b) condensation of the steam discharged from the ST in the condenser from 6-1; (c) reversible adiabatic pumping process of condensate from condensing to the HRSG pressures, 1-2; and (d) heat addition at constant pressure in the steam generator (SG) to raise feedwater to saturation temperature, evaporate it, and superheat it from 2-3. In reheat steam cycle, the steam leaving the HP section returns to the SG from 4-5 for further heating before being admitted to the IP cylinder. Reheat is sometimes necessary to raise the steam dryness fraction at the turbine exit than the minimum of 0.88 required by the industry to avoid the blades pitting and raise the efficiency of the LP cylinder.

The use of GTCC to produce both EP gives high-energy utilization factor (UF), up to 80 %, where

$$\text{UF} = \left(\text{Work output} + \text{process heat}\right)/\text{fuel heat supplied}$$

The GTCC is usually used for baseload operations because of its high efficiency. The HRSG can have single-, double-, or triple-pressure stages. The HRSG of single and double-pressure stages and their temperature distribution are shown in Figure 17b. A bottoming steam cycle using double-pressure steam HRSG is shown in Figure 18a. A triple-pressure stage HRSG is shown in Figure 18b. Several differences exist between the steam PP cycle using conventional steam generator (SG) (Figure 19a) and steam cycle in the GTCC (Figure 19b) using HRSG of the GT. The ST plant in Figure 19a has 300 MW electric power (EP) output capacity, using reheat cycle where steam leaving the HP cylinder is reheated in the SG before its introduction to the IP cylinder. This cycle has five closed feed heaters and one open feed heater (deaerator), and the steam flow rate leaving the condenser is 197.86 kg/s, about 76 % that of throttling condition (261.1 kg/s).

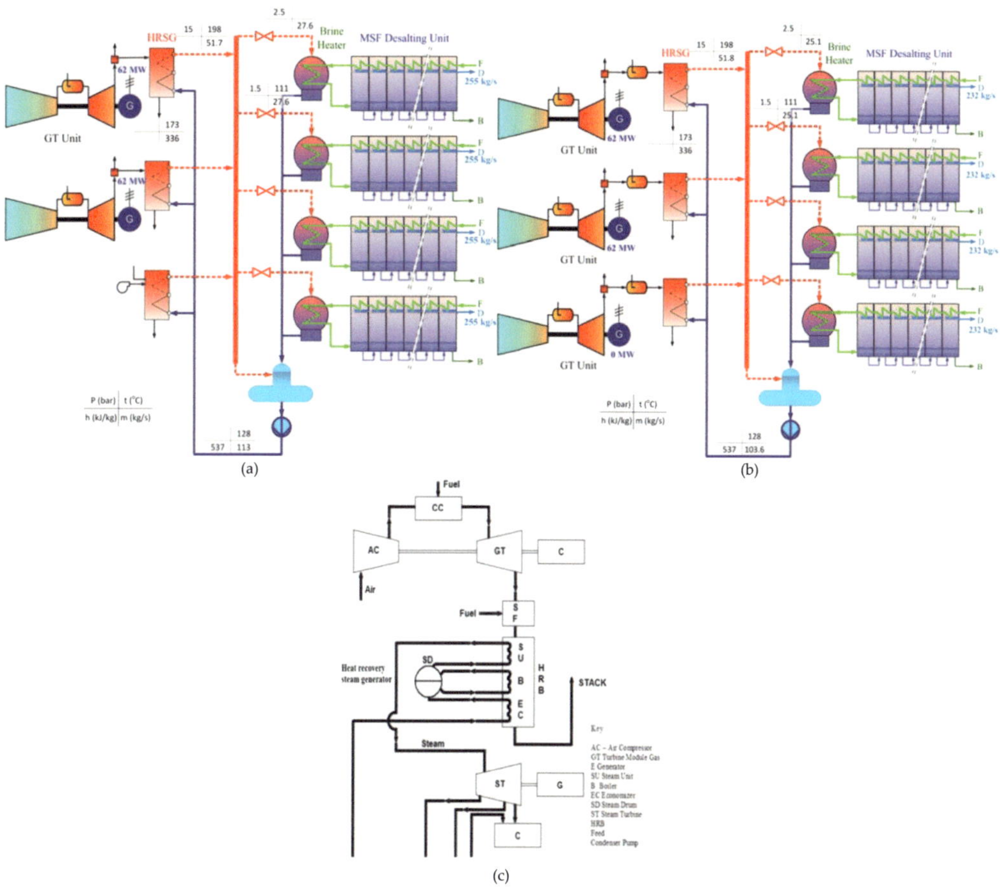

Figure 15. a: Two GTs conned to two HRSGs generating steam to operate two MSFs and using standby boiler [21] b: Three GTs conned to two HRSGs generating steam to operate three MSFs and using auxiliary burners to operate the HRSG [21] c: GTCC with GT, HRSG, and steam turbine (ST)

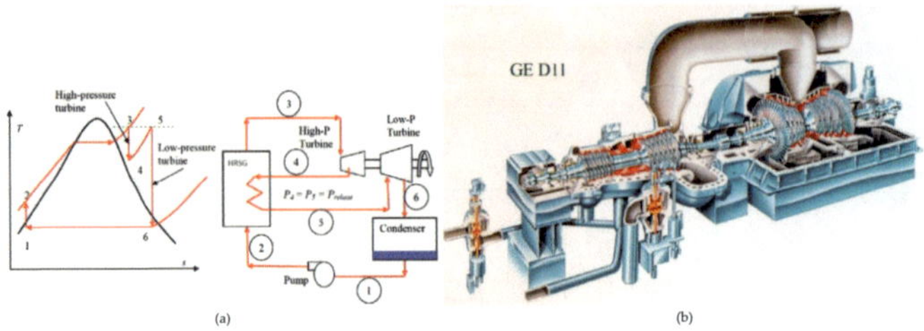

Figure 16. a: Schematic steam turbine (Rankine) using high-pressure (HP), intermediate-pressure (IP), and low-pressure (LP) cylinders b: Steam turbine with HP, IP, and LP cylinders

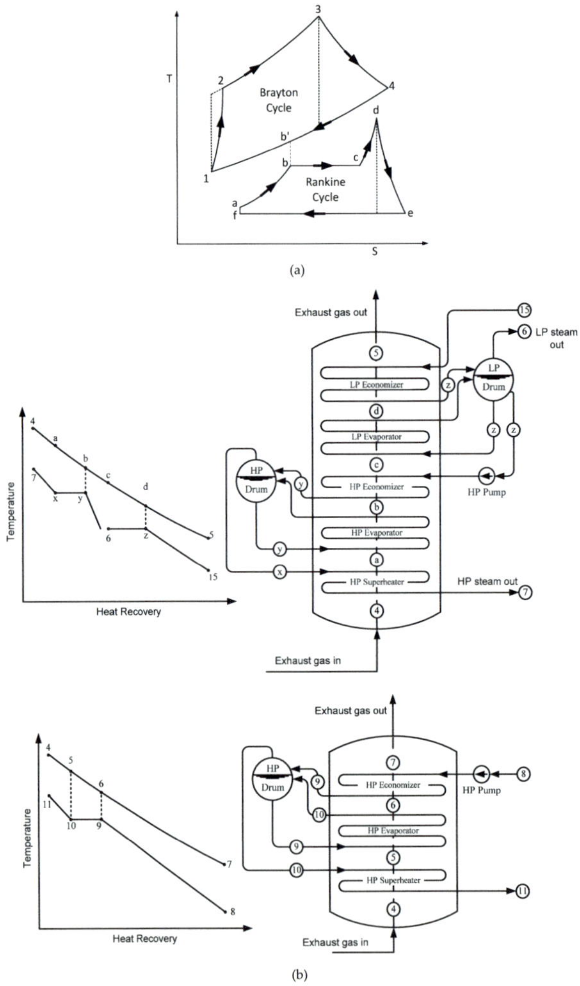

Figure 17. a: Upper GT turbine cycle (Brayton) and bottom Rankine steam cycle. b: HRSG of dual pressure in the middle and single pressure at the bottom steam stages and their temperature distribution [20]

http://dx.doi.org/10.5772/60209

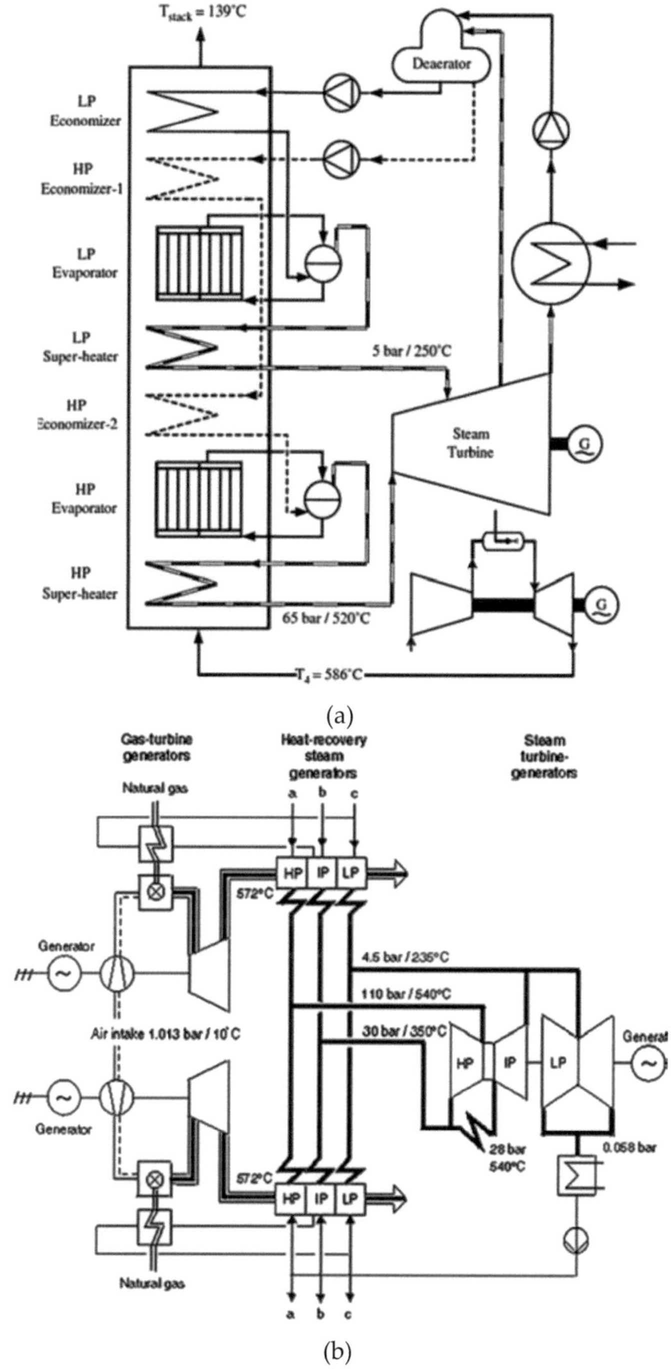

Figure 18. a: Bottom steam cycle with dual-pressure steam stages HRSG [1] b: GTCC using triple-pressure steam HRSG [23]

The ST cycle shown in Figure 19b and with data given in Table 3 is also a reheat cycle of 275 MW output capacity. It utilizes the hot gases leaving three GTs of 164 MW of EP output each. Contrary to the cycle in Figure 19a, the cycle using the HRSG has no feed heaters as all feedwater heating is done in the HRSG, and thus, the steam flow rate leaving the condenser is 438.8 kg/s, about 123.3 % that of throttling condition (355.3 kg/s), and the steam to the ST is admitted from the three stages HRSG to the ST at three points (pressures); see Figures 19a,b .

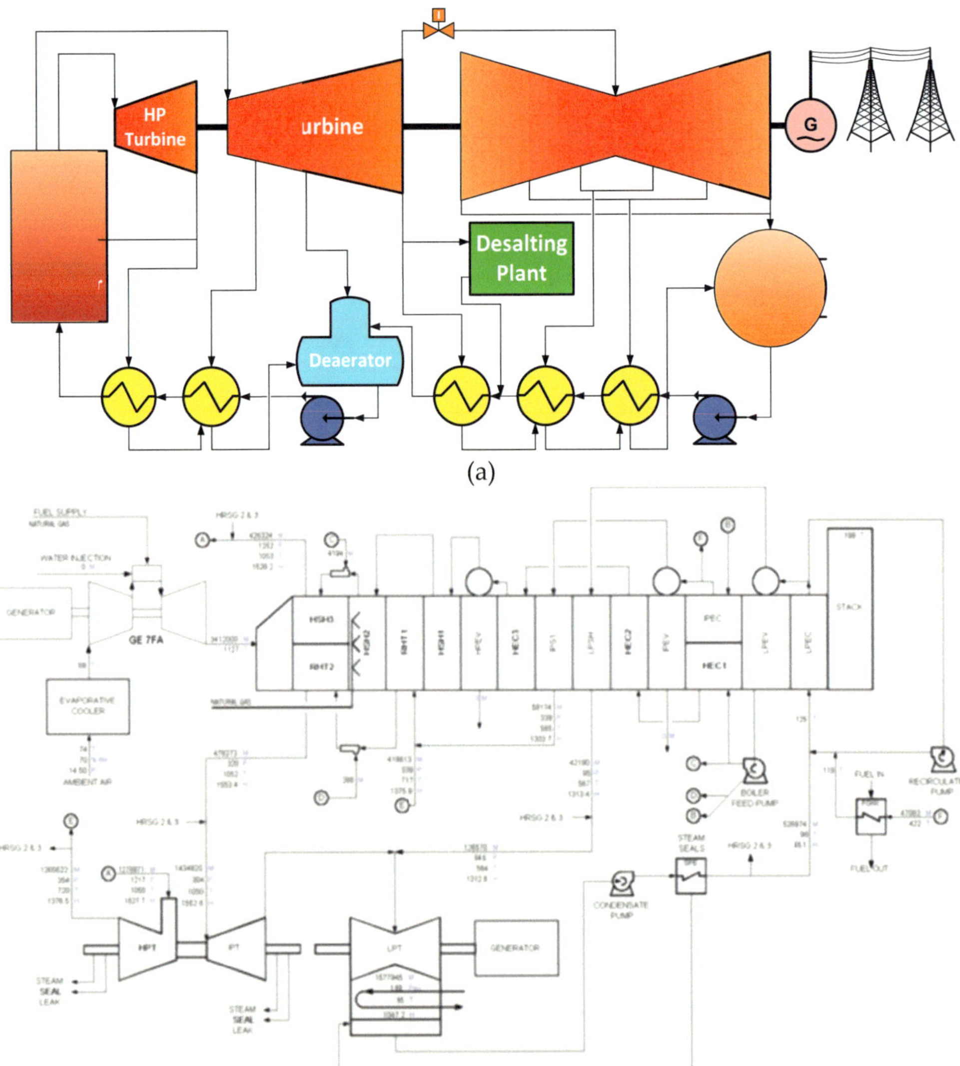

Figure 19. a: Conventional steam power plant operated by fuel-operated steam generator. b: Combined GTCC using 3 GT, 3 HRSG, and one steam turbine [5]

Case	Case 1: 74 °F ambient temperature, unfired
Ambient temperature, °F, relative humidity 70 %	
GT capacity	3×164.1 MW
GT fuel	NG
Total GT output	492.3 MW
Steam turbine output	275.55 MW
Total gross EP output	767.850
Auxiliary power	18.45 MW
Net EP output	794.4 MW
Gross GT heat rate, BTU/kWh (LHV), Btu/kWh	9,310
Gross GT fuel consumption (LHV)	4,583.3
NPHR, Btu/kWh, (LHV)	6,116
Net plant efficiency, % (LHV)	44.8

Table 3. Data of the GTCC given in Figure 19b [5]

3.2. Steam turbines in GTCC

The steam turbine in the GTCC can be extraction-condensing steam turbine (ECST) (Figure 20a) or back pressure steam turbine (BPST) (Figure 20b). In the ECST, steam is expanded from inlet pressure (say at 100 bar) and high temperature (up to 538 °C) to the condenser pressure (about 10 kPa) below atmospheric pressure. As steam expands, its pressure and temperature decrease, while its specific volume and its volumetric flow rate increase. This requires increasing the blade length of the turbine as steam expands to accommodate the increased volumetric steam flow (Figure 20c). In large-scale steam turbines, the steam volumetric flow is limited by the size of the turbine last stages (see Figure 16b), and this can enforce the use of double-flow condensing steam turbine where the last stage flow is divided between two rows of blades.

In BPST, the steam exits the turbine at the pressure required by the process to be heated as desalination, say 2–3 bar and is higher than that in the end condenser of the ECST cycle, say at 10 kPa. Condensation of discharged steam in industrial processes provides process heat needed for desalination, heating, absorption cooling, or any other processes.

The steam expansion in the ST is usually represented on the enthalpy-entropy (called Mollier chart) as turbine line shown in Figure 21a. For an adiabatic process, the change in enthalpy Δh is equal to the specific work, w per kg of flowing steam. The steam line on the *h-s* diagram would be a vertical line in reversible (ideal) expansion. The entropy increases during expansion in actual adiabatic process on a Mollier chart. The end point of the irreversible process still lies on that constant-pressure line corresponding to the exhaust pressure. Figure 21a shows that an increase in entropy during expansion decreases the work output, since the change $\Delta h(actual)$

is less than *Δh(isentropic)* as the isentropic efficiency defined by: η(isentropic) = Δh(actual)/ Δh(isentropic) <1.

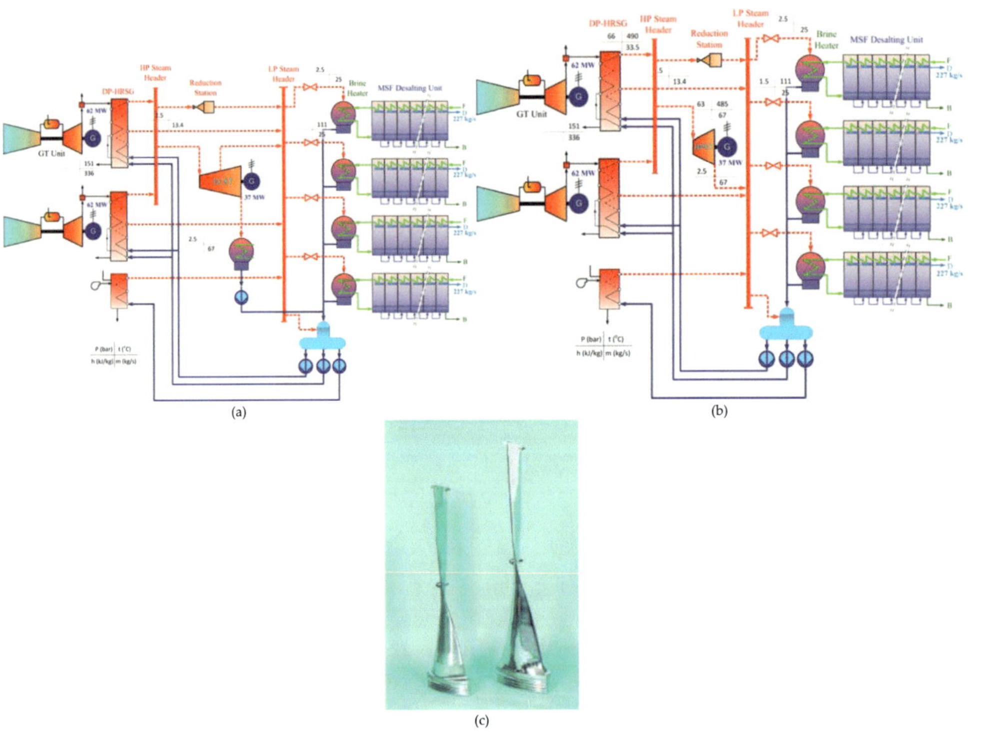

Figure 20. a: GTCC with GT, HRSG, and using extraction condensing steam turbine (ECST) [21] b: GTCC with GT, HRSG, and using back pressure steam turbine (BPST) and standby boiler [21] c: The new last stage buckets come in two sizes – 85 cm and 1.21 m [24]

One of the main concerns in the design of the ST is its exhaust size selection discharging to the condenser. Lowering the condenser pressure allows more expansion of the steam in the ST, i.e., more decrease in the enthalpy *Δh* that is transferred to work. However, decreasing the pressure increases the steam specific volume, thus increasing the steam velocity and increasing the kinetic energy loss of the steam as it leaves the turbine to condenser at almost zero velocity. Figure 21a shows that for the turbine line ABC on the h-s diagram, the exit steam dryness fraction is about 0.84, which is less than 0.88 and not acceptable. Once reheating is done, line ED, the dryness fraction increases to 0.92, which is acceptable. Figure 21b illustrates the exhaust loss curve for a condensing steam turbine. The exhaust area for a particular application should provide a balance between exhaust loss and capital investment in turbine equipment.

Some of the GTCC mount the GT and ST on the same shaft (Figures 22a, b). Since the steam turbine comes to operation after heating up the whole steam cycle, a freewheel clutch is installed between the steam turbine and the generator to prevent the GT from spinning up the steam turbine in a cold steam cycle. Due to the freewheel clutch, the shafts of the gas turbine

http://dx.doi.org/10.5772/60209

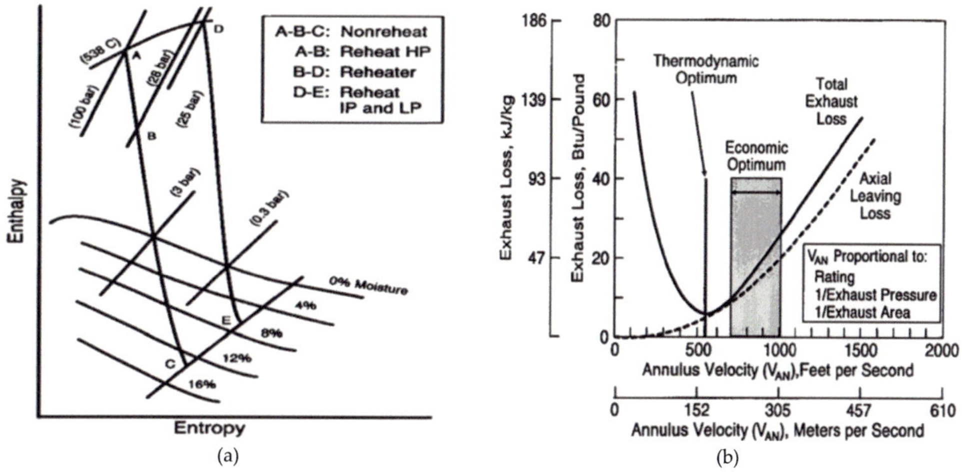

Figure 21. a: Enthalpy-entropy diagram for a steam turbine [25] b: Illustrative exhaust loss curve [25]

and the steam turbine are spinning up separately, which prevents them from reaching speed ranges that would cause dangerous resonance frequencies. As soon as the boiler is heated up to operation temperature, the control valve is opened and the steam turbine provides its part of power to drive the generator [23].

Figure 22a shows the ST mounted on the same shaft of the GT and both use the same generator, one GT and one ST of single- and double-flow LP cylinders as developed by GE. In addition, the steam turbine can be combined with single GT but with separate shafts or several gas turbines and one ST and several shafts as shown in Figures 22a, b.

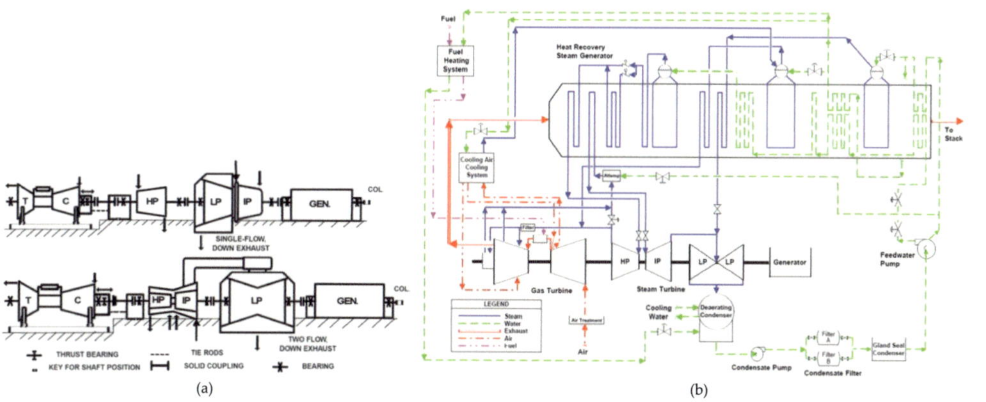

Figure 22. a: S107H and S109H single-shaft steam turbine and GT (STAG) equipment configuration [26] b: STAG 107H/109H cycle diagram [26]

3.3. Cogeneration steam turbine

Steam can be extracted from ST for processing heat by using a nonautomatic extraction ST that has openings in the turbine casing for steam extraction, with no means for controlling the pressure of the extracted steam. Steam can also be extracted from an automatic extraction steam turbine with openings in the turbine casing for extraction and means for directly regulating the steam flow to the next turbine stages after extraction opening. Automatic extraction turbines are used when there is a need for process steam at specific pressure between turbine inlet and outlet pressures, as in the case of desalination. There is simultaneous control of the desired extraction steam pressure and turbine speed, even though the demand for extraction steam and the power requirements of the driven load may vary over a wide range. Also an induction-extraction ST that can admit and exhaust steam. In extraction condensing steam turbine (ECST), the steam or part of it exits the turbine at a given pressure and may further be used. The 300 MW steam turbine operating in Kuwait provides full steam demand to two MSF desalting units of 7.2 MIGD each when the turbine EP load varies between 300 and 75 MW.

In Kuwait CPDP, the MSF unit gain ratio defined by desalted water (DW) output to heating supply S (i.e., D/S) has a typical value of 8, and the steam pressure at extraction point to the MSF at full load is 3.5 bar and is throttled to the pressure required by the MSF of 2 bar. When the turbine load is lowered, the steam pressure throughout the turbine is also lowered and reaches about 2 bar at the MSF extraction point when the turbine load is 25 % of the 300 MW nominal load. So, a throttling valve between the extraction point and the MSF is installed to keep the pressure to the MSF plant at 2 bar (Figure 19a). If the steam at the extraction point is less than 2 bar, extraction to the MSF is stopped. In this case, if the MSF can work directly from the high pressure steam supply to the turbine after being throttled and desuperheated (Figures 23a, b).

Steam condensation in the DP provides the steam latent heat as the heating source to the DP. The specific work produced by expanding steam from throttling condition of P_1 and T_1 to the condenser pressure P_o and T_o is represented by the area encircled by ABCDA in Figure 24a and the area BEFC, represents the specific rejected heat. When steam is extracted at P_3 to the DP, the specific work per kg of steam is represented by AGHD in Figure 23b, and the area GBCH is the work loss for each kg extracted to the DP. In the Kuwaiti plant, the steam to the MSF unit is extracted from crossover pipe between the intermediate-pressure (IP) and the LP cylinders (Figure 23b). So, the ratio of power to water outputs in CPDP varies as the EP load is always variable and cannot be stored, while water depends on the demand and available storage capacity. So, the EP and DW production ratio is not always constant or matched together.

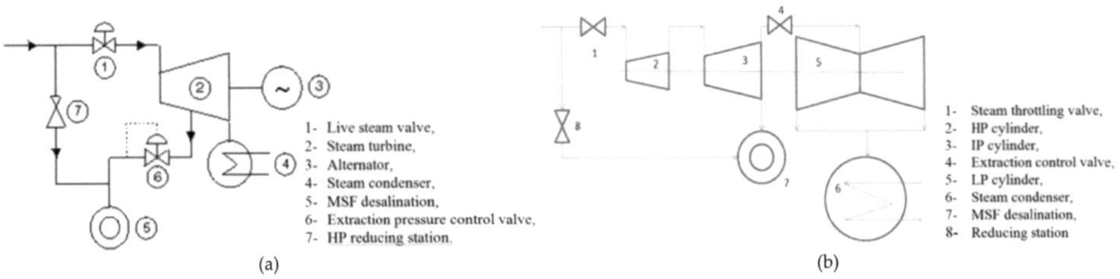

Figure 23. a: Case 1. HP-LP cylinder features for steam extraction from turbine casing [28] b: Case 2. HP-LP cylinder features for steam extraction from crossover pipe between IP and LP cylinders [28]

http://dx.doi.org/10.5772/60209

Steam condensation in the DP provides the steam latent heat as the heating source to the DP. The specific work produced by expanding steam from throttling condition of P_1 and T_1 to the condenser pressure P_o and T_o is represented by the area encircled by ABCDA in Figure 24a and the area BEFC, the rejected heat. When steam is extracted at P_3 as in Figure 24b to the DP, the specific work per kg of steam is represented by AGHD in Figure 24b, and the area GBCH is the work loss for each kg extracted to the DP.

It is noticed here that in BPST, the steam flow to the turbine depends on the turbine load, and thus, the steam discharged to the DP is slave to the turbine load. So, BPST is usually used in baseload operation, and steam to the DP can be supplied from HP steam line, which is very expensive.

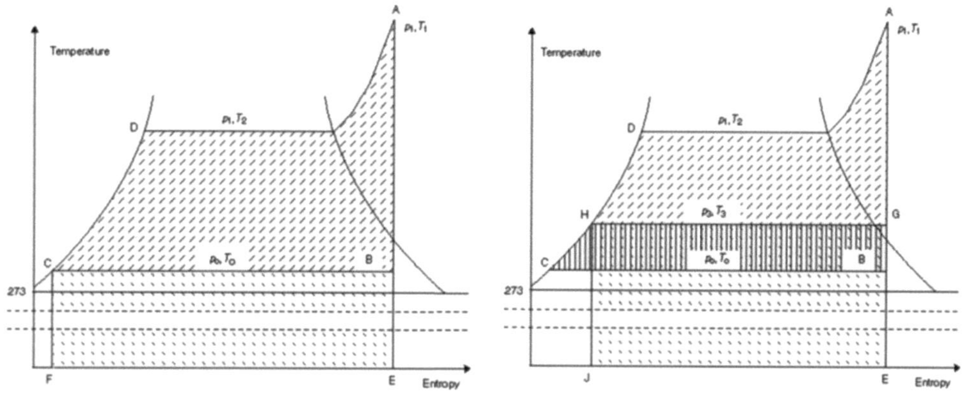

(a) (b)

Figure 24. Enthalpy-entropy diagram for a steam cycle of (a) power-only plant and (b) dual-purpose plant [23]

3.4. Heat Recovery System Generator (HRSG)

The HRSGs utilize the hot gases leaving the GT to generate steam that can be used to operate thermally driven desalting plants or steam turbines bottoming power cycle. The HRSG can be unfired, supplementary fired or called post-fired (PF), and fully fired. The HRSG can be horizontal or vertical (Figures 25a–d). As given before, the HRSG can have single-, dual-, or triple-pressure level type. The single-pressure stage HRSG has low efficiency, compared to dual-pressure HRSG. In single-pressure HRSG, high efficiency is attained by lowering the stack temperature, and this requires lowering the steam pressure. Lowering the steam pressure lowers the steam cycle efficiency. In dual-pressure designs, lowering stack temperatures would only decrease the first (low)-stage pressure while leaving the second state conditions approximately unchanged. A design parameter of the HRSG is the pinch point (pp), which is the temperature difference between the gas leaving the boiling section and generated steam saturation (or boiling) temperature. The choice of high pp increases the mean temperature difference between the hot gases and water and reduces the heat transfer area but decreases to a certain extent the HRSG efficiency. The low-pressure (LP) generated steam in dual-pressure HRSG can feed the steam turbine at a suitable point or it may be used as process steam for industrial applications (drying, desalination, absorption refrigeration, etc.).

In CPDP, electricity and process heat for desalination are simultaneously produced regardless of gas turbine load; supplementary firing or post-firing (PF) is usually used. In Ras Laffan B CPDP, a very flexible plant design was developed with PF to allow very high thermal power input (maximum 280 MWth) to cope with a wide operational range of GT electrical power and steam production for electricity or desalinated water production. The power island having a total capacity of 1025 MW is equipped with three V943A gas turbines with bypass stack to allow open-cycle operation, three HRSGSs equipped with double PF firing, and two 200 MW range backpressure steam turbines; steam from the power island is fed to four desalination units supplied by Doosan for a total water production of 273,000 m^3 per day. Each GT has 310 MW power output at generator terminals, 39.8 % efficiency, and 750 kg/s exhaust gas mass flow rate at 576 °C exhaust gas temperature.

The HRSGs are of the horizontal gas flow, top supported, natural circulation type, with single-pressure stages, and two-staged supplementary firing. The HRSG steam parameters at full GT load are pressure = 85.4 bar and temperature = 563 °C, 636 t/h nominal, and 703 maximum steam flow.

The post-firing modified the steam flow as follows: first firing increased the steam flow to nominal 110 t/h and maximum 145 t/h, and second firing increased the steam flow rate to nominal 150 t/h and 170 t/h.

4. Cost allocation in CPDP utilizing GTCC

This section develops a mathematical model to evaluate the performance of a typical CPDP using typical GTCC plant and how this performance is affected by parameters such as ambient temperature, compression ratio, air-to-fuel ratio, turbine inlet temperature, and stack temperature. The fuel consumed by the GT is allocated to each of the products (EP and DW) on the basis of the first and second laws of thermodynamics [30].

Figure 26 shows a schematic diagram of a typical CPDP using GTCC, which is considered as reference plant considered here. The plant's design is based on the data given in Table 4 and 50 °C ambient summer temperature and 600 °C temperature of exhaust gases leaving the GT.

No. of units	Fuel type	LHV	Gross output	Ambient temp.	Humidity	Pressure
3	NG	47,806 kJ/kg	215.5 MW	50 °C	30 %	1.013 bar
HRSG, type: natural circulation			Desalination: MSF	Steam turbine, ST: BPST		
No. of HRSG	Integral type deaerator	HRSG blowdown	No. of units and capacity	No. of ST	Gross capacity, MW	Cooling SW temperature
3	3	1 %	3×15 MIGD	1	215.7 MW	°C

http://dx.doi.org/10.5772/60209

No. of units	Fuel type	LHV	Gross output	Ambient temp.	Humidity	Pressure
GTCC						
Gross GTCC output, MW			Net GTCC output, MW			
862.2 MW			819.7 MW			

Table 4. Technical specifications of Shuaiba North GTCC power-desalination plant Gas turbines *GE912FA*

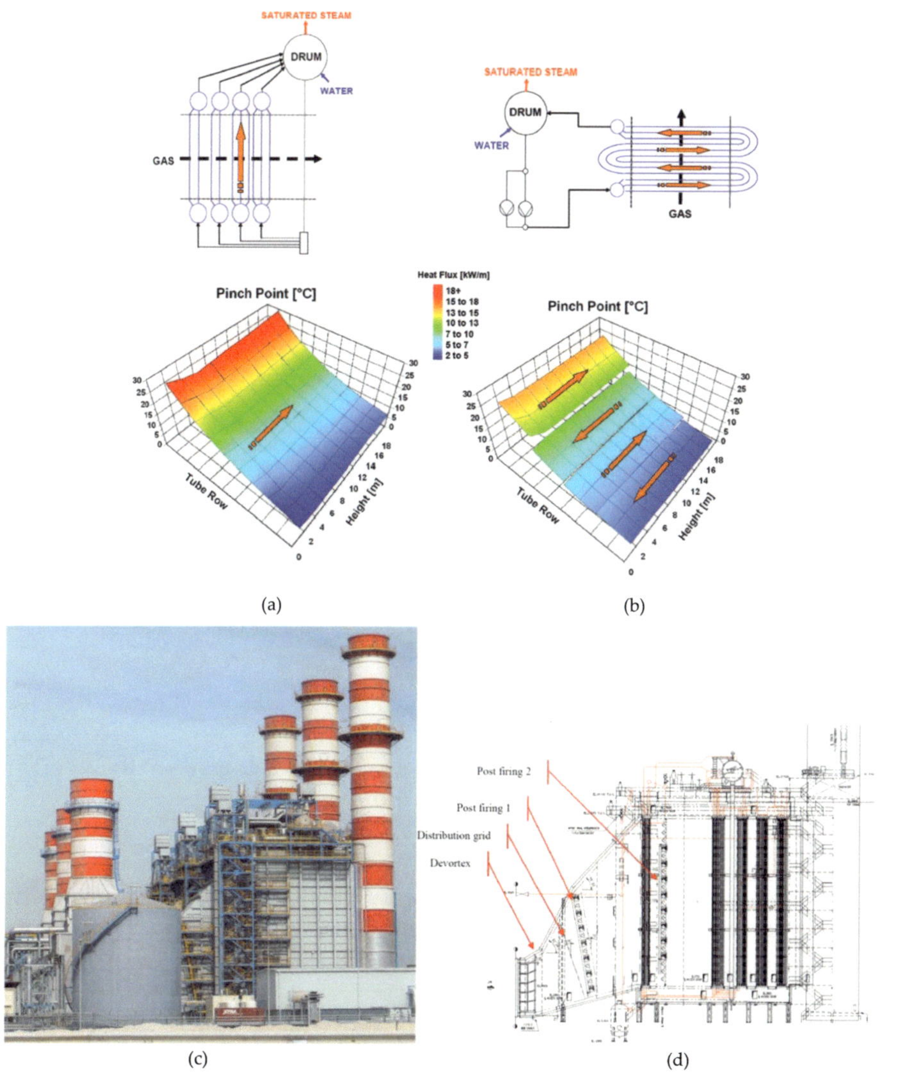

Figure 25. (a) Vertical and (b) horizontal, HRSG LPEVA arrangement and pinch diagram Figure 25c: Ras Laffan B HRSGs [29] Figure 25d: Ras Laffan longitudinal section, single pressure [29]

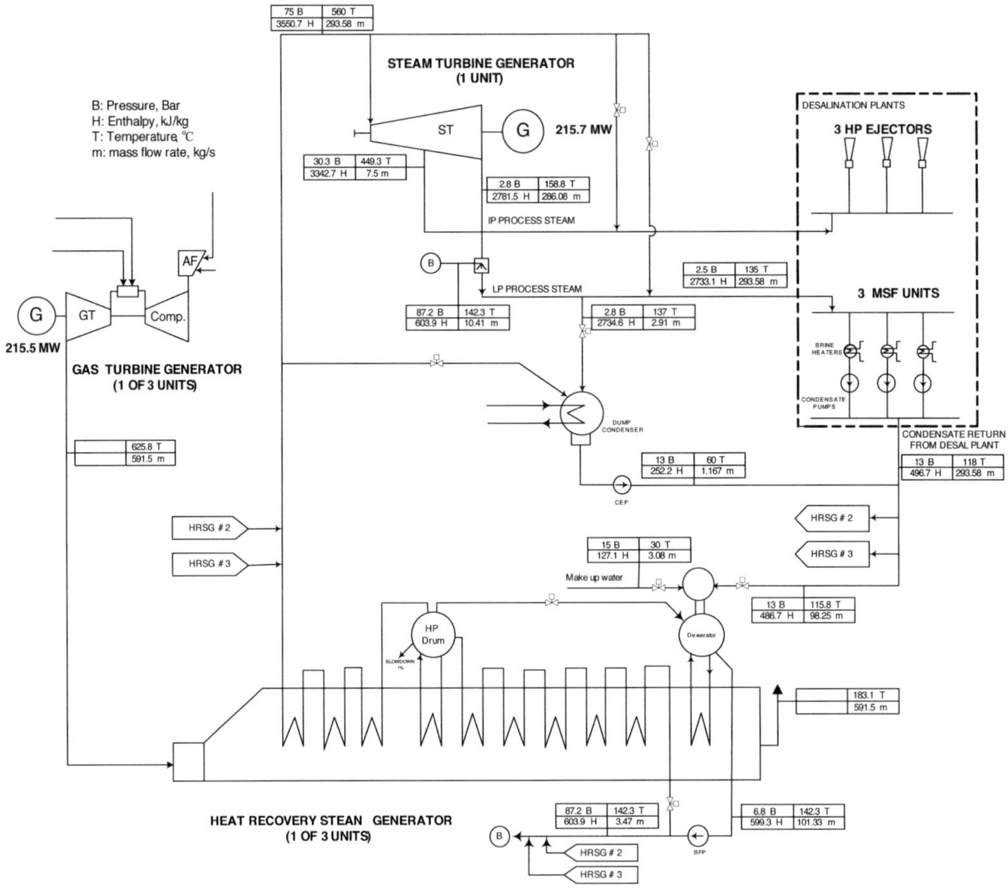

Figure 26. Mass and heat balance diagram of Shuaiba North GTCC power-desalination plant

4.1. Energy analysis

An energy analysis, based on the first law of thermodynamics, is given as follows.

4.1.1. Gas Turbine (GT) cycle

The GT cycle data give 625 °C exhaust gases exit temperature, 50 °C ambient temperature, and 215.5 MW power output for each GT. The isentropic efficiency is 0.85 for the compressor and 0.9 for the turbine. The mechanical efficiency is 0.998 for the turbine and 0.995 for the compressor.

The used fuel is NG having 47.806 MJ/kg low heating value (LHV) and 12.897 kg/s (45 t/h) flow rate. The airflow rate to each GT is 578.62 kg/s. The air-to-fuel ratio (*A*/*F*) is then 44.86; and the exhaust gases flow rate from each GT is 591.52 kg/s.

The compressor work is

$$W_c = m_a \times \frac{(h_2 - h_1)}{\eta_{mc}} = 578.62 \times \frac{(759.1 - 323.6)}{0.995} = 253.2\ MW$$

$$Q_f = m_f \times LHV = 12.897 \times 47{,}806 = 616.55\ MW$$

The heat gain by the air in is the combustion chamber

$$Q_{in} = m_a \times (h_3 - h_2) = 578.62 \times (1724 - 759.1) = 558.31\ MW$$

The turbine work output is

$$W_t = m_g \times (h_3 - h_4)\eta_m = 591.52 \times (1724 - 930.8) \times 0.998 = 468.25\ MW$$

The GT power output is

$$W_{GT} = (W_t - W_c) = 468.25 - 253.2 = 215.05\ MW$$

The gross GT cycle efficiency based on LHV is

$$\eta_{GT} = \frac{W_{GT}}{Q_f} = \frac{215.05}{616.55} = 0.35$$

4.1.2. Heat Recovery Steam Generator (HRSG)

There are three GTs, three HRSGs, and only one ST. One third (1/3) of feedwater from the steam cycle returns to each HRSG, and the heat gained by this feedwater is equal to that lost by the exhaust gases, then,

$$m_g \times C_p \times (T_4 - T_{stack}) = m_s \times (h_s - h_f)$$

where T_4 is the exhausted gas temperatures at the GT exit; T_{stack} is the HRSG stack exit; C_p is the gases' specific heat (~1.11 kJ/kg°C); h_s and h_f are the specific enthalpies in kJ/kg of superheated steam leaving the HRSG and feedwater entering the HRSG, respectively; and m_s is the steam flow rate from each HRSG. The temperature profile of the hot gases and steam-water temperature in the HRSG is shown in Figure 27.

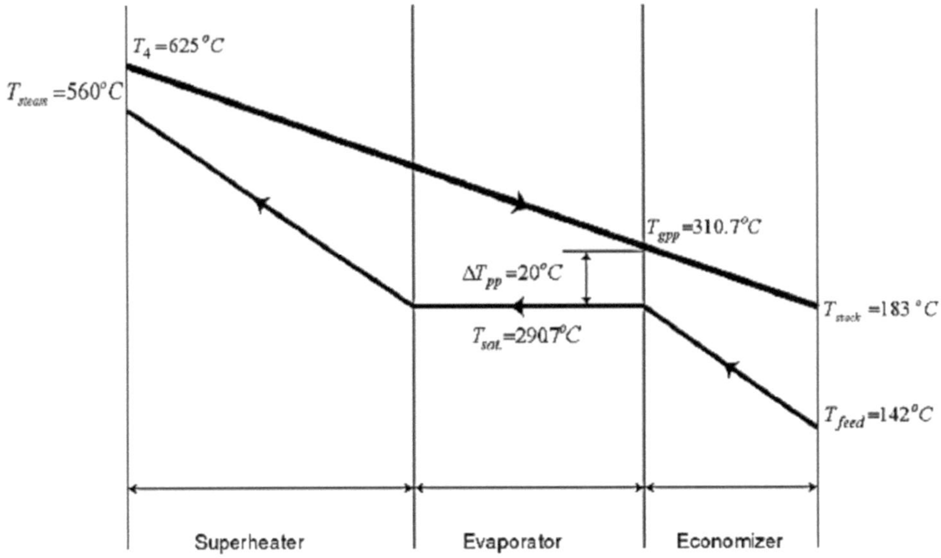

Figure 27. Gas and steam-water temperature profile of the HRSG

The superheated steam temperature T_s at the HRSG exit is determined by the terminal temperature difference (T_4 – T_s) with typical value in the range of 50 °C. The pinch point temperature (pp) difference is defined by the minimum temperature difference between the hot gases Tp and the steam saturation temperature, pp = (T_p – T_{sat}), say equal to 20 °C.

For the reference plant, m_g = 591.5 kg/s, T_4 = 625.8 °C, and T_{stak} = 183 °C, and thus the heat loss from the hot gases is

$$m_g \times C_p \times \left(T_4 - T_{stack}\right) = 591.5 \times 1.11 \times \frac{625 - 183}{1000} = 290.7\ MW$$

The feedwater is heated from its inlet feed temperature to saturation liquid temperature T_{sat}, evaporated to saturated steam, and then superheated to T_s.

The steam leaving the three HRSGs is directed to the ST at mass flow rate, $3m_s$ = 1056.9 t/h (293.58 kg/s) or m_s = 97.86 kg/s from each HRSG. The heat gain in the

HRSG by water, QHRSG is

$$Q_{HRSG} = m_s \times \left(h_s - h_f\right) = 97.86 \times \frac{3550.7 - 599.3}{1000} = 288.82\ MW$$

3*ms* = 1056.9 t/h (293.58 kg/s) or *ms* = 97.86 kg/s from each HRSG. The heat gain in the

HRSG by water, is Q_{HRSG}

This is almost equal the heat loss by hot gases; and the heat input to the steam cycle, Q_s, in from the three HRSGs is

$$Q_{s,in} = 3Q_{HRSG} = 3 \times 288.82 = 866{,}46\ MW$$

4.1.3. Steam cycle

The steam leaving the three HRSGs is directed to a back-pressure steam turbine (BPST). The steam discharged from the BPST enters the brine heaters of three MSF units.

The throttling condition of the steam inlet to the turbine is 75 bar pressure, 560 °C temperature, and 3,550.7 kJ/kg specific enthalpy. A small part of the expanded steam is extracted from the BPST to operate the steam ejectors of the MSF, at 30.3 bar, 449.3 °C, and 3,342.7 kJ/kg enthalpy, whereas the balance continues to expand and is exhausted to the three MSF units at 2.8 bar, 158.8 °C, and 2,781.5 kJ/kg enthalpy. This steam is desuperheated before entering the MSF units to 2.5 bar, 135 °C, and 2,733.1 kJ/kg enthalpy. The power generated by the BPST is

$$W_{st} = \left[\sum m_{in} h_{in} - \sum m_{out} h_{out}\right] \times \eta_{mech}$$
$$= (293.58 \times 3350.7\ - 7.5 \times 3342.7\ - 286.07 \times 2781) \times 0.99 / 1000\ = 219.4\ \mathrm{MW}$$

The power consumed by the steam cycle pumps in the steam cycle is negligible except that of the boiler feedwater pump (BFP), which can be calculated as follows:

$$W_{BFP} = \frac{v_f \times (P_{BFP} - P_{6.8})}{\eta_{BFP}} = 0.001408 \times (82200\ - 680) / (0.8 \times 1000) = 4.3\ \mathrm{MW}$$

The net power generated by the steam turbine is 215.1 MW, and the heat gained by the water in the three HRSGs is 866.46 MW. This gives the ST efficiency as

$$\eta_{st} = \frac{(W_{ST} - W_{BFP})}{Q_{in}} = \frac{(219.4 - 4.3)}{866{,}46} = 0.248$$

This efficiency underestimates the performance of the ST cycle. It does not account for the benefit gained by the steam leaving the ST to the MSF to produce desalted seawater and not expanded to an end condenser.

4.1.4. Desalination units

The heating steam mass flow rate to each MSF unit is m_s = 97.86 kg/s (one third of the steam discharged from the ST plus water used for its desuperheating). Each MSF unit produces desalted seawater (DW) at the rate D = 15 MIGD (789 kg/s). This gives gain ratio (GR=D/S) equal to mass of DW/heating steam = 789/97.75 = 8.06

The heat consumed by each MSF unit is

$$Q_{de} = m_s \times \left(h_{d,in} - h_{d,ex}\right) = 97.86 \times \frac{2733.1 - 496.7}{1000} = 277\ MW$$

where $h_{d,in}$ and $h_{d,ex}$ are the enthalpy of steam entering the MSF brine heater and its return condensate, respectively.

So, the heat consumed for each 1 m^3 of desalted water is q_d = = (218.86×1000)/789 = 277 MJ/m^3.

It is more rational to express the heat supplied to the MSF by its real value in terms of mechanical equivalent energy. The turbine work loss due to discharging its steam to brine heater of MSF unit and not expanding to an end condenser can be calculated; if this steam was expanded in low-pressure (LP) turbine to condenser pressure at 10 kPa and dryness fraction of 0.9, its enthalpy would be 2,345.5 kJ/kg, and the produced work is

$$W_{de} = m_s \times \left(h_{MSF} - h_{cond}\right) = 97.86 \times \frac{2781 - 2345.5}{1000} = 42.7\ MW$$

This 42.6 MW is equivalent mechanical work W_{de} to the heat Q_{de} = 218.86 MW supplied to each MSF unit.

Another small amount of steam is extracted from the steam turbine at higher pressure to operate the steam ejectors of each MSF plant at 2.5 kg/s flow rate, 30.3 bar pressure, 449.3 °C temperature, and 3,342.5 kJ/kg enthalpy. If this steam was expanded in a turbine to the condensing pressure of 10 kPa and 90 % dryness fraction, its enthalpy would be 2,345.5 kJ/kg, and its work output is

$$Q_{ejector} = m_{ejector} \times \left(h_{ejector} - h_{cond}\right) = 2.5 \times \frac{3342.5 - 2345.5}{1000} = 2.4925\ MW$$

So, the work loss by the steam supplied to one 15 MIGD (789 kg/s) is

$$W_{th} = W_{de} + W_{ejector} = 42.6 + 2.5 = 45.1\ \text{MW},$$

and specific work loss is equal to

http://dx.doi.org/10.5772/60209

$$45{,}100\ \text{kW}/780\ (\text{kg/s}) = 57.16\ \text{kJ/kg} = 15.9\ \text{kWh/m}^3.$$

Since the pumping energy of the MSF is in the range of 4 kWh/m^3 (14.4 kJ/kg), the total equivalent mechanical energy (counting for pumping and thermal energy) to produce 1 m^3 of desalted water is

$$W_{eq} = W_{th} + W_{pumping} = 15.9 + 4 \cong 20\ \text{kWh/m}^3 = 72\ \text{kJ/kg}$$

So, the equivalent mechanical energy required to produce DW at the rate of 45 MIGD (2,367 kg/s) can be calculated as follows:

$$Weq = 72 \times 2367 / 1000 = 170.424\ \text{MW}$$

This 170.424 MW consists of 34.1 MW for pumping energy and 136.34 MW for thermal energy.

The pumping energy of the BFP as well as for the MSF should be subtracted from the total power output of the turbines to become

$$\text{Net power output} = 3\ W_{GT} + W_{ST} - W_{pump} - W_{BFP}$$
$$= (3 \times 215.02) + 215.1 - 34.085 - 4.3 = 821.77\ \text{MW}$$

4.1.5. Total cycle

The fuel energy consumed by the three GT units is

$$Q_{f,t} = 3 \times Q_f = 3 \times 616.5 = 1849.5\ \text{MW}$$

The total power output from the GTCC is 3 $W_{GT} + W_{ST} = 645 + 215.1 = 960.1$ MW and the GTCC overall efficiency $\eta_t = 960.1/1849.5 = 0.465$.

Again, this efficiency underestimates the performance of the GTCC, since it does not account for the heat gained by the 3 MSF units.

Another term is usually considered and known as the utilization factor (UF):

$$UF = \frac{3\ W_{GT} + W_{ST} + Q_{de}}{Q_{f,t}} = \frac{645 + 215.1 + 218.86}{1849.5} = 0.82$$

where $Q_{f,t}$ is the only heat added to the three GT cycles.

In fact, the UF overestimates the performance of the CPDP since it adds the work by both GTs and ST (high-quality energy) to heat supplied to the MSF units 3 Q_{de} (low-quality energy).

A new modified total efficiency η_{mf} is

$$\eta_{mf} = \frac{3W_{GT} + W_{ST} - W_{pump} + W_{de}}{Q_{f,t}} = \frac{(645 + 215.1 - 34,1 + 170.43)}{1849.5} = 0.54$$

So, an energy balance of the given CPDP using GTCC and MSF units shows that $Q_{f,t}$ is supplied to the overall system, which is mainly converted partially to the following main items:

a. Total work output by 3 $W_{GT} + W_{ST} - W_{pump} - W_{BFP}$=3×215+ 215.1- 34.1- 4.3 = 821.7 MW (44.4 % net efficiency).

b. Heat added to 3 MSF units, $Q_{des,t}$ =3 × Q_{de} = 3×218.86 = 656.58MW (35.51 % of total heat input).

c. The heat rejected to the environment in the form of hot gases leaving the three stacks of the HRSGs is

$$Q_{staks} = 3 \times m_g \times C_p \times \left(T_{stack4} - T_e\right) = 591.5 \times 1.11 \times \frac{183.1 - 50}{1000} = 262.2\ MW$$

where *Te* is the ambient temperature.

The balance will be unaccounted energy losses,

$$\begin{aligned} Q_{losses} &= Q_f - 3W_{GT} - W_{ST} - 3Q_{de} - W_{stack} = \\ &= 1849.4 - 645 - 215.1 - 656.58 - 262.2 = 70.52\ MW \end{aligned}$$

Figure 28 shows the energy balance of the given GTCC and MSF units.

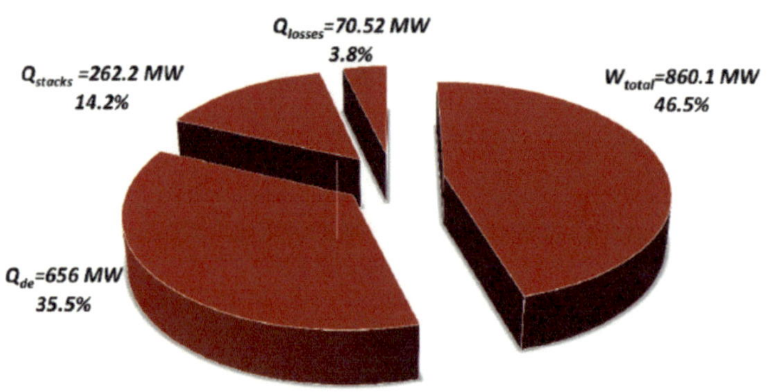

Figure 28. Energy balance of the given GTCC and MSF units (see online version for colors)

http://dx.doi.org/10.5772/60209

Q_{stack} is the largest heat loss and is accounted for about 14 % of the heat input.

The high stack temperature (183 °C) is due to the high feedwater temperature returning back from the MSF desalting units at 142 °C.

If GTCC is chosen without desalting plant, lower feedwater temperature is chosen and the stack temperature when NG without sulfur content was used; T_{stack} could be in the order of 100 °C.

5. Exergy analysis

An exergy analysis, based on the second law of thermodynamics, is conducted here for the cases considered before.

5.1. Compressor

The exergy destruction (irreversibility) in the compressor can be presented as follows:

$$I_{comp} = W_{comp} - A_{comp}$$

where A_{comp} is the increase of flow availability in the air stream across the compressor and equal to

$$A_{comp} = m_a \times [(h_2 - h_1) - T_e(s_2 - s_1)]$$

The second law efficiency of the compressor is expressed as

$$\varepsilon_{comp} = \frac{A_{comp}}{W_{comp}}$$

5.2. Combustion Chamber (cc)

The main exergy loss (or destruction) of the GT cycle occurs in the combustion chamber (cc) of the GT cycle. An exergy balance in the combustion chamber gives $E_f = E_3 - E_2 + I_{cc}$

The E_f, E_2, E_3, and i_{cc} are the exergies of fuel input, compressed air inlet, combusted gas exit, and exergy destructed (irreversibility), respectively.

E_f is almost equal to the mass fuel flow×high heating value (HHV), HHV = 55,530 kJ/kg

For each GT, the value of ($E3 - E2$) can be approximated by

$$E_3 - E_2 = m_g \times \frac{\left[\left(h_3 - h_2\right) - T_e\left(s_3 - s_2\right)\right]}{1000} = 426.196\,\text{MW}$$

Thus, the exergy destruction I_{cc} and combustion chamber second law efficiency ε_{cc} are calculated as

$$I_{cc} = E_f - \left(E_3 - E_2\right) = 716.17 - 426.19 = 290\ MW$$

ε_{cc} = Exergy gain/Exergy input =426.19/716.17 =0.595

This is much lower than the combustion chamber energy efficiency, η_{cc}, based on the thermodynamics first law and is usually assumed equal to 0.99.

5.3. Gas turbine

An exergy balance around the GT cycle gives

$$E_1 + E_f = E_4 + W_{GT} + I_{GT}$$

E_1 *and* E_4 are the exergy of air inlet to and gases leaving from the GT, respectively, and I_{GT} is the exergy destruction in the GT cycle. The values of these terms are calculated as

$$E_{4-1} = m_g \times \left[\left(h_4 - h_1\right) - T_e\left(s_4 - s_1\right)\right] / 1000 = 159.244\ \text{MW}$$

Since W_{GT} = 215 MW, the exergy destruction in the GT cycle is

$$I_{GT} = E_f - \left(E_4 - E_1\right) - \left(W_{GT}\right) = 341.4\ \text{MW}.$$

This I_{GT} (341.4 MW) includes the energy destruction in the combustion chamber (Icc = 290 MW) and the balance = 51.4 MW is the exergy destruction in the cycle components, other than the combustion process.

The exergy difference utilized to produce the W_{GT} is ($E3 - E4$) = 241.5 MW.

For the three GTs, this exergy difference is 3 ×241.5 = 724.5 MW.

http://dx.doi.org/10.5772/60209

So, the effectiveness of the GT, ε_{GT} (without combustion chamber losses), is

$$\varepsilon_{GT}\left(\text{without combustion energy losses}\right) = 215/241.5 = 0.89$$

and the exergy destruction in the turbine, compressor due to friction is

$$I_{fric} = E_3 - E_4 - W_{GT} = \left(241.5 - 215\right) = 26.5\text{MW}.$$

The effectiveness of the GT cycle, when used as simple GT cycle is

$\varepsilon_{GT}(total) = W_{GT}/E_f = 215/716.17 = 0.3$ which is the same with gross efficiency based on HHV.

When GTCC is used, the exergy of the exhaust gases E_4 is utilized to generate steam and operate steam turbine, and ε_{GT} in a GTCC is

$$\varepsilon_{GT}\left(in\ GTCC\right) = \frac{W_{GT} + (E_4 - E_1)}{E_F} = \frac{215 + 159.2}{716.17} = 0.523$$

5.4. Heat Recovery Steam Generator (HRSG)

In heat recovery steam generator (HRSG), the heat of hot gases leaving the GT is transferred to feedwater in deaerator, economizer, evaporator, and superheater and the heat transfer in the HRSG.

An exergy balance around the HRSG gives

$$\Delta E_g = \Delta E_w + I_{HRSG}$$

where ΔEg is the exergy loss by the hot gases which is equal to the exergy gain by the water ΔEw plus exergy destruction in one HRSG, I_{HRSG}

$$\Delta E_g = m_g \times [\left(C_p \times \left(T_4 - T_{stack}\right) - T_e\left(s_4 - s_{stack}\right)\right] = 129.108\ MW$$

$$\Delta E_w = m_w \times \left[\left(h_s - h_f\right) - T_e\left(s_s - s_f\right)\right] = 116.438\ MW$$

Then, $I_{HRSG} = (\Delta E_g - \Delta E_w) = 12.67\ MW$, and the effectiveness of HRSG is $\varepsilon_{HRSG} = \frac{\Delta E_w}{\Delta E_g} = 0.9$

So, the exergy difference gained by water in 3 HRSG = 3 ×116.438 = 349.314 MW, and this is the exergy input to the steam cycle including the three MSF units.

5.5. Steam turbine cycle

The exergy difference across the ST cycle, ⊗*EST*, is equal to (*Esi* – *Ese*), where *Esi* and *Ese* are the exergy of the steam inlet to the turbine and steam outlet to the MSF units, respectively.

$$\Delta E_{ST} = 3 \times m_s \times \left[\left(h_{si} - h_{se}\right) - T_e\left(s_{si} - s_{fe}\right)\right] =$$
$$= 3 \times 97.86 \times \left[\left(3550.7 - 2781.5\right) - 323\left(6{,}95 - 7.2\right)\right] / 1000 = 249.531\ MW$$

where s_{si} and s_{se} are the specific entropy of steam at the turbine inlet and outlet, respectively.

Then the exergy loss in the ST is

$$I_{ST}\left(loss\right) = \Delta E_{ST} - W_{ST} = 249.531 - 215.1 = 34.431\text{MW}$$

5.6. Desalination system

The exergy difference between the discharged steams from the turbine to the condensate from the brine heaters of one MSF unit, ΔEde is as follows:

$$\Delta E_{de} = m_{sd} \times \left[\left(h_{d,in} - h_{d,e}\right) - T_e\left(s_{d,in} - s_{de}\right)\right] =$$
$$= 97.86 \times [\left(2783.5 - 496\right) - \frac{323\left(7.2 - 1.52\right)}{1000} = 39.34\ MW$$

where m_{sd} and $(h_{d,in} - h_{d,e})$, $s_{d,in}$, and s_{de} are the steam flow rate to each MSF desalting unit, its specific enthalpy difference between the steam inlet, and its condensate exit from the desalting unit, specific entropy at steam inlet, and specific entropy of its condensate at the exit, respectively.

So, an exergy balance of the given CPDP using GTCC and MSF units shows that there are unaccounted losses due to steam extracted at moderately high pressure to operate the steam ejectors of the MSF units, and others and can be calculated as follows:

$$E_{unacct.} = E_{f(total)} - W_{GT(total)} - \Delta E_{GT(total)} - \Delta E_{HRSG(total)} - W_{ST} - \Delta E_{de(total)} - DE_{ST} - E_{rej} \quad (50)$$
$$E_{unacct.} = 2148.5 - 646.5 - 1024.2 - 38 - 215.7 - 134.56 - 33.83 - 38.85 = 16.85\ MW$$

5.7. Exergy distribution of the overall GTCC

Exergy balance of the given CPDP using GTCC and MSF units is conducted for the whole GTCC cycle.

The fuel exergy of the fuel supplied for the three GTs is

http://dx.doi.org/10.5772/60209

$$E_{f(total)} = 3 \times m_f \times HHV = 2148.5\ MW$$

This fuel exergy is used to produce the power output power from the three GT= 3× 215 = 646.5 MW.

So, the exergy loss from the GT cycles is

$$\Delta E_{GT(total)} = 3 \times I_{GT} = 3 \times 341.4 = 1024.2\ MW$$

The exergy destruction in the three HRSG can be calculated as follows:

$$\Delta E_{HRSG(total)} = 3 \times\ I_{HRSG} = 3 \times 12.67 = 38\ MW$$

The fuel exergy is utilized to produce output power from the steam turbine is

$$W_{ST} = 1 \times 215.7 = 215.7\ MW$$

The fuel exergy used to produce 45 MIGD from 3 MSF units is

$$\Delta E_{de(total)} = 3 \times \Delta E_{de} = 3 \times 39.34 = 134.56\ MW$$

The exergy destruction in the steam turbine is I_{ST} =33.83 MW and exergy loss to environment through the HRSG stacks can be calculated as follows:

$$E_{rej} = 3 \times m_g \times \left[\frac{\left(h_{stuck} - h_1\right) - T_1\left(S_{stuck} - S_1\right)}{1000} \right] =$$
$$= 3 \times 591.3 \times \left[\frac{\left(458.1 - 323.6\right) - 323\left(6.125 - 5.776\right)}{1000} \right] = 38.85\ MW$$

There are unaccounted losses due to steam extracted at moderately high pressure to operate the steam ejectors of the MSF units, and others and can be calculated as follows:

$$E_{unacct.} = E_{f(total)} - W_{GT(total)} - \Delta E_{GT(total)} - \Delta E_{HRSG(total)} - W_{ST} - \Delta E_{de(total)} - DE_{ST} - E_{rej}$$
$$= 2148.5 - 646.5 - 1024.2 - 38 - 215.7 - 134.56 - 33.83 - 38.85 = 16.85\ MW$$

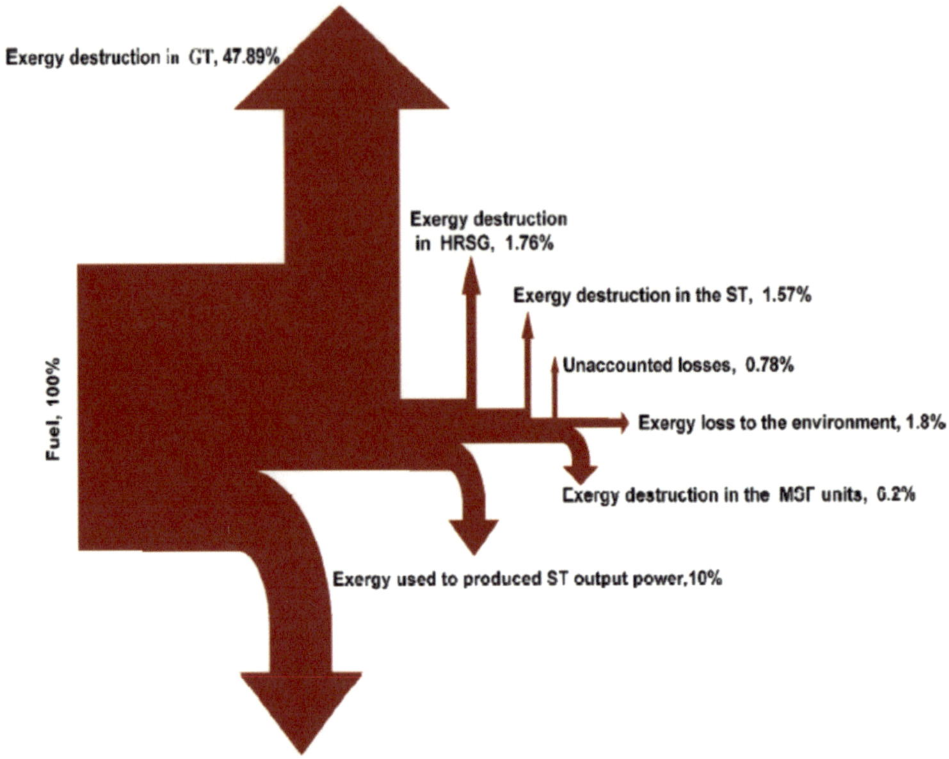

Figure 29. Grassmann diagram of the GTCC system (see online version for colors)

6. Fuel allocation between the EP and DW production

There are two methods to allocate the fuel between the EP and DW production, while the second is exergy method.

6.1. Work loss method

The first method is the work loss method. As mentioned, there is work loss due to discharging steam to the MSF units instead of its expansion to the condensing turbine. The CPDP outputs are
3 W_{GT} = 646.5 MW by the three GT, W_{ST} = 215.7 MW by steam turbine, and thermal energy input to the 3 MSFs, $3Q_{de}$ =3×218.68 = 656 MW. It was showed that $3Q_{de}$ causes the loss (or equivalent to) 3 W_{de} = 136.34 MW. The MSF units consume pumping energy at the rate of $W_{(pumping)}$= 34.1 MW, which should be deducted from the total power output.

So, the fuel charged to desalination to the total fuel supply should be

http://dx.doi.org/10.5772/60209

$$\text{Fuel}_{\text{energy to desal.}} = \left(\frac{W_{pumping} + 3W_{de}}{3W_{GT} + W_{ST} + 3W_{de}} \right) \times Q_f =$$

$$= \left(\frac{34.1 + 136.34}{646.5 + 215.7 + 136.34} \right) \times 1847.1 = 304.77 \text{ MW}$$

The specific fuel energy charged to produce 1 m^3 can be calculated as follows:

$$\text{Spesific}_{\text{fuel energy to desal.}} = \left(\frac{\text{Fuel}_{\text{energy to desal.}} \times 24 \times 60 \times 60}{45 \text{ MIGD} \times 4550 (\frac{m^3}{day})} \right) = 128 \frac{MJ}{m^3}$$

The fuel charged to produce the net power output can be calculated as

$$\text{Fuel}_{\text{energy to power}} = Q_f - \text{Fuel}_{\text{energy to desal.}} = 1847.1 - 304.77 = 1542.3 \text{ MW}$$

6.2. Exergy method

The aim of combining the 3 MSF units with the GTCC is to supply these units with its heat needs, 3 Q_{de} =3×218.68=656 MW. The exergy difference across the three MSF units is 3Δ E_{de} = 134.56 MW and represents the exergy consumed by the desalting system.

The pumping work $W_{(pumping)}$= 34.1 MW and work loss due to extraction of steam to the MSF steam ejector (2.4925 MW) should be added to 3Δ E_{de} to become 171.15 MW

This almost the same work was charged to the desalting units in the method of lost work, and there is no need to repeat the share of desalting in the fuel energy again.

It is clear that both methods give very close results, but the first method is easier and understandable by practitioner engineers.

6.3. Desalinated water cost

Since all combined cycle power plants in Kuwait were dual fuel (i.e., can be operated either by natural gas or heavy oil), the cost of desalinated water is evaluated in this section based on the current oil and natural gas prices. Hence, the desalinated water produced by this plant will be estimated based on two different types of fuel as follows:

The oil price is 60 $/bbl and the low heating value of the oil is LHV_{oil} = 42229 kJ/kg; so, the energy content in 1 barrel of oil (density of 900 kg/m^3) can be calculated as follows:

1 barrel=0.159 $m^3 \times 900 \frac{kg}{m^3} \times 42299 \frac{kJ}{kg}$ =6. 04 GJ; so, the oil price per GJ will be $9.33/GJ

$$\text{The desalinated water cost} = \text{Spesific}_{\text{fuel energy to desal.}} \frac{GJ}{m^3} \times \text{fuel price} \frac{\$}{GJ}$$

$$\text{The desalinated water cost} = 0.128 \frac{GJ}{m^3} \times 9.33 \frac{\$}{GJ} = 1.272 \frac{\$}{m^3}$$

When the gas price is \$2/MMBTU which is equivalent to 1.895 \$/GJ and LHV_{NG}= 47806 kJ/kg.

$$\text{The desalinated water cost} = 0.128 \frac{GJ}{m^3} \times 1.895 \frac{\$}{GJ} = 0.243 \frac{\$}{m^3}$$

The following assumption is assumed for the analysis: steady state operation.

7. Sensitivity analysis

The developed equations were used to evaluate the performance of the reference CPDP using the GTCC in this section. A simplified schematic diagram of plant is shown in Figure 30, while state point conditions of the model are given in Table 5.

The model was tested against the available data of Al-Shuaiba CCPP, and the results showed good agreements as shown in Table 4. This model can also be used for simulation and/or parametric studies of the plants in order to evaluate its performance. A sensitivity analysis is carried out to investigate the effects of some combined cycle parameters on the overall efficiency, specific fuel energy to desalination, as well as the desalinated water cost. The selected parameters are ambient air temperature, compression ratio, and air-to-fuel ratio, turbine inlet temperature, and stuck temperature.

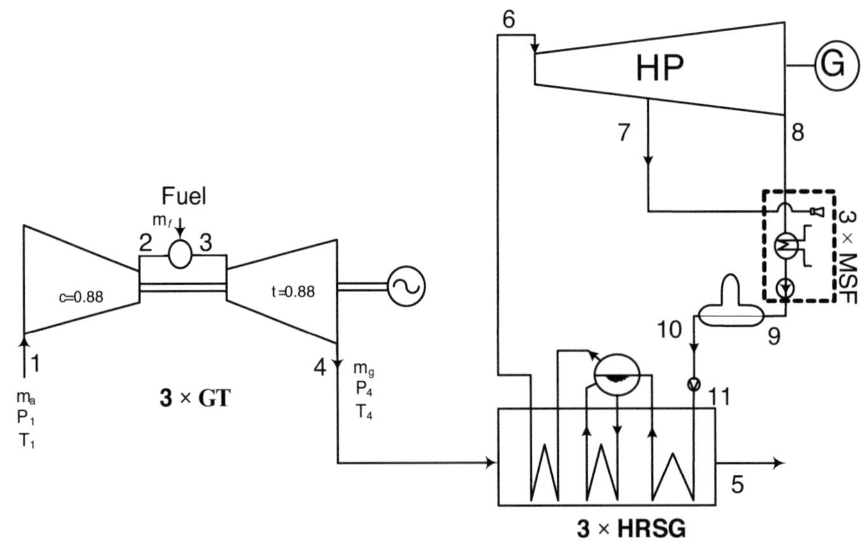

Figure 30. A simplified schematic diagram of Al-Shuaiba combined cycle power plant model

Point	Mass, kg/s	Pressure, Bar	Temp. °C	Enthalpy, kJ/kg	Entropy, kJ/kg	Exergy, kJ/kg
1	578.4	1.013	50	323.6	5.776	-1398
2	578.4	17.73	478.4	768.9	5.827	-967.6
3	591.3	17.73	1297	1720	6.678	-269.6
4	591.3	1.013	625	930.8	6.845	-1109
5	3 × 591.3	1.013	183	458.1	6.125	-1367
6	293.58	75	560	3520	6.941	1481
7	3 × 2.5	30.3	449.3	3342	7.076	1233
8	286.08	2.8	158.8	2781	7.155	648.4
9	293.58	13	118	495.4	1.503	47.39
10	293.58	6.8	142.3	598.7	1.761	73.81
11	293.58	87.2	142.3	603.9	1.753	81.5

Table 5. State point conditions of the model

Model	Actual	Operating and design conditions
		Gas turbine, 3 GT
578.62	578.62	Air mass flow rate, kg/s
44.86	44.86	Air-to-fuel ratio, A:F
17.5	NA	Pressure ratio, r_p
1297	NA	Turbine inlet temperature, °C
625	625	Exhaust gas temperature, °C
88.3	NA	Compressor isentropic efficiency, %
85	NA	Turbine isentropic efficiency, %
215.6	215.5	Power output, MW
35	NA	Gas turbine cycle efficiency, %
		Steam turbine, 1 ST
293.58	293.58	Steam mass flow rate, kg/s
75	75	High turbine inlet pressure, bar
560	560	High turbine inlet temperature, °C
2.8	2.8	Inlet pressure to MSF, bar
158.8	158.8	Inlet temperature to MSF, °C
215.5	215.7	Power output, MW
24.88	NA	Steam turbine cycle efficiency, %
		Heat recovery steam generator, 3 HRSG
6.8	6.8	Feedwater pressure, bar
142.6	142.3	Feedwater temperature, °C
20	20	Pinch point temperature difference, °C
183	183.1	Stack temperature, °C
		Desalination unit, 3 MSF
8.06	8	Gain ratio

Model	Actual	Operating and design conditions
789	789	Distillate output, kg/s
2.5	2.5	Inlet pressure to MSF, bar
135	135	Inlet temperature to MSF °C
2.5	2.5	Steam flow rate to ejector, kg/s
30	30	Inlet pressure to ejector, bar
450	449.3	Inlet temperature to ejector, °C
		GTCC
46.66	46.7	Combined cycle efficiency, %

Table 6. Mathematical model results against actual plant

The effect of ambient air temperature on the fuel allocation between the electric power and desalinated seawater production is presented in Figure 31. It shows that as the ambient air temperature increases, the allocated fuel to the electric power decreases, and this led to increase the allocated fuel to desalination. Figure 32 shows the effect of ambient temperature on the combined cycle efficiency at different compression ratios. It is clear that the cycle efficiency is the highest at maximum pressure ratio and minimum ambient temperature. On the other hand, the effect of air-to-fuel ratio is limited as shown in Figure 33.

Figure 34 shows the effect of ambient and turbine inlet temperatures (TIT) on the specific fuel energy to desalination. It shows that the specific fuel energy to desalination increases at high ambient temperatures, while it decreases at higher turbine inlet temperatures.

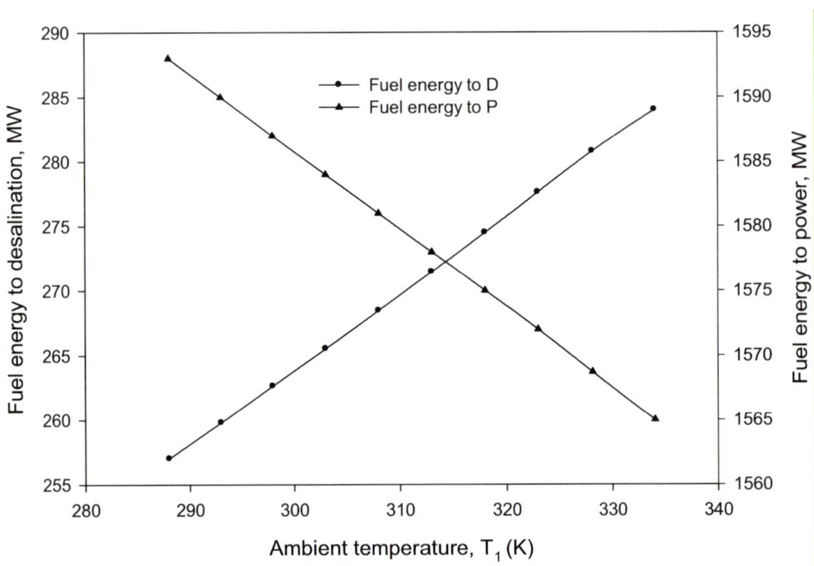

Figure 31. The effect of air ambient temperature to fuel allocation between the electric power and desalinated seawater production

http://dx.doi.org/10.5772/60209

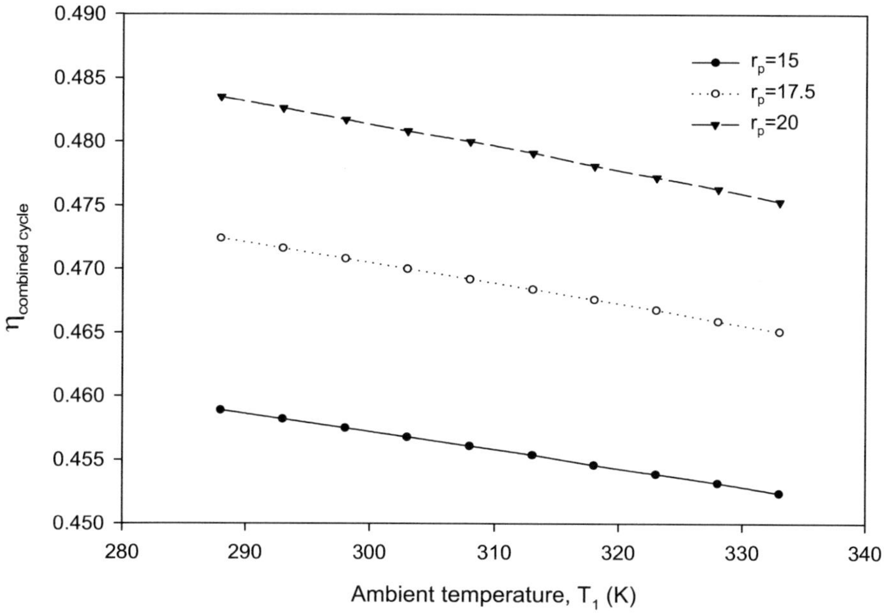

Figure 32. The effect of ambient temperature on the combined cycle efficiency at different compression ratios

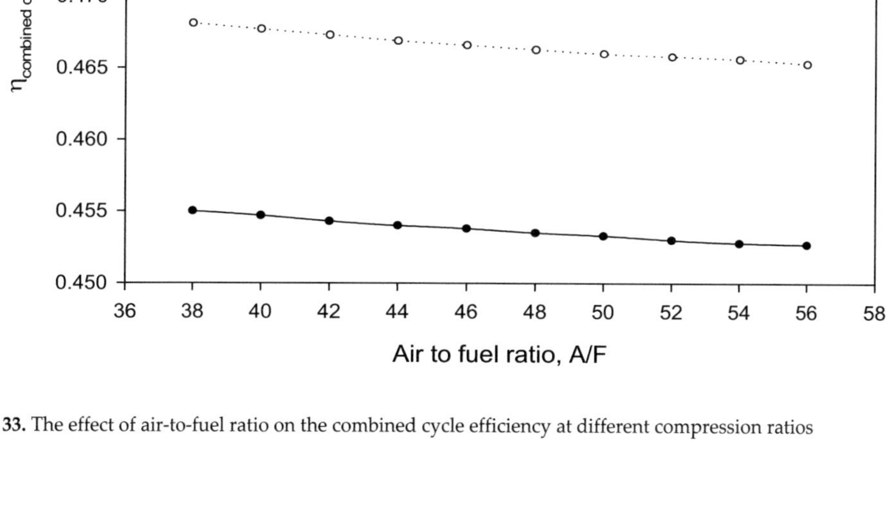

Figure 33. The effect of air-to-fuel ratio on the combined cycle efficiency at different compression ratios

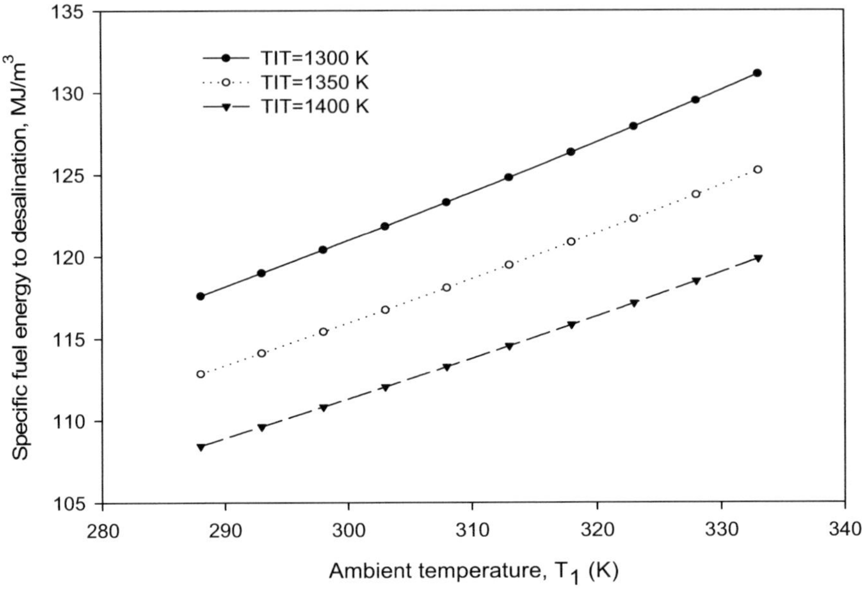

Figure 34. The effect of ambient temperature on the specific fuel energy to desalination at different turbine inlet temperatures

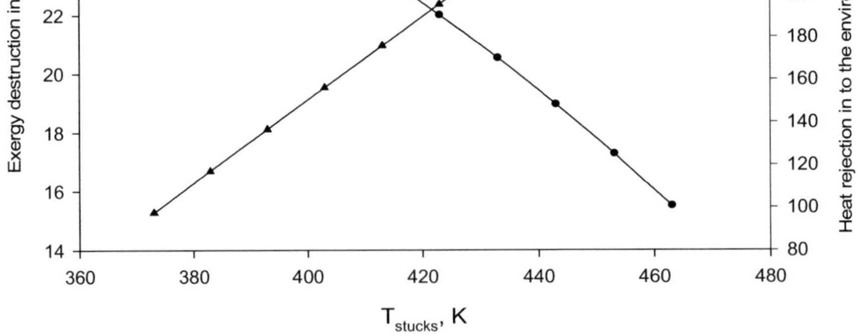

Figure 35. The effect of stack temperatures on the heat rejection into environment as well as on the HRSG exergy destruction

On the other hand, the effect of stack temperatures on the heat rejection into environment as well as on the exergy destruction of HRSG is depicted in Figure 35. It is clear that as the stack temperatures increases, the heat rejection into the environment will increase and consequently the exergy destruction in HRSG will decrease.

The main real problem by this plant is the high stack temperature of 183 °C because the high feedwater returns from the MSF desalting units at 135 °C. If the stack temperature is reduced to 110 °C, the heat gained by the water in the HRSG temperature is increased to 16.5 %, and the ST output would increase to 251.32 MW. The difference between this ST output and actual output of 215.7 by the ST or 35.62 should be charged as well to the desalting process. Again for net efficiency of 44 %, the fuel energy of 80.955 MW should be added to the 188.8 MW calculated before, or the fuel charged for thermal energy would be 269.75 or 113 MJ/m^3.

Although the effect of steam turbine inlet pressure on the specific fuel energy to desalination is negligible, it will slightly be affected by the types of fuel as shown in Figure 37.

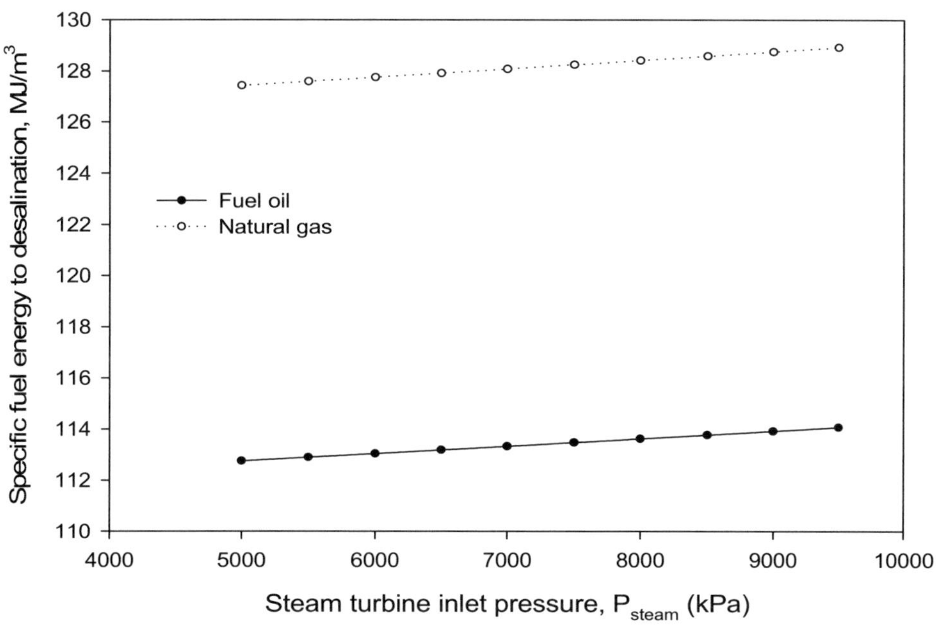

Figure 36. The effect of steam turbine inlet pressure on the specific fuel energy to desalination at different types of fuel

Figure 38 shows the effect of ambient temperature on the specific fuel energy to desalination when different types of fuel are used. Since there is a large difference in price per unit of energy between oil and natural gas, the cost of desalinated water is strongly affected by the fuel type as shown in Figure 39.

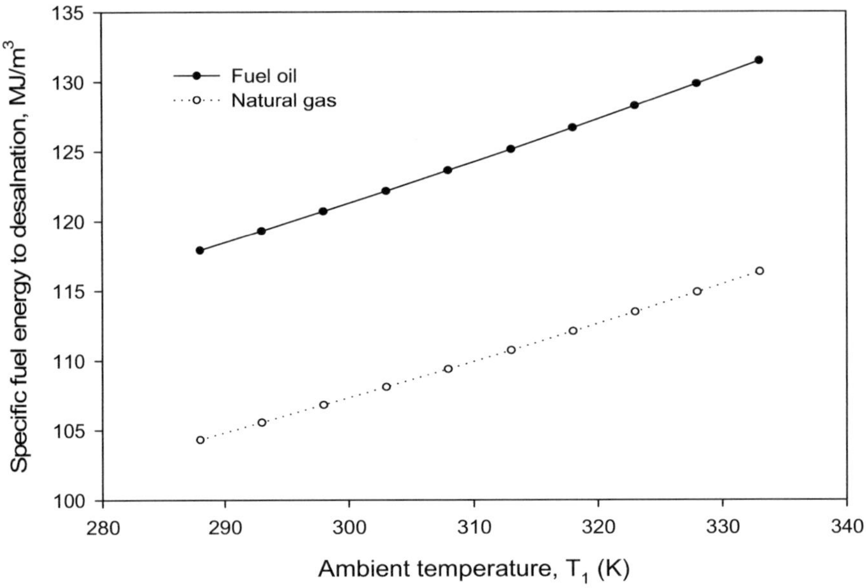

Figure 37. The effect of ambient temperature on the specific fuel energy to desalination at different types of fuel

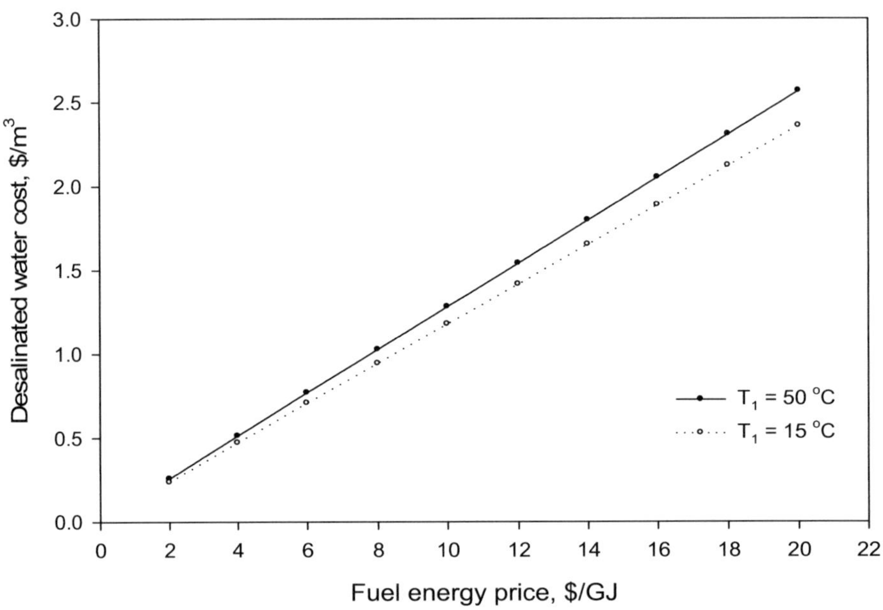

Figure 38. The effect of fuel energy price on the desalinated water cost at different ambient temperatures

http://dx.doi.org/10.5772/60209

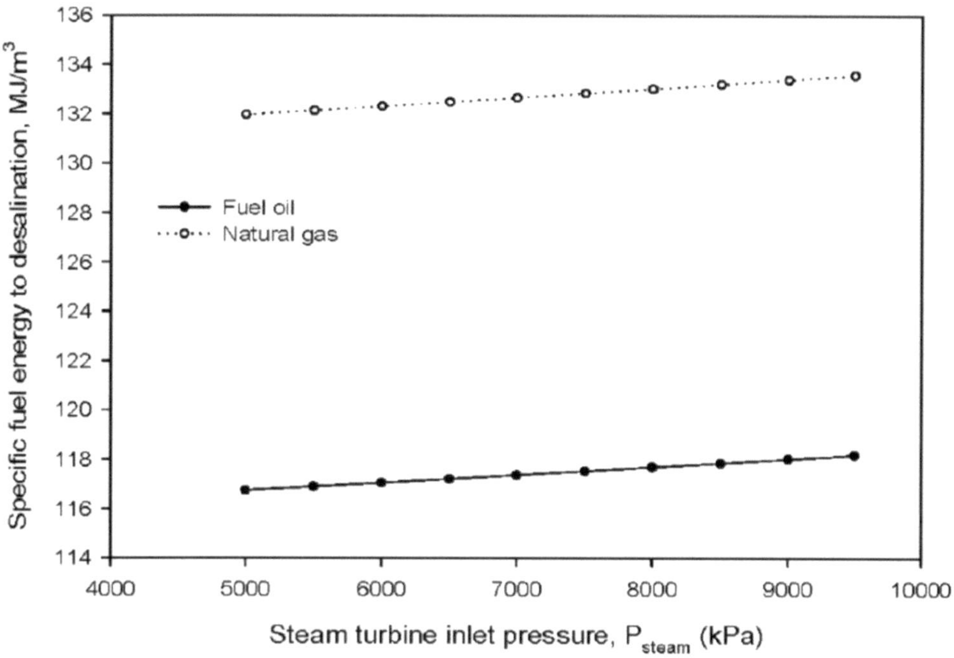

Figure 39. The effect of steam turbine inlet pressure on the fuel energy allocated to DW at different types of fuel

8. Conclusion

A general overview on the CPDP using GTCC is presented including description and analysis of the GTCC component. Energy and exergy analyses, based on the first and second laws of thermodynamics, respectively, were conducted on CPDP using GTCC connected with MSF desalting system. The concept of work loss due to exhausting steam from the ST at higher pressure and temperatures compared to end condenser condition was introduced and calculated. The exergy at different points of both GT and ST cycles and HRSG and the exergy destructions in several components were calculated. The main exergy loss was found in the GT combustion chambers. The fuel energy allocation between the desalting process and power production was conducted, based on the work loss and exergy methods. Both methods gave almost the same results. The main problem detected from the design of the given plant the high stack temperature of 183 °C of the HRSG to match that of high feedwater returning from the MSF desalting units at 135 °C. In GTCC using condensing turbines and NG with no sulfur, the typical HRSG stack temperature is 100 °C. The decrease of the stack temperature is reduced from 183 °C to 100 °C, which would increase the heat gain by the HRSG and ST work output about 19 %. This means that ST work output would be 256.6 MW. The difference between this ST output and actual output of 215.7 by the ST or 40.9 MW should be charged to also to the desalting process. Again, for net efficiency of 44 %, the fuel energy of 92.94 MW should be added to the 188.8 MW calculated before, or the fuel charged for thermal energy would be 281.82 MW or 119 MJ/m^3.

Sensitivity analysis shows that the pressure ratio, inlet air temperature, turbine inlet temperature, and stack temperature have a significant role in the combined cycle performance. It shows also that the cost of desalinated water is strongly affected by the fuel type because there is a large difference in price per unit of energy between oil and natural gas.

Nomenclature

a	specific exergy, kJ/kg
A	stream availability, mass flow rate×specific exergy, kW or MJ
B	brine flow, kg/s
BH	brine heater
BPST	back pressure steam turbine
CPDP	cogeneration power-desalting plant
D	distillate output flow rate, kg/s or MIGD
DP	desalting plant
DW	desalted seawater
ECST	extraction condensing steam turbine
EP	electric power
F	seawater feed flow rate, kg/s
FH	feed heater
GCC	Gulf Cooperation Countries
GR	gain ratio, distillate D per heating steam S, D:S
GT	gas turbine
GTCC	gas-steam turbine combined cycle
h	specific enthalpy, kJ/kg
H	heat rate (3,600/h), kJ/kWh
HHV	fuel high heating value, kJ/kg
HP	high pressure
HRSG	heat recovery steam generator
IP	intermediate pressure
kWh	3,600 kJ/s
L	latent heat of vaporization, kJ/kg

http://dx.doi.org/10.5772/60209

LHV	fuel low heating value, kJ/kg
LP	low pressure
m	mass flow rate, kg/s
MED	multi-effect distillation system
ME-TVC	multi-effect thermal vapor compression desalting system
MIGD	million imperial gallons per day (4,546 m^3/day or 52.616 kg/s)
MSF	multistage flash desalination
MVC	mechanical vapor compression desalting system
MW	megawatts
net	net output
P	pressure, kPa or bar
PP	power plant
PR	performance ratio, kg of distillate/2,330 kJ of heat
Q	rate of heat addition or rejection
RO	reverse osmosis
Sd	steam supply to desalting plant
SG	steam generator
SPD	single-purpose desalting plant
SPP	separate power plant
ST	steam turbine
SWRO	seawater reverse osmosis
TC	thermal compressor
TTD	terminal temperature difference
TVC	thermal vapor compression desalting system
v	specific volume, m^3/kg
W	power output
Wcp	cycle work
Wd	equivalent work of heat supply to desalting units
Wnp	net power output
Wp	pumping work of desalting unit

Greek letters

ε	second law efficiency or effectiveness
ηb	boiler efficiency
ηis	isentropic efficiency of pump or turbine

Subscripts

b	boiler or brine
bd	blowdown stream
cw	cooling seawater
d	desalting unit, discharge vapor, or distillate
e	extracted steam
e	environment
f	saturated liquid properties, feed heaters, or fuel
g	saturated vapor properties
pp	power plant
r	reheat

Author details

M.A. Darwish, H.K. Abdulrahim*, A.A. Mabrouk and A.S. Hassan

*Address all correspondence to: habdelrehem@qf.org.qa

Qatar Environment and Energy Research Institute, Qatar Foundation, Doha, Qatar

References

[1] Mohamed A. Darwish, Hassan K. Abdulrahim, Anwar B. Amer, On better utilization of gas turbines in Kuwait, Energy 33 (2008) 571–588

[2] John Xia, Rick Antos, (W501F) 3 Million hours fleet operational experience, POWER-GEN International 2006 – Orlando, FL, November 28-30, 2006, http://www.energy.siemens.com/co/pool/hq/energy-topics/pdfs/en/gas-turbines-power-plants/9_SGT65000F.pdf

[3] Armin Städtler, Meeting the Middle East Energy Demand with the Proven 8000H Series, Power-Gen Middle East, Abu Dhabi, 2014-10-12, http://www.energy.siemens.com/nl/pool/hq/energy-topics/technical-papers/2014-10-12_PGME_8000H.pdf

[4] Olav Bolland, Thermal power Generation, (2014), http://folk.ntnu.no/obolland/pdf/kompendium_power_Bolland.pdf

[5] Bob Shepard, Gas turbines technologies for electric generation, Power Plant Primer - Combustion Turbines - IEEE Mississippi Section, http://www.ieeems.org/Meetings/presentations/MS3-ASME%20Gas%20Turbine%20Technologies%20Presentation.ppt

[6] LM2500+ Aeroderivative Gas Turbine Package (29 MW), https://www.ge-distributed-power.com/products/power-generation/15-to-35-mw/lm2500

[7] Magdalena Milancej, Advanced Gas Turbine Cycles: Thermodynamic Study on the Concept of Intercooled Compression Process, Diploma thesis, Institut für Thermodynamik und Energiewandlung Technische Universität Wien and Institute of Turbomachinery, Technical University of Lodz, http://publik.tuwien.ac.at/files/pub-mb_3689.pdf

[8] Ivan Sigfrid, Investigation of a prototype industrial gas turbine combustor using alternative gaseous fuels, Ph. D. Thesis, Division of Thermal Power Engineering, http://lup.lub.lu.se/luur/download?func=downloadFile&recordOId=3972178&fileOId=3972217

[9] Turbine Inlet Air Cooling, From Wikipedia, the free encyclopedia, http://en.wikipedia.org/wiki/Turbine_Inlet_Air_Cooling

[10] Frank J. Brooks, GE gas turbine performance characteristics, http://www.muellerenvironmental.com/Documents/GER3567H

[11] Bob Omidvar, Gas Turbine Inlet Air Cooling System, http://www.albadronline.com/oldsite/books/49_GasTurbineInlet.pdf

[12] Gas turbine inlet air fogging, http://www.meefog.com/wp-content/uploads/br-gt-gasturbine.pdf

[13] Claire M. Soares, Gas turbines in simple cycle and combined cycle applications, http://www.netl.doe.gov/File%20Library/Research/Coal/energy%20systems/turbines/handbook/1-1.pdf

[14] Multistage Axial Compressors, section 12.4, http://web.mit.edu/16.unified/www/FALL/thermodynamics/notes/node92.html

[15] Ernesto Benini (2010). Advances in Aerodynamic Design of Gas Turbines Compressors, Gas Turbines, Gurrappa Injeti (Ed.), ISBN: 978-953-307-146-6, InTech, Available from: http://www.intechopen.com/books/gas-turbines/advances-in-aerodynamic-design-of-gas-turbines-compressors

[16] Scott Samuelsen, Conventional Type Combustion, http://www.netl.doe.gov/File%20Library/Research/Coal/energy%20systems/turbines/handbook/3-2-1-1.pdf

[17] The Jet Engine, Rolls Royce, ISBN 0 902121 2 35, http://www.amazon.com/The-Jet-Engine-Rolls-Royce/dp/0902121049

[18] Mohammad Nazri Mohd. Jaafar, Azeman Mustafa, Madya Hamidon Musa, Wan zaidi Wan Omar, Mohd. Zamri Yusoff, Kamsani Abdul Majid, Mohamad Shaiful Ashrul

Ishak, Development of low NOx liquid fuel burner, Faculty of Mechanical Engineering, Universiti Teknologi Malaysia 2005, http://eprints.utm.my/539/1/LAPORAN_AKHIR_IRPA_74069.pdf

[19] AP 42, Fifth Edition, Volume I, Chapter 3: Stationary Internal Combustion Sources, Stationary Gas Turbines, http://www.epa.gov/ttn/chief/ap42/ch03/final/c03s01.pdf

[20] Ueli Honegger, Gas turbine combustion modeling for a parametric emissions monitoring system, M.Sc. thesis, Kansas State University, 2007, https://krex.k-state.edu/dspace/bitstream/handle/2097/371/UeliHonegger2007.pdf?sequence=1

[21] Bassam G. Jabboury and Mohamed A. Darwish, Performance of gas turbine co-generation power desalting plants under varying operating conditions in Kuwait, Heat Recover), Systems & CHP Vol. 10, No. 3, pp. 243-253, 1990

[22] Bassam G. Jabboury and Mohamed A. Darwish, The effect of the operating parameters of heat recovery steam generators on combined cycle/sea-water desalination plant performance, Heat Recover)' Systems & CHP, Vol. 10, No. 3, pp. 255-267, 1990

[23] J. Dastych, Pickhardt, H. Unbehauen, Control schemes of cogenerating power plants for desalination, in Process Instrumentation, Control and Automation, from Encyclopedia of Desalination and Water Resources, Eolss Publishers, Paris, France, [http://www.desware.net] [Retrieved January 5, 2015]

[24] Steam turbine technology HEATs up, Power Engineering International, http://www.powerengineeringint.com/articles/print/volume-11/issue-3/features/steam-turbine-technology-heats-up.html

[25] M. Boss, Steam turbines for STAG combined-cycle power systems, http://site.ge-energy.com/prod_serv/products/tech_docs/en/downloads/ger3582e.pdf

[26] SSS CLUTCH, Key to combined cycle flexibility, http://www.sssclutch.com/power-generation/combinedcycle/attachments/NR9905_2.pdf

[27] R.W. Smith, P. Polukort, C.E. Maslak, C.M. Jones, B.D. Gardiner, Advanced Technology Combined Cycles, http://site.ge-energy.com/prod_serv/products/tech_docs/en/downloads/ger3936a.pdf

[28] J.P. Ninan and B. Khan, SPECIAL DESIGN ASPECTS OF CO-GENERATION UNITS, in Thermal Power Plants and Co-generation Planning, from Encyclopedia of Desalination and Water Resources, Eolss Publishers, Paris, France, [http://www.desware.net] [Retrieved January 5, 2015]

[29] G. Volpi, G. Silva, R. Piasente, Ansaldo Caldaie experience in HRSG design developments. http://www.ansaldoboiler.it/prodotti/generatori-di-vapore-a-recupero/%3Faid%3D747%26sa%3D1

[30] M.A. Darwish, Anwar Bin Amer, Cost allocation in cogeneration power–desalinationplant utilising gas/steam combined cycle (GTCC) in Kuwait, *Int. J. Exergy, Vol. 14, No. 3, 2014* 275

Chapter 7

Adsorption Cycle and Its Hybrid with Multi-Effect Desalination

Muhammad Wakil Shahzad, Kyaw Thu, Ang Li,
Azhar Bin Ismail and Kim Choon Ng

Additional information is available at the end of the chapter

http://dx.doi.org/10.5772/60400

Abstract

Adsorption (AD) cycle is recently pioneered for cooling and desalination applications. For water treatment, the cycle can be used to treat highly concentrated feed water, ranging from seawater, ground water, and chemically laden waste water. This chapter presents a review of the recent development of AD cycle and its hybridization with known conventional cycles such as the MED and MSF. We begin by looking at the basic sorption theory for different adsorbent–adsorbate pairs, namely the silica gel–water and the zeolite–water pairs. Under the IUPAC categorization, there are six types of isotherm behavior that capture almost all types of adsorbent–adsorbate behaviors and many isotherm correlations have been developed to described their uptake patterns, namely the Henry, Langmuir, Toth, etc. We have recently developed a correlation that can universally capture all six types of isotherms of IUPAC and it requires only four regression coefficients.

We present also the basic AD cycle for seawater desalination as well as its hybridization with known conventional thermally driven cycles. We present the performances of the AD pilot which was powered by renewable solar thermal input. Owing to thermodynamic synergy between the thermally driven cycles, the AD cycle is combined with the robust multi-effect distillation cycle to improve the water production yields. The hybrid cycle is called the "MED+AD" or MEDAD in short. With hybridization, it allows the bottom-brine temperature of the MED to operate below ambient temperature, as low as 5°C, in contrast to the conventional MED which is limited by the ambient, resulting in a quantum increase of distillate production by two to three times. We demonstrate this efficiency improvement in a pilot comprising a three-stage MED and AD plant and the top-brine temperature is maintained at 70°C. Lastly, we present the concept of exergy to apportion the operating cost of fuel input in a cogeneration plant where both electricity and water are being produced, and this concept gives an operating cost of fuel input to be lower than the conventional enthalpic approach.

Keywords: Thermal Desalination, Adsorption Desalination (AD), Exergy Analysis, Multi-Effect Distillation (MED), Hybrid Cycles

1. Introduction

Fresh water is a precious entity which is needed for economic development of every sector of a country such as agriculture, industrial, and domestic sectors. The persistent quest of economic development and exponential population growth in many countries has intensified the water demand which is projected to grow annually at a rate of 3–4% [1–3]. Even though 70% of the earth is covered by water, most of the part is in nonpotable state due to its salinity, either in the form of brackish, waste, and seawater [3–7]. About 20% of the world population is living below the acute water poverty level of 500 m^3 per capita per year [8], risking their health due to poor water quality and substandard sanitation. The demand for fresh water in 2030 is expected to increase up to 6,900 billion cubic meters (Bm^3), as compared to total available water resources of 4,500 Bm^3 [9]. The fresh water supply and demand scenario is shown in Figure 1, and it is predicted that the sustainable natural water cycle of our earth cannot meet the projected future water demand.

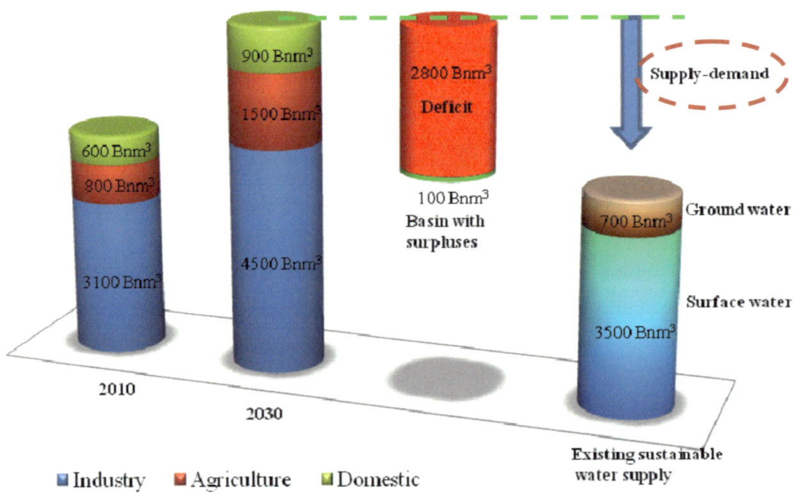

Figure 1. Fresh water supply demand gap: current and future estimates

http://dx.doi.org/10.5772/60400

In 2012, installed desalinated water capacity was 72 Mm^3/day (mcmd) and it is estimated to increase to 98 mcmd by 2015 as reported in the literature [10]. Almost half of total desalination capacity is installed in the Gulf Cooperation Council (GCC) countries. In the 1960s and 1970s, many GCC countries relied heavily on the ground water supply for all sectors such as agriculture, industrial, and domestic sectors. However, the ground water resource in these countries has depleted rapidly over the decades due to excessive water extraction and insufficient aquifer recharge. Despite a higher desalination market share in GCC, the fresh water availability is dropping rapidly to below the acute water poverty level of 500 m^3 per capital per year, caused by an exponential population growth and higher GDP thrust.

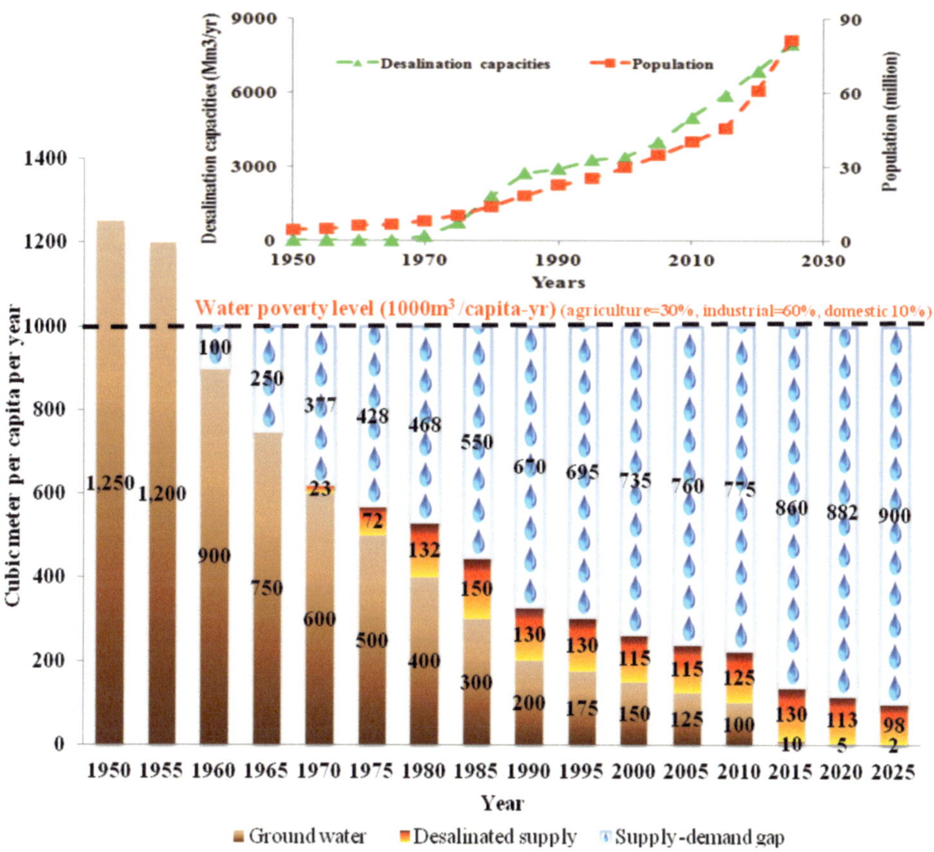

Figure 2. Fresh water availability per capita in GCC countries

GCC countries population, fresh water availability, and desalination scenario are shown in Figure 2, spanning from the early decades in 1950 to the future years up to 2025 [11–17]. Seawater desalination was started in GCC countries in the 1960s, but significantly desalinated water supply was observed in the past two decades, contributing to the overall water consumption share. Increasing trend of desalination capacities has resulted in higher energy consumption, more than 25% of total energy production in GCC countries. It is estimated that, with contracted desalination facilities, the total primary energy requirement will grow up to 255.45 GWh in 2025 with 55% increment as compared to 112.47 GWh in 2000 and CO_2 emission will be doubled in 2025 as compared to 60 ktonne/year in 2000 as shown in Table 1. The income loss attributed to desalination was 3.6 billion USD in 2000 and this loss is expected to rise to 31 billion USD in 2025 as calculated by considering yearly fuel price in terms of US$/bbl as shown in Table 1.

Year	Specific energy utilization for desalination (GWh_pe)	CO_2 production on the basis of fuel (0.527 tonne/MWh) (ktonne/yr)	Displaced income due to desalination (1 bbl = 1628kWh), (million USD)
1970	6.32	3.3	517.2
1975	25.19	13.3	2066.0
1980	62.29	32.8	5107.6
1985	92.79	48.9	5103.7
1990	98.26	51.8	3930.6
1995	109.92	57.9	2748.0
2000	112.47	59.3	3598.9
2005	131.93	69.5	7256.2
2010	163.96	86.4	13116.6
2015	191.21	100.8	21989.5
2020	221.97	117.0	26636.7
2025	255.45	134.6	31164.5

Table 1. Specific energy consumed and CO_2 footprint in GCC countries due to desalination processes

Water production by desalination processes has significant effect on the energy requirement and environment. The intricate nexus between water, energy, and environment has encouraged scientists and engineers to innovate desalination methods with better energy efficiency and environment-friendly processes. Although RO processes are energy-efficient and dominantly used, they have certain limitations with respect to local conditions. For example, the frequent maintenance issues from high operating pressure; water quality problems in term of residuals of boron, chlorides, and bromides; and severe fluctuations in the seawater intake quality are some of the challenges faced by the RO membranes. In the GCC region, frequent occurrence of harmful algae blooms (HABs) that may contain neuroparalytic and diarrhetic toxins which can pass through the pores of membranes lead to health problems. Large fluctuations in the feed water quality have direct implication to the operation and maintenance costs of RO plants [18–20]. Owing to the uncertainty of RO operation, thermal desalination

http://dx.doi.org/10.5772/60400

methods are deemed as the dominant processes employed in desalination market in the Middle East countries, more than 65% of installed capacities.

An innovative solution to strengthen the thermally driven and yet robust multi-effect desalination (MED) is its integration with the heat-driven adsorption desalination (AD) cycle. AD cycle is an emerging low-cost desalination system that requires only low temperature waste heat or solar energy to operate the cycle. The hybridization of both cycles, called MEDAD, extend the range of downstream (last stage) temperature of conventional MED system typically from 40°C to 5°C. The additional number of stages enhances the water production of the MED cycle by 2 - 3 fold at the same top-brine temperatures. In addition, AD integration cycle has the following advantages: (i) it increases inter-stage temperature differential of each MED stages due to the lowering of the bottom-brine temperature; (ii) it helps to scavenge the ambient energy in last part of the MED stages where the latent energy is further recycled; (iii) the AD cycle utilizes only low temperature waste heat; (iv) it has almost no major moving parts; (v) it reduces the chances of corrosion and fouling due to high concentration exposed to low temperature (5°C) in the last stages; (vi) it produces additional cooling effect from last stages of MED operating below ambient temperature; and (vii) significant increase in system performance. The basics of adsorption phenomenon, AD cycle, and its hybrids and desalination processes economics are presented in detail in this chapter.

2. Basic adsorption phenomena

The understanding of an AD cycle is predicated on the availability of basic equilibria–vapor uptake or isotherms of an adsorbent–adsorbate pair at assorted pressures and temperatures. From literature, all isotherms of adsorbent–adsorbate pairs can be categorized into six types (IUPAC) and they are described by many types of empirical and semi-empirical models. The simplest adsorption isotherm model is the classical Langmuir model [21] where it assumes a homogeneous surface with a monolayer vapor uptake where all adsorbent surfaces contain a uniform charged energy. Each pore vacant site is assumed to be filled by a single vapor molecule, forming a single sorption event. Invoking the rate of gas molecules filling the adsorption sites ($\frac{d\theta}{dt}$), as given by Ward and co-workers [22–25], the expression of the Langmuir isotherm model can be derived as follows:

$$\frac{d\theta}{dt} = K'\left[\exp\left(\frac{\mu_g - \mu_\alpha}{RT}\right) - \exp\left(\frac{\mu_\alpha - \mu_g}{RT}\right)\right] \tag{1}$$

where μ is the chemical potential in kJ/mol and subscripts g and a are the gaseous and adsorbed phases, T is the absolute temperature, and K' is the dimensionless constant. The isotherm, θ, can be obtained by integrating the rate equation over the energy level from E_c to infinity and in this simple case, when $\rightarrow 0$ then $\theta \rightarrow 0$, and as P is large, θ approaches 1.

$$\theta = \frac{K\exp\left(\frac{\varepsilon}{RT}\right)P}{1+K\exp\left(\frac{\varepsilon}{RT}\right)P} \tag{2}$$

A real solid surface of an adsorbent contains geometrical roughness that is formed during its formation or activation process and the energetic heterogeneity has to be accounted for. The gas molecules experienced varying potential at adsorption sites of uneven energy levels and the surfaces are subdivided into infinitesimal pieces. Thus, the total adsorption of a heterogeneous surface is the sum of the product of the sorption event or surface coverage, $\tilde{\theta}(\varepsilon_i)$ and its probability energy distribution, $\chi(\varepsilon_i)$, i.e.,

$$\theta = \sum_i \chi(\varepsilon_i)\cdot\tilde{\theta}(\varepsilon_i) = \int_0^\infty \chi(\varepsilon)\cdot\tilde{\theta}(\varepsilon)\mathrm{d}\varepsilon \tag{3}$$

where the Langmuir model can be applied with a probability function over the entire surface being summed to unity, i.e.,

$$\int_0^\infty \chi(\varepsilon)\cdot\mathrm{d}\varepsilon = 1 \tag{4}$$

Applying a condensation approximation at moderate temperature, the local surface coverage can be simplified to a step-like dirac-delta function as shown in Figure 3, and expressed below;

$$\lim_{T\to 0}\tilde{\theta}(\varepsilon) = \begin{cases} 0 \text{ for } \varepsilon < \varepsilon_c \\ 1 \text{ for } \varepsilon \geq \varepsilon_c \end{cases} \tag{5}$$

where, ε_c relates to the adsorption energy in equilibrium conditions. Hence, Equation 3 can be simplified to

$$\begin{aligned}\theta &= \int_0^{\varepsilon_c} 0\cdot\chi(\varepsilon)d\varepsilon + \int_{\varepsilon_c}^\infty 1\cdot\chi(\varepsilon)d\varepsilon \\ &= \int_{\varepsilon_c}^\infty \chi(\varepsilon)d\varepsilon\end{aligned} \tag{6}$$

A solution of $\chi(\varepsilon)$ to represent the site energy allocation for a real adsorbent surface is obtained by approximating it to a continuous probability distribution function. Depending on the adsorbent surface energy characteristics during adsorption interactions, symmetrical or asymmetrical Gaussian functions are usually assumed, as illustrated in Figure 4.

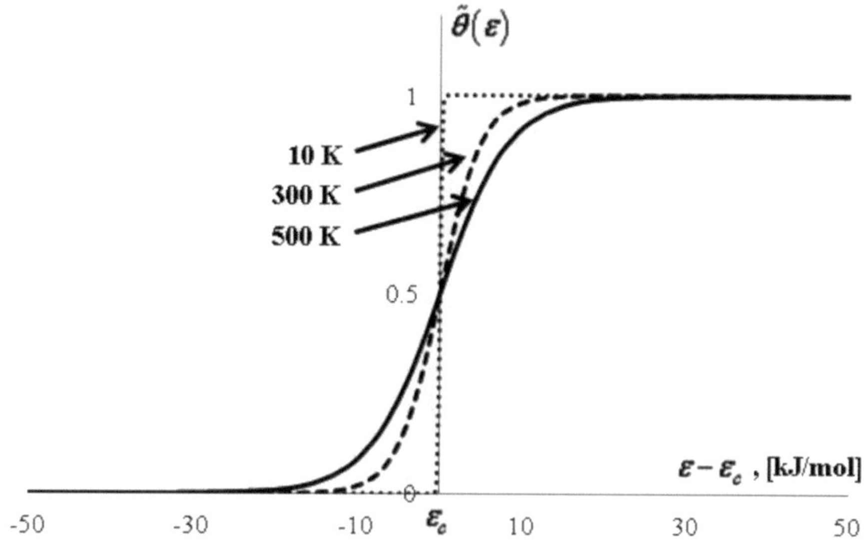

Figure 3. Step-like profile yielded from Equation 5 at moderate temperatures

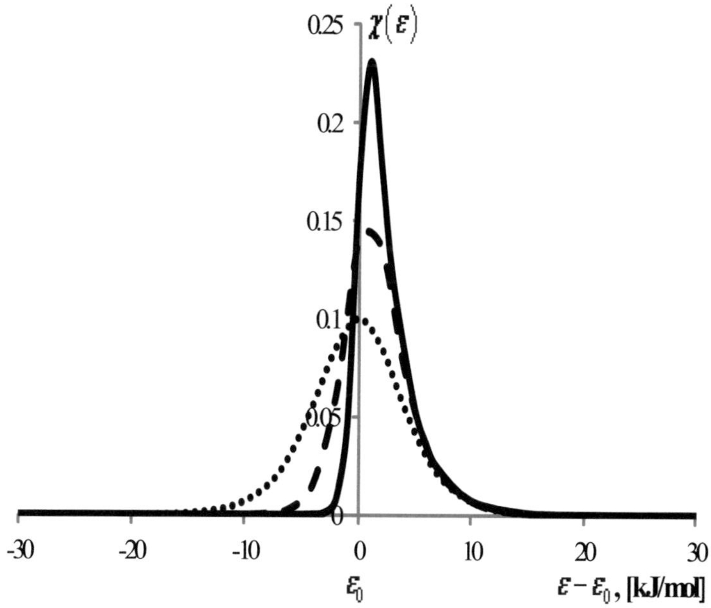

Figure 4. An illustration of symmetrical or asymmetrical Gaussian function to represent the adsorption site energy distribution for classical isotherm models

The details of EDFs, $\chi(\varepsilon)$, for the classic Langmuir–Freundlich [26], Dubinin–Astakhov [27], Dubinin–Raduskevich [28], and Tóth isotherm models are outlined in Table 2, and integrating

EDFs from the cut-off energy ε_c to ∞, the exact correlation of these isotherm models can be obtained.

Model Name	Adsorption Site Energy Distribution $\chi(\varepsilon)$	Isotherm Equation
Langmuir–Freundlich	$\chi(\varepsilon)=\dfrac{\frac{1}{c}\exp\left(\frac{\varepsilon-\varepsilon_0}{c}\right)}{\left[1+\exp\left(\frac{\varepsilon-\varepsilon_0}{c}\right)\right]^2}$	$\theta=\dfrac{\left[KP\exp\left(\frac{\varepsilon_0}{RT}\right)\right]^{\frac{RT}{c}}}{1+\left[KP\exp\left(\frac{\varepsilon_0}{RT}\right)\right]^{\frac{RT}{c}}}$
Dubinin–Astakhov		$\chi(\varepsilon)=\dfrac{r(\varepsilon-\varepsilon_1)^{r-1}}{E^r}\exp\left[-\left(\frac{\varepsilon-\varepsilon_1}{E}\right)^r\right]$ $\theta=\exp\left[-\left(\frac{RT}{E}\ln\frac{P_0}{P}\right)^r\right]$
Dubinin–Radushkevich	$\chi(\varepsilon)=\dfrac{2(\varepsilon-\varepsilon_1)}{E^2}\exp\left[-\left(\frac{\varepsilon-\varepsilon_2}{E}\right)^2\right]$	$\theta=\exp\left[-\left(\frac{RT}{E}\ln\frac{P_0}{P}\right)^2\right]$
Tóth	$\chi(\varepsilon)=\dfrac{\frac{1}{RT}\left[\exp\left(\frac{\varepsilon-\varepsilon_3}{RT}\right)\right]^t}{\left\{1+\left[\exp\left(\frac{\varepsilon-\varepsilon_3}{RT}\right)\right]^t\right\}^{\frac{t+1}{t}}}$	$\theta=\dfrac{KP\exp\left(\frac{\varepsilon_3}{RT}\right)}{\left\{1+\left[KP\exp\left(\frac{\varepsilon_3}{RT}\right)\right]^t\right\}^{\frac{1}{t}}}$

Table 2. Assorted forms of adsorption site energy distribution functions (EDF) for the Langmuir-Freundlich, Dubinin-Astakhov (DA), Dubinin-Raduskevich (DR) and Tóth isotherm models

2.1. Universal site-Energy probability Distribution Function (EDF)

In the recent development of adsorption isotherm theory, Li [29] proposed a universal model that was able to fit all types of isotherms, as specified in Equation 7. Type I to V patterns at various temperatures could be directly captured by the equation with four regression parameters.

$$\frac{q}{q_0}=\frac{A\phi\exp\left(\beta\frac{P}{P_s}\right)\frac{P}{P_s}+C\frac{P}{P_s}}{\left\{1+\phi\exp\left(\beta\frac{P}{P_s}\right)\frac{P}{P_s}\right\}^t} \tag{7}$$

and,

$$\beta=\exp\left(\frac{E_c}{RT}\right) \tag{8}$$

http://dx.doi.org/10.5772/60400

$$A = \frac{\left[1+\phi \exp(\beta)\right]^{t} - C}{\phi \exp(\beta)} \tag{9}$$

where the alphabet ϕ, C are constants, t is surface heterogeneity factor, and E_c denotes the characteristic energy of the adsorbent–adsorbate pair. These four parameters are required to calculate in the regression process. The rest of the letters have their usual means.

Using the unified adsorption isotherm framework, a universal adsorption site energy distribution function (EDF) was proposed, which relates directly to their isotherm types, and the proposed EDF fitted well with the statistical rate theory of adsorption. The EDF yielded a single peak asymmetrical distribution for Type I which was similar to that for the classical LF, DA, DR, and Tóth isotherm models. The EDF is given as below;

$$\chi(\varepsilon) = \frac{\beta^*}{RT}\left[\phi \exp(\beta^*)\frac{\beta^*}{\beta}+1\right]^{-t-1} \left\{ \begin{array}{l} \left[A\phi \exp(\beta^*)\frac{(\beta^*+1)}{\beta}+\frac{C}{\beta}\right] \\ \left[1-\phi \exp(\beta^*)\frac{\beta^*}{\beta}(t-1)\right] \\ -Ct\phi \exp(\beta^*)\left(\frac{\beta^*}{\beta}\right)^2 \end{array} \right\} \tag{10}$$

where the variable β^* is a function of the adsorption site energy ε and it is expressed as

$$\beta^* = \exp\left(\frac{2E_c - \varepsilon}{RT}\right) \tag{11}$$

From the data available from literature and the proposed Eqs. 9 and 10, the assorted isotherms as categorized by the IUPAC can be successfully captured succinctly by these equations using only four coefficients of regression, and these energy distribution functions and isotherms are depicted in Table 3. For each type of isotherm, the corresponding energy distribution functions (EDFs) have been developed.

The adsorbent–adsorbate interaction is the key in the design of adsorption cycle, which can be functionalized to adopt to warmer ambient temperatures, particularly for operation in the summer period of semi-arid or desert regions. Figures 5(a) and (b) depict two major types of useful adsorbents in water uptake: The former is silica gel Type 3A, suitable for an AD cycle operating below 33°C ambient such as tropical weather conditions. The latter figure depicts the isotherms of zeolite (Z0-alumina phosphate oxide, $Al_XPh_YO.nH_2O$). It has properties that can be tailored for adsorption/desorption at ambient temperatures up to 50°C (corresponding to desorption at 3.8–4.5 kPa (60°C isotherm) and adsorption at 2–2.8 kPa (40°C isotherm). The thermos-physical properties of both adsorbents are tabulated in Table 4, showing the BET surface-pore areas of 600–800 m^2/g.

Types of Isotherms (IUPAC Categorization)		Adsorbate–Adsorbent Pair / References
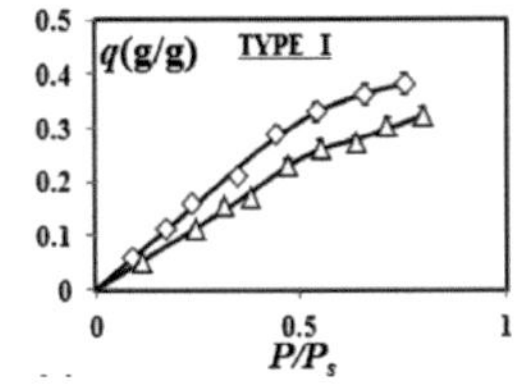	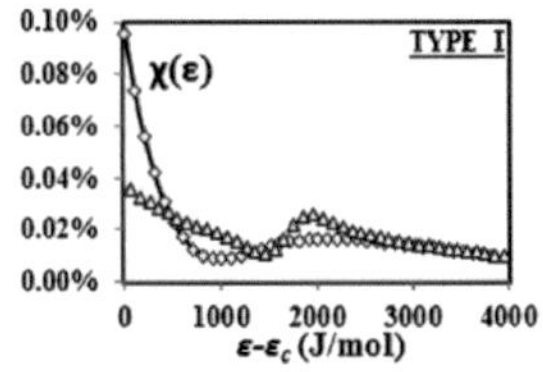	Type I Water–silica gel Type RD Type I Water–silica gel Type A Qiu J. Characterization of silica gel–water vapor adsorption, MEng Thesis, 2004, NUS
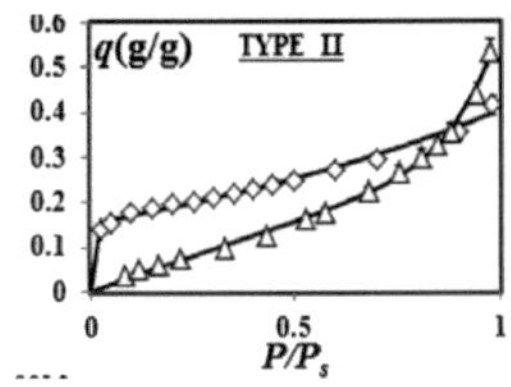	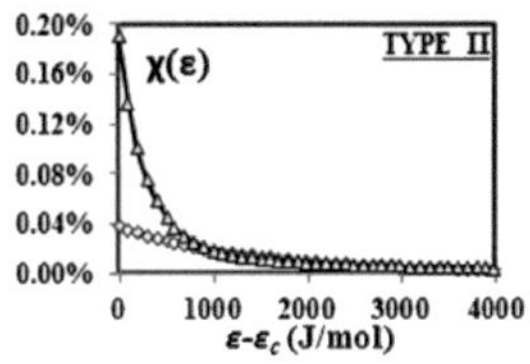	Type II ◇ Water–boehmite Type II Δ Water–polyvinyl pyrrolidone 1. Wang, S-L, Johnston CT, David L, White JL, Stanley LH. Watervapor adsorption and surface area measurement of poorly crystalline boehmite. J Colloid Interface Sci 2003;260(1):26–35. 2. Zhang, J, Zografi G. The relationship between "BET" and "free volume"-derived parameters for water vapor absorption into amorphous solids. J Pharm Sci 2000;89(8):1063–72.
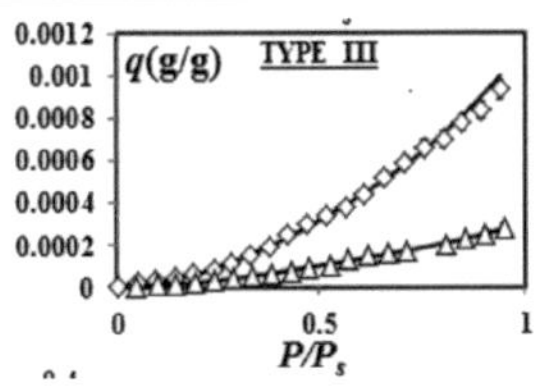	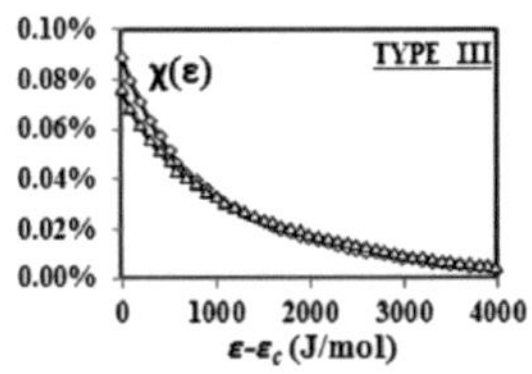	Type III ◇ Water-sctivated carbon Type III Δ Water–carbon S-W nanotube Kim P, Agnihotri S. Application of water-activated carbon isotherm models to water adsorption isotherms of single-walled carbon nanotubes. J Colloid Interface Scie 2008;325(1):64–73.
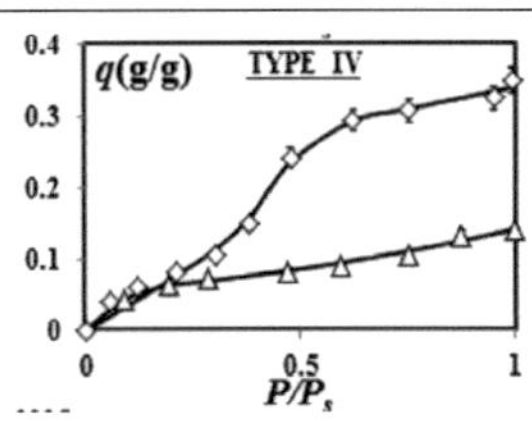 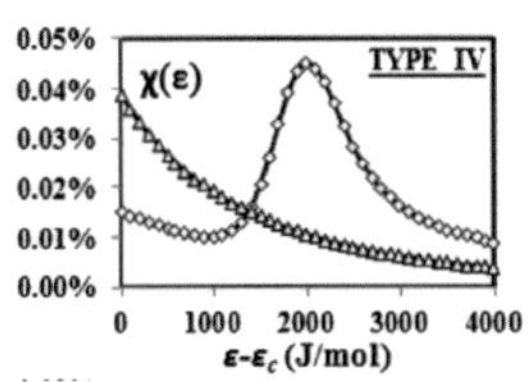		Type IV ◇ Water–boehmite Type IV Δ Water–polyvinyl pyrrolidone Bansal RC, Dhami TL. Surface characteristics and surface behaviour of polymer carbons—II: adsorption of water vapor. Carbon 1978;16(5):389–95.

http://dx.doi.org/10.5772/60400

Types of Isotherms (IUPAC Categorization)	Adsorbate–Adsorbent Pair / References
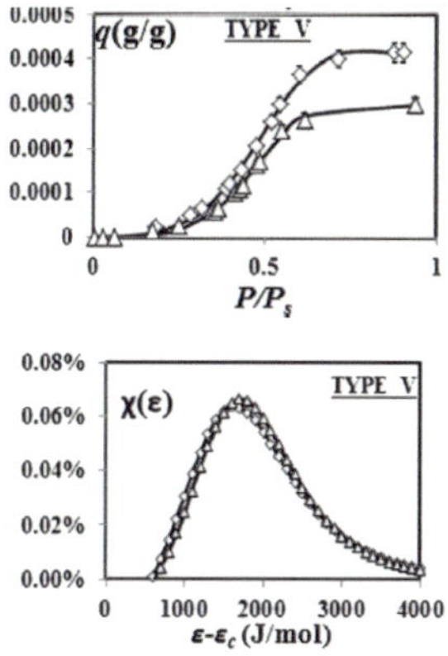	Type II ◇ Water–boehmite Type II Δ Water–polyvinyl pyrrolidone Lagorsse S, Campo MC, Magalhaes FD, Mendes A. Water adsorption on carbon molecular sieve membranes: experimental data and isotherm model. Carbon 2005;43(13):2769–79.

Table 3. A summary of the energy distribution functions and isotherms as categorized by the IUPAC

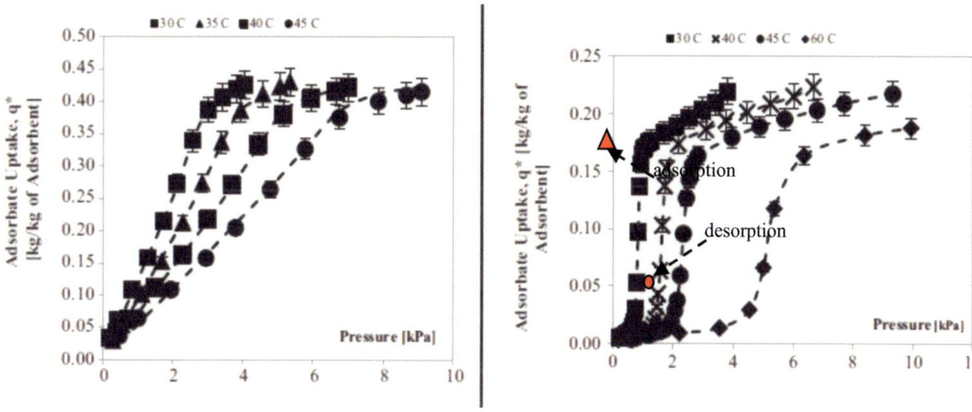

Figure 5. (a) shows the isotherms of water-silica gel Type 3A at four temperatures: 30°C to 45° C at increasing pressure up to 10 kPa, (b) depicts the isotherms of Zeolite (Z01- Alumina Phosphate Oxide) from 30° C to 60° C.

Properties	Silica Gel Type 3A	Zeolite FAM Z01
BET surface area [m^2/g]	680	147.3
Porous volume [ml/g]	0.47	0.071
Apparent density [kg/m^3]	770	600–700
Thermal conductivity [W/m.K]	0.174	0.113 (30°C)
Heat of adsorption (H_2O) [kJ/kg of H_2O]	2800	3110 (25°C)
Specific heat capacity [kJ/kg.K]	0.921	(30°C)

Table 4. Thermo-physical properties of the silica gel type 3A and Zeolite (Z01)

3. Design of AD batch-operated cycle

There are five main components of AD system namely: (i) evaporator, (ii) adsorption and desorption reactor beds, (iii) condenser, (iv) pumps, and (v) pretreatment facility. The detailed process diagram is shown in Figure 6. For the batch-operated AD cycle, it involves two main processes.

3.1. Adsorption-assisted-evaporation

In which the vapors generated in AD evaporator are adsorbed on the pore surface area of adsorbent. The heat source is circulated through the tubes of evaporator and seawater is sprayed on the tube bundle. It is observed that the evaporation is initiated by heat source, but during adsorption process the high affinity of water vapor of adsorbent drops the evaporator pressure and contribute in evaporation. The AD evaporator operation temperature can be controlled by heat source temperature that is normally circulated in terms of chilled water. The AD evaporator can operate at a wide range of chilled water temperature varying from 5°C to 30°C to produce the cooling effect as well at low temperature operation. The vapor adsorption process continues until the adsorbent bed reaches a saturation state.

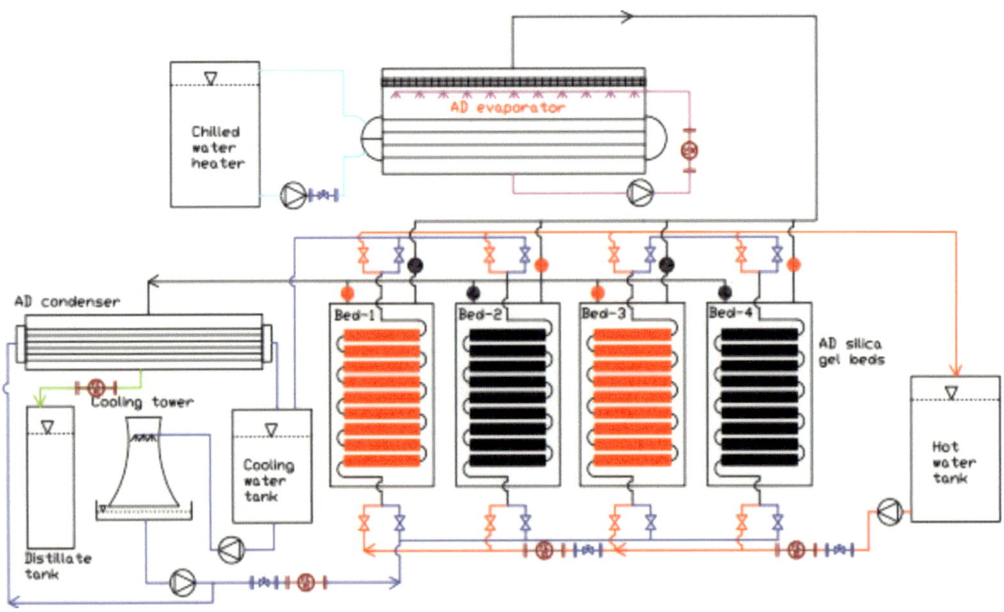

Figure 6. Detailed schematics of an adsorption desalination cycle. The circle (filled) dots are the vapor valves and the pair of triangles refers to the liquid valves. The range of vapor pressures in evaporator and condenser are 1-2 kPa and 5 to 7 kPa

http://dx.doi.org/10.5772/60400

3.2. Desorption-activated-condensation

In which saturated adsorbent is regenerated using the low-grade industrial waste heat or renewable energy (desorption temperature varies from 55°C to 85°C) and desorbed vapors are condensed in a water-cooled condenser and collected as a distillate water.

It can be seen that two useful effects produced by AD cycle are the cooling effect by the first process "adsorption-assisted-evaporation" and fresh water production by converting the seawater by second process "desorption-activated-condensation". Useful effects which are cooling and water production can be produced simultaneously by introducing the multi-bed technique.

In multi-bed AD system, the operation and switching technique is used. During operation, one or pair of adsorbent reactor beds undergo the adsorption process and at the same time one or pair of adsorbent reactor beds execute the desorption process. The time for adsorbent reactor beds operation, either adsorption or desorption, depends on the heat source temperature and silica gel quantity packed in a bed. Prior to changing the reactor duties, there is a short time interval called switching in which the adsorber bed is preheated whilst the desorber bed is precooled to enhance the performance of cycle. In AD cycles, the operation (adsorption and desorption) and switching processes are controlled by automated control scheme that can open and close the respective bed hot/cold water valves. During switching operation, all vapor valves are closed so that there is no adsorption or desorption taking place.

4. MED-AD hybrid cycle

MEDAD is a hybrid of two thermal systems namely multi-effect desalination system and adsorption cycle. The main components of this novel thermal hybrid system are (1) multi-effect distillation (MED) system, (2) adsorption desalination (AD) cycle, and (3) auxiliary equipments. In this hybridization system, the last stage of the MED is connected to adsorption beds of AD cycle for the direct vapor communication to adsorption beds. Figure 7 shows detailed flow schematic of MED plant combined with AD system.

Adsorption-based desalination is investigated by many researchers [30–45] and reported that optimal specific daily water production (SDWP) for four bed scheme is about 4.7 kg/kg silica gel. The first adsorption desalination plant was installed in the National University of Singapore (NUS) which consists of four silica gel beds. Ng et al. investigated the processes using chilled water at assorted temperature and demonstrated that the specific water production of the system [46, 47]. They also introduced and patented a novel hybrid desalination method "MEDAD cycle" that is a combination of conventional MED and AD cycle [48, 49]. This novel desalination cycle can mitigate the limitations of conventional MED system to increase the system performance. This combination allows the MED last stage to operate below ambient temperature typically at 5°C as compared to traditional MED at 40°C. This not only reduces the corrosion chances but also increases the distillate production to almost 2 – 3 fold as compared to traditional MED systems.

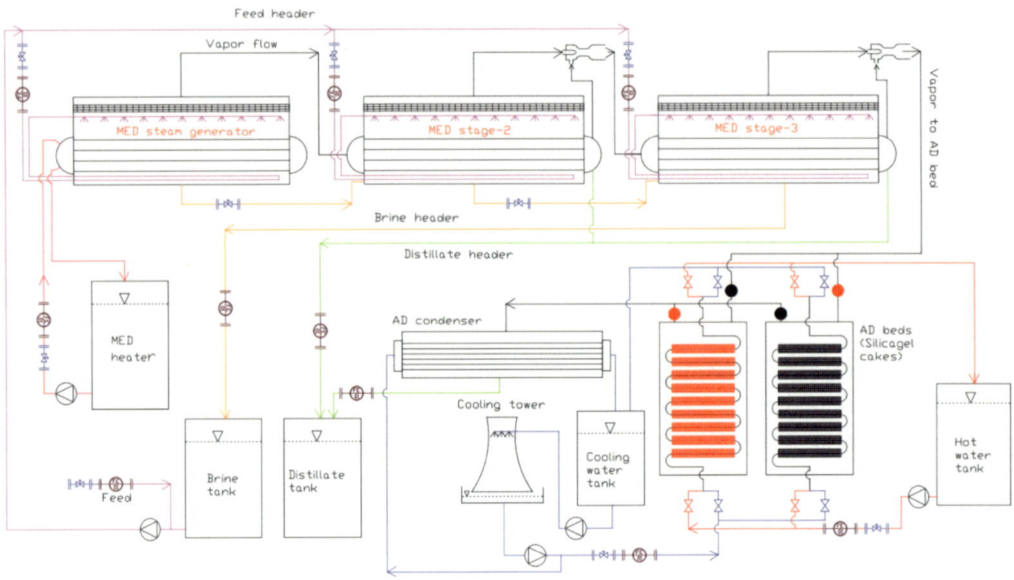

Figure 7. Detailed P& ID diagram for an experimental MED facility of nominal heating capacity of 12 kW. The first stage comprises a water-fired steam generator whilst the last stage (3rd stage) is open to a water-cooled cooling tower

4.1. MED-AD simulation

MEDAD cycle simulation have been conducted [50–53] and presented in Table 5. The simulation is based on a fully transient model and the predicted results are compared with conventional MED system. It is observed that at same input parameters such as a top-brine-temperature (TBT), water production can achieve up to two fold increase when the hybridized MEDAD is compared with the MED cycle alone.

Equation	Remarks
Modeling equations for steam generator	
$[(M_{hw}.Cp_{hw})]\frac{dT_{hw}}{dt}=(\dot{m}_{hw}h_{f,Thw,in})-(\dot{m}_{hw}h_{f,Thw,out})-h_{in,o}.A_{in,i}(T_{hw}-T_{tube,i})$	Energy balance for the hot water flowing inside the tubes of brine heater.
$[(M_{HX,i}.Cp_{hx,i})]\frac{dT_{tube,i}}{dt}=h_{in,i}.A_{in,i}(T_{hw,i}-T_{tube,i})-h_{out,i}.A_{out,i}(T_{tube,i}-T_{v,i})$	Energy balance for metal tubes.
$\frac{dM_{b,i}}{dt}=\dot{m}_{f,i}-\dot{m}_{b,i}-\dot{m}_{v,i}$	Mass balance for the seawater inventory in the

Equation	Remarks
	evaporator side of the brine heater.
$Q_{in}=h_{o,i}A_i(T_t-T_{v,i})[(M_{b,i}.Cp_{b,Tb})+(M_{HX,i}.Cp_{HX})]\frac{dT_i}{dt}=(\dot{m}_{f,i}h_{f,Tf})-(\dot{m}_{b,i}h_{f,Tb})-(\dot{m}_{v,i}h_{g,Tv})+Q_{in,i}$	Energy balance for the evaporator side of the steam generator.
$M_{b,i}\frac{dX_{b,i}}{dt}=(\dot{m}_{f,i}X_{f,i})+(\dot{m}_{b,i}X_{b,i})-(\dot{m}_{v,i}X_{v,i})$	Material/ concentration balance
$Nu=\frac{h_{in,i}d_{in,i}}{K_{tube,i}}=0.023\mathrm{Re}_l^{0.80}\mathrm{Pr}_l^{0.40}$	Convective heat transfer coefficient equation
$R_{wall,i}=\frac{\ln\left(\frac{d_{out,i}}{d_{in,i}}\right)}{2.\pi.K_{tube,i}.L_{tube,i}}$	Tube wall resistance
$h_o\left(\frac{v^2}{k^3g}\right)^{1/3}=0.0007\,Re^{0.2}\,Pr^{0.65}q^{\prime\prime 0.4}$	Falling film evaporation heat transfer coefficient, Han and Fletcher's correlation
$U_iA_i=\frac{1}{\frac{1}{h_{in,i}A_{in,i}}+R_{wall,i}+\frac{1}{h_{out,i}A_{out,i}}}$	Overall heat transfer coefficient
Modeling equations for intermediate stages	
$[(M_{l,i+1}.Cp_{l,Tcond})]\frac{dT_{cond,i+1}}{dt}=[\dot{m}_v h_{fg,Tv}]_i-[h_{in}.A_{in}(T_{cond}-T_{tube})]_{i+1}$	Energy balance for the condenser side of the ith effect
$[(M_{HX,i+1}.Cp_{hx,i+1})]\frac{dT_{tube,i+1}}{dt}=h_{in,i+1}.A_{in,i+1}(T_{cond,i+1}-T_{tube,i+1})-h_{out,i+1}.A_{out,i+1}(T_{tube,i+1}-T_{v,i+1})$	Energy balance for tube metal
$\frac{dM_{b,i+1}}{dt}=\dot{m}_{f,i+1}-\dot{m}_{b,i+1}-\dot{m}_{v,i+1}$	Brine inventory balance
$Q_{in,i+1}=h_{out,i+1}A_{i+1}(T_{t,i+1}-T_{v,i+1})[(M_{b,i+1}.Cp_b)+(M_{HX,i+1}.Cp_{HX,i+1})]\frac{dT_{i+1}}{dt}=(\dot{m}_{f,i+!}h_{f,Tf})-(\dot{m}_{b,i+1}h_{f,Tb})-(\dot{m}_{v,i+1}h_{g,Tv})+Q_{in,i+1}$	Energy balance for evaporator side
$M_{b,i+1}\frac{dX_{b,i+1}}{dt}=(\dot{m}_{f,i+1}X_{f,i+!})-(\dot{m}_{b,i+1}X_{b,i+1})-(\dot{m}_{v,i+!}X_{v,i+!})$	Material/ concentration balance

Equation	Remarks
$$Nu=\frac{h_{in,i+1}L_{i+1}}{K_{tube,i+1}}=0.728\left[\frac{gh_{fg,Tcond}\rho_{l,Tcond}(\rho_l-\rho_v)_{Tcond}K^3_{l,Tcond}}{\mu_{l,Tcond}d_i(T_{v,i+1}-T_{tube,i+1})}\right]^{1/4}$$	Nusselt film condensation correlation for the calculation of the heat transfer coefficient inside the condenser tubes
$$U_iA_i=\frac{1}{\frac{1}{h_{in,i}A_{in,i}}+R_{wall,i}+\frac{1}{h_{out,i}A_{out,i}}}$$	Overall heat transfer coefficient equation
MED last stage connected with AD beds	
$$[(M_{b,n}.Cp_b)+(M_{HX,n}.Cp_{HX,n})]\frac{dT_n}{dt}=(\dot{m}_{f,n}h_{f,Tf})-(\dot{m}_{b,n}h_{f,Tb})-(M_{sg}h_{g,Tv})\frac{dq_{ads}}{dt}+Q_{in,n}$$ $$Q_{in,n}=h_{out,n}A_n(T_{t,n}-T_{v,n})$$	Energy balance for the condenser side of the ith effect

Table 5. MEDAD modeling equations

Figure 8 shows the transient temperature profiles of a MEDAD cycle. It can be seen that last stages of MED are operating below ambient temperature due to hybridization. It can also be noticed that MED last stages profiles are affected by cyclic AD operation.

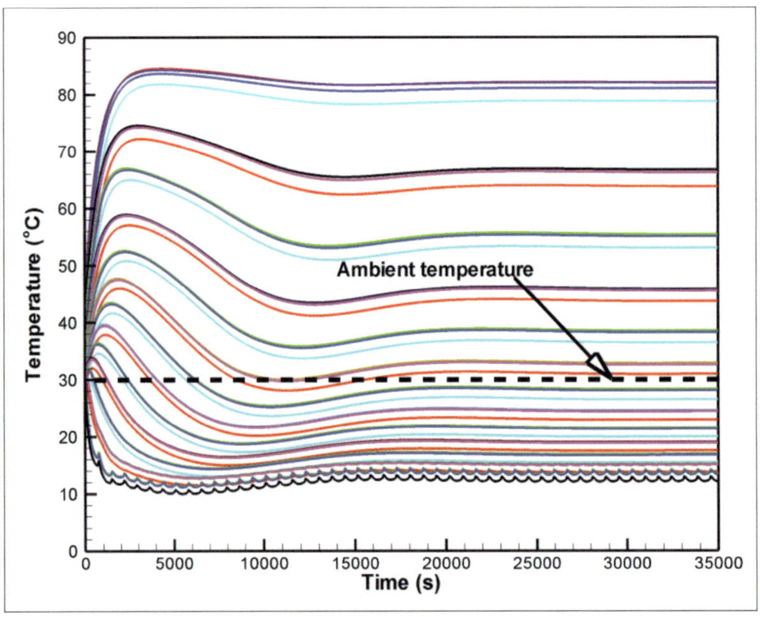

Figure 8. MEDAD cycle components temporal temperature profiles

http://dx.doi.org/10.5772/60400

Figure 9 mapped the performance parameters of hybrid MEDAD cycle (concentration, GOR, PR, and WPR). It can be observed that the batched-mode water vapor uptake by the coupled AD cycle dominates the performance of the cycle. The performance of the MEDAD cycle with different additional effects with the conventional eight effect MED cycle as baseline cycle is studied in terms of water production rate.

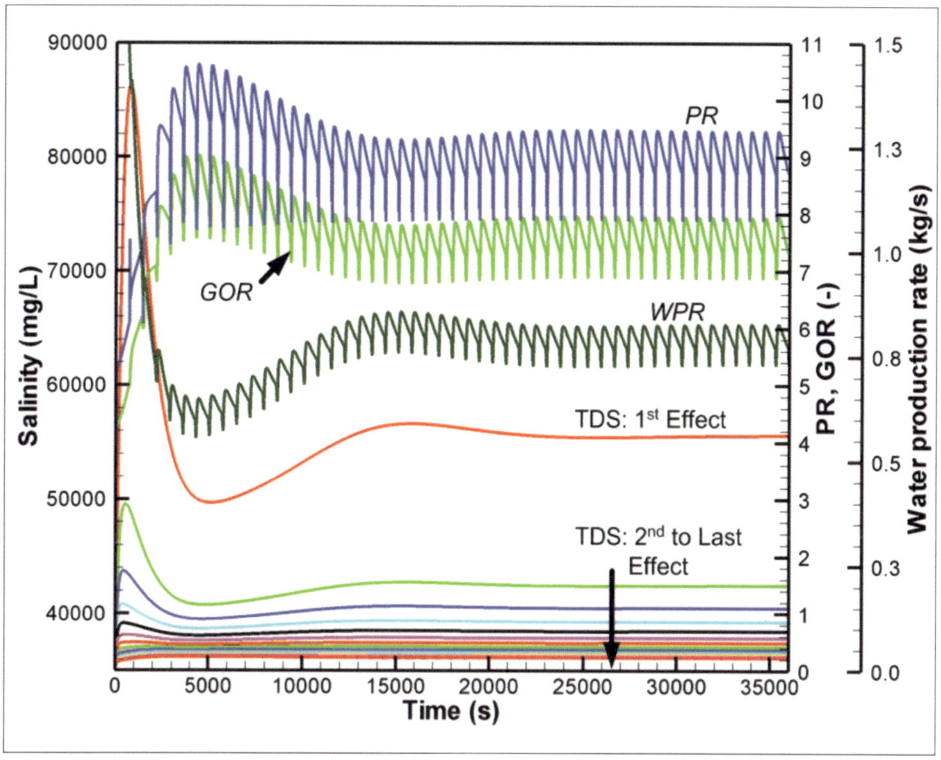

Figure 9. MEDAD hybrid cycle performance (concentration, GOR, PR and WPR)

Figure 10 shows the quantum jump in the water production rate of the proposed MEDAD cycle. Another aspect of this hybridization is that the desorbed water vapor from the AD cycle can be recycled back into the MED system for further energy recovery.

4.2. MED-AD experimentation

A three-stage MED system is designed [54], fabricated, and installed in NUS as shown in Figure 11. In MED stages, vapor emanation from feed seawater is achieved by falling film-evaporation process. Evaporation energy is recovered by series of reutilization of vapor condensing energy in successive stages of those produced in preceding stages. Process of vapor production and energy recovery by condensation continues until the last stage of MED. The vapors from the last stage are then directed toward AD beds where they are adsorbed on the adsorbent surface. Adsorbent high affinity for water vapor drops the pressure and hence the saturation temper-

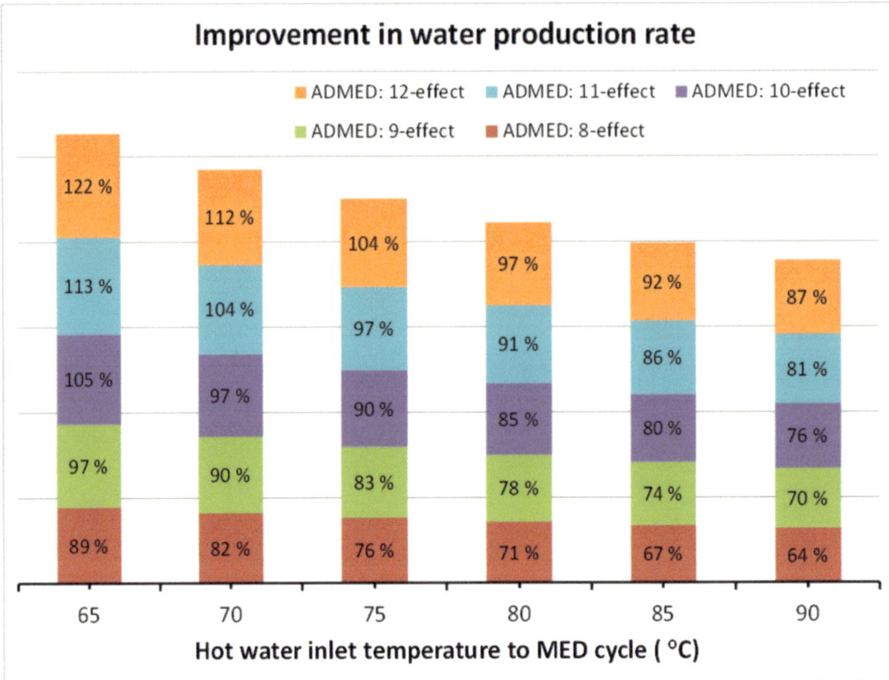

Figure 10. Percentage improvement in water production rate by the MEDAD cycle

ature of last stages falls below ambient, typically up to 5°C. It is observed that this drop in pressure and temperature of last stage also affected the operational parameters of the few preceding stages.

Figure 11. MEDAD system installed in NUS

http://dx.doi.org/10.5772/60400

Experiments are conducted in two steps. In first part, system is operated as a conventional MED at assorted heat source temperature ranges from 38°C to 70°C. In second part, experiments are conducted as a hybrid MEDAD system at assorted heat source temperature ranges from 15°C to 70°C and results are compared with conventional MED system.

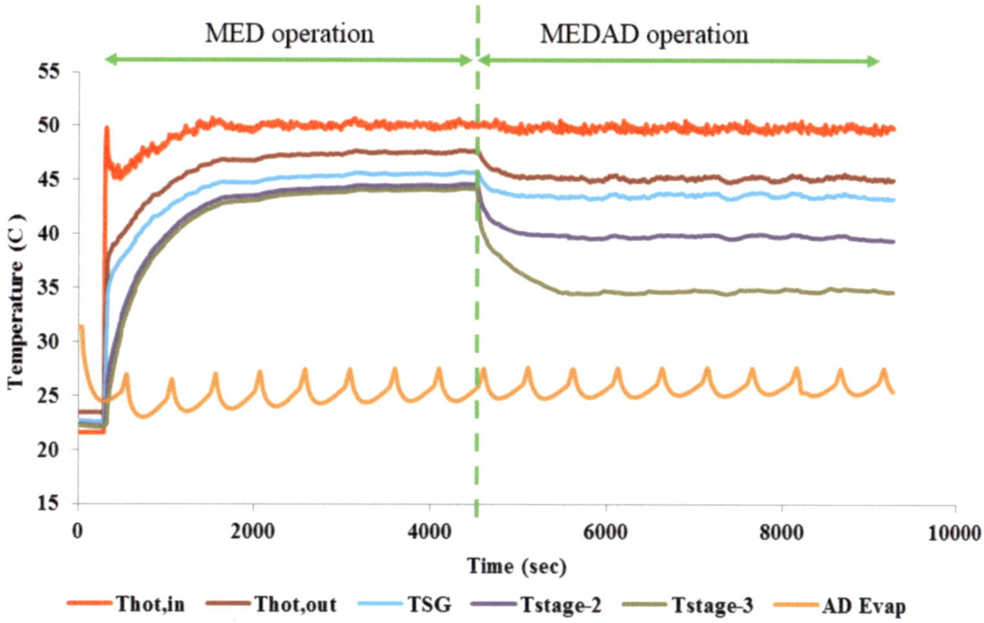

Figure 12. MEDAD components temperature profiles at 38°C heat source temperature

Figure 12 shows the instantaneous temperatures of MED and MEDAD components at a heat source temperature of 38°C. It can be seen that steady-state events (minimum temperature fluctuations) occur after 1 hour from start-up and experiments for distillate collection are continued for 4 to 5 hours. It is noticed that the inter-stage temperature difference (ΔT) is more than twice per stage as compared to the conventional MED stages. This is attributed to the vapor uptake by the adsorbent of AD cycle, resulting in the increase of vapor production. The MEDAD cycle yields a stage ΔT from 3°C to 4°C as compared to 1°C or less in the case of MED alone. Table 6 shows the comparison of MEDAD and MED components steady-state temperature values.

Figure 13 shows the distillate production trace at heat source 38°C from MED stages, AD condenser and combined. The batch operated AD production can be seen clearly. At the start of desorption, the production is higher and it drops with time to zero during the switching period while MED stages production is quite stable. Small fluctuations in MED water production may be due to the fluctuations in the spray of the feed that affect the condensation rate. It can be seen that hybridization boost water production 2 – 3 fold as compared to conventional MED system at same TBT. Water production profiles are similar as explained in simulated results.

Heat source temperature (°C)	MED SG	MEDAD SG	MED Stage-2	MEDAD Stage-2	MED Stage-3	MEDAD Stage-3	MED condenser	AD evaporator
70	63.9	54.4	62.8	50.8	62.3	46.7	56.6	26.1
65	59.6	50.7	58.7	47.0	58.1	42.5	53.9	25.8
60	55.4	49.2	54.4	45.7	53.8	41.3	49.4	25.5
55	51.5	48.5	50.5	44.8	49.9	40.8	45.8	25.1
50	47.4	44.6	46.7	41.6	46.2	37.9	42.7	23.2
45	43.9	41.1	43.4	38.6	43.0	34.9	40.3	22.4
40	38.7	36.6	38.4	34.8	38.1	31.6	36.1	21.0
38	37.4	35.4	37.2	33.5	37.0	30.3	35.3	18.2
Operating limit of Conventional MED. The lower operational points are from MEDAD Hybrids								
35	-	30.9	-	27.9	-	23.5	-	16.1
30	-	26.0	-	23.7	-	19.1	-	12.2
25	-	22.1	-	19.9	-	16.3	-	10.1
20	-	18.1	-	14.2	-	11.3	-	7.1
15	-	13.2	-	11.5	-	9.4	-	5.6

Table 6. A comparison of conventional MED and hybrid MEDAD systems components temperatures at different heat source temperatures

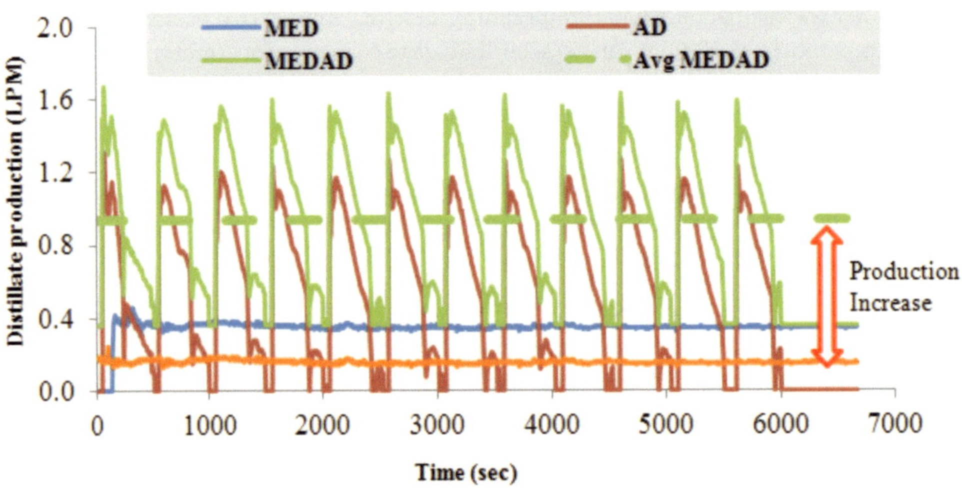

Figure 13. Conventional MED and hybrid MEDAD cycle water production profiles at 38°C heat source temperature

http://dx.doi.org/10.5772/60400

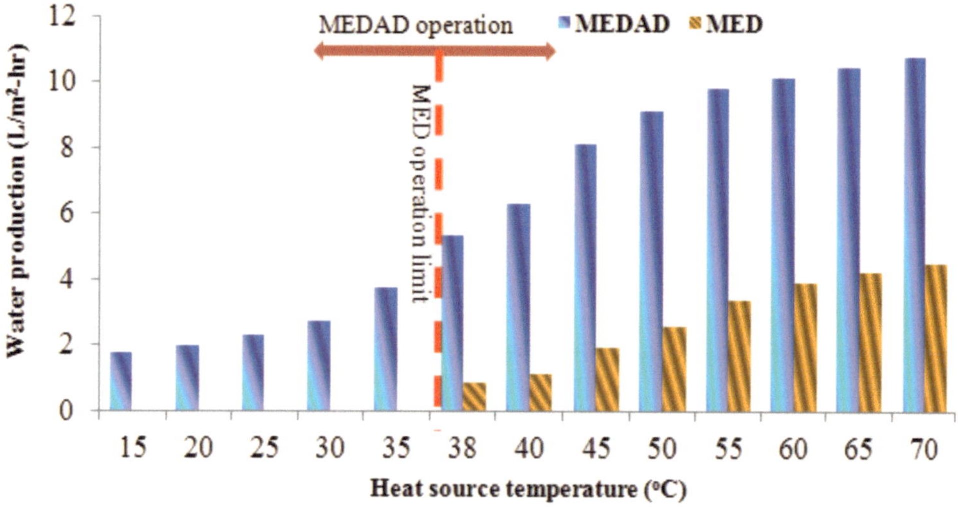

Figure 14. MED and MEDAD steady state water production at different heat source temperatures

Figure 14 shows the comparison of water production of MED and hybrid MEDAD cycles at assorted heat source temperatures. Quantum increase in water production (two- to threefold) can be observed at all heat source temperatures. These results have good agreement with simulation results. It can also be seen that in conventional MED system last stage temperature is limited to 38°C due to condenser operating with cooling water from cooling tower. While in the case of hybrid MEDAD, the last stage temperature can be as low as 5°C because there is no condenser and last stage is connected to AD beds for vapor adsorption. This higher overall operational gap in proposed hybrid MEDAD cycle helps to insert more number of stages (up to 19 stages) as compared to conventional MED system (about 4–6 stages). More number of stages increases the vapor condensation heat recoveries and hence the water production at same top brine temperatures.

5. Exergy analysis for operational cost apportionment

A computation model is developed for the cogeneration plant where the properties of expanding steam, such as enthalpy (h) and entropy (s) at a given temperature and pressure, are computed for the key states in the schematic diagrams of "PP", "PP+MED," and "PP +MEDAD" cycles. The approach employed here is to calculate the total exergy changes or destruction across the inlet and exit sections of the equipment. For example, the exergy associated with the turbines is the sum of all contributions from (a) the HP-T unit at the same mass flow rate across it, (b) the LP-T unit until the extraction point, and (c) the exergy changes after the point of steam extraction, i.e., $E_{T,1-2}=E_{1-a}+E_{b-1'}+E_{1'-2}$. On the basis of total exergy available across whole system, proportion utilized by power plant and desalination system is calculated.

For this comparison study, an assorted range of bled-steam is extracted at low pressure but the mass flow rates are varied to cover the expected practical operational ranges, typically from 10% to 50% of the total flow. At 20% bled-steam from the LP-T, the ratios of power-to-water for exergy and energetic analyses are found to be 95.7%:4.3% (exergy) and 72.2%:27.8% (energetic) methods, respectively. This implies that if the 20% steam were to be continuously consumed by the low pressure turbines (LP-T), its work contribution from LP_T would be insignificant, i.e., a maximum of 4.3%. However, the energetic value of bled-steam accounts for a disproportionate share of 27.8% due primarily to the high latent heat content of the low pressure steam. This ratio of energy-to-exergy shares of the total working steam is found to be 4 – 7 fold higher, descending with the larger amount of bled-steam as shown in Figure 15. The higher effectiveness of working steam incurred at the MED cycle is attributed to the better thermodynamic matching of steam's latent energy.

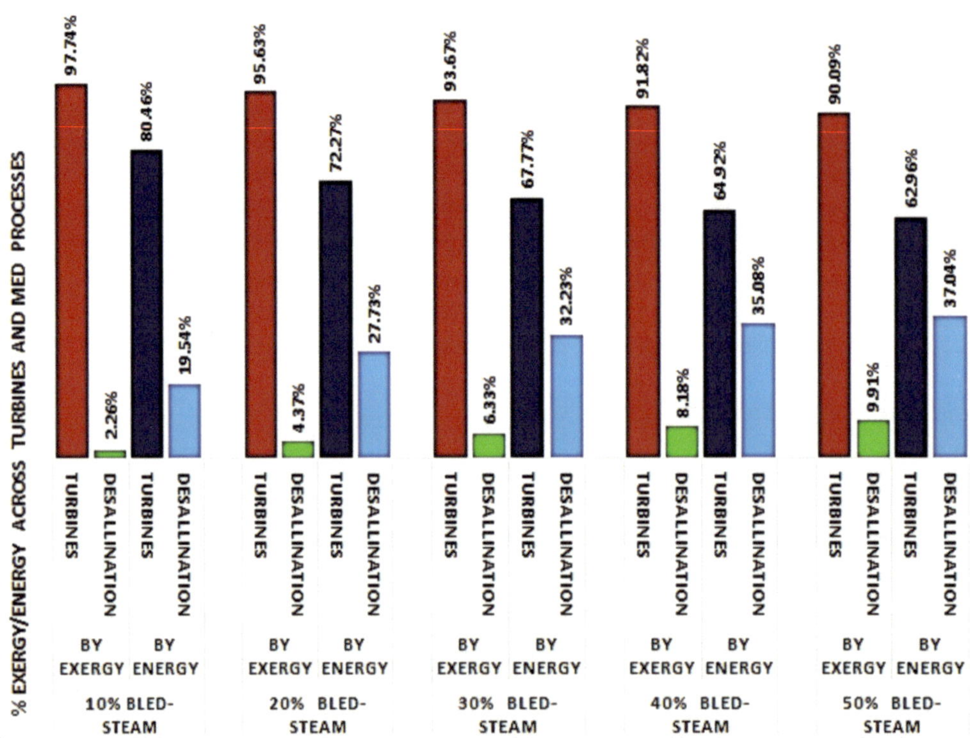

Figure 15. Exergetic and energetic proportions at different percent of steam extraction

On the basis of the above analysis for primary fuel cost and with data from the published literature [55, 56], the life-cycle cost (LCC) of water production is compared for all capital expenditure (Capex) and operation expenditure (Opex), across all proven industrial processes,

http://dx.doi.org/10.5772/60400

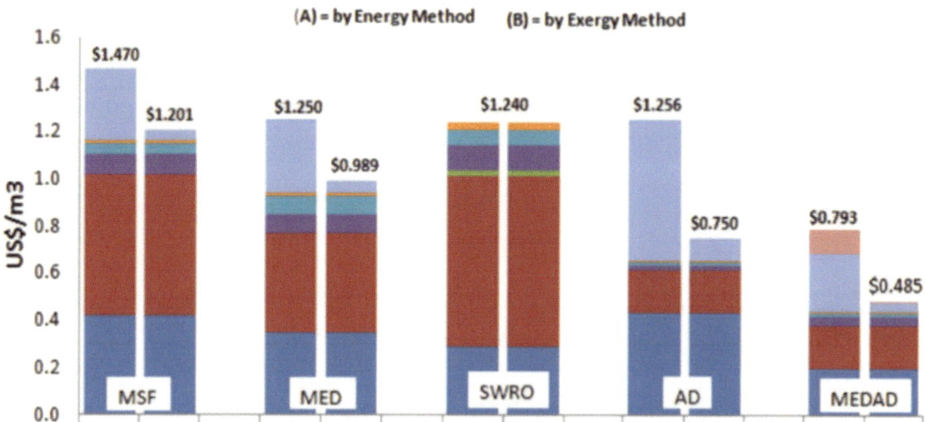

	MSF A	MSF B	MED A	MED B	SWRO A	SWRO B	AD A	AD B	MEDAD A	MEDAD B
Hybrid / AD thermal	0.000	0.000	0.000	0.000	0.000	0.000	0.000	0.000	0.103	0.005
Thermal	0.310	0.049	0.310	0.049	0.000	0.000	0.600	0.095	0.248	0.039
Parts	0.010	0.010	0.010	0.010	0.030	0.030	0.011	0.011	0.012	0.012
Chemicals	0.050	0.050	0.080	0.080	0.070	0.070	0.010	0.010	0.010	0.010
Labour	0.080	0.080	0.080	0.080	0.100	0.100	0.020	0.020	0.042	0.042
Membranes	0.000	0.000	0.000	0.000	0.030	0.030	0.000	0.000	0.000	0.000
Electrical	0.600	0.600	0.420	0.420	0.720	0.720	0.185	0.185	0.180	0.180
Capital cost	0.420	0.420	0.350	0.350	0.290	0.290	0.430	0.430	0.197	0.197

Cost of electricity is taken at US$0.15 /kWh, the exergy ratio for Turbine to MED is 0.9563 to 0.0437,
Increase in water production by MEDAD is 2 folds of the conventional MED.

Figure 16. A comparison of life-cycle unit water cost for various desalination methods

as shown in Figure 16. Exergy factor calculated above is utilized only for thermal and electricity cost calculations. Energy based analysis has good agreement with GWI [56] data. It can be seen that by LCC, unit water production cost is highest for PP+MSF which amounts to US$ 1.201/m^3, whilst the lowest unit cost is the PP+MEDAD method which is only at US$ 0.485/m^3 and this unit cost is even lower than the LCC of reverse osmosis (RO) plants.

6. Summary

Recent developments in adsorption theory, adsorption desalination (AD), and conventional MED desalination cycles have been reviewed in this chapter. We highlight the key role of AD cycles which can be hybridized with the proven cycles such as the MED cycle, exploiting the thermodynamic synergy between the thermally driven cycles that significantly improve the water production yields. Experiments were conducted in a lab-scale pilot MEDAD and confirmed the excellent synergetic effects that boosted the water production up to two- to threefold over the conventional MED. We believe that if the hybrid MEDAD cycles are well optimized and operated, it can achieve high GOR and the projected LCC of water production can be lowered to as low as US$ 0.485/m^3.

7. Abbreviation

MED; Multi-effect desalination

AD; Adsorption desalination

RO; Reverse osmosis

SDWP; Specific daily water production

EDF; Energy distribution function

GOR ; Gain output ratio

SG; Steam generator

PR; Performance ratio

WPR; Water production ratio

TBT; Top brine temperature

PP; Power plant

HP-T; High pressure turbine

LP-T; Low pressure turbine

LBT; Lower brine temperature

LCC; Life cycle costing

PDF; Probability distribution function

Symbols	
P	Pressure [kPa]
P_0	Saturation pressure [kPa]
R	Universal gas constant [kJ/kmol.K]
T	Temperature [K]
t	Surface heterogeneity for Tóth [-]
Greek Symbols	
γ	Dubinin–Astakhow power factor [-]
ε	Adsorption site energy [kJ/kmol, or kJ/kg]
ε_c	Equilibrium adsorption site energy [kJ/kg]
$\varepsilon_{1\ to}\ \varepsilon_3$	Reference energy [kJ/kmol]
θ	Adsorption uptake [kg/kg of adsorbent]

http://dx.doi.org/10.5772/60400

θ	Local adsorption uptake [kg/kg of adsorbent]
μ	Chemical potential [kJ/kmol]
$\chi(\varepsilon)$	Energy distribution function [-]
Subscript	
α	Adsorbed phase
g	Gaseous phase

Acknowledgements

The authors wish to thank National Research Foundation (NRF), Singapore (grant WBS no. R-265-000-399-281), and King Abdullah University of Science & Technology (KAUST) (Project no. 7000000411) for financial support for MED plant at the National University of Singapore.

Author details

Muhammad Wakil Shahzad[1*], Kyaw Thu[2], Ang Li[2], Azhar Bin Ismail[1] and Kim Choon Ng[2]

*Address all correspondence to: muhammad.shahzad@kaust.edu.sa

1 King Abdullah University of Science & Technology, Water Desalination & Reuse Center (WDRC), Thuwal, Saudi Arabia

2 Department of Mechanical Engineering, National University of Singapore, Singapore

References

[1] Sources of fresh water, UNESCO module-7 (http://amper.ped.muni.cz/~miler/aktivity/unesco/Science%20and%20Technology%20Education/07_Sources+of+fresh+water.pdf)

[2] http://sciencenordic.com/earth-has-lost-quarter-its-water

[3] Earth's water supply (http://gen.uga.edu/documents/water/Earth.pdf)

[4] National oceanic and atmospheric administration, United State department of commerce. (http://oceanservice.noaa.gov/facts/wherewater.html)

[5] The encyclopedia of earth (http://www.eoearth.org/view/article/152861/)

[6] http://www.universetoday.com/65588/what-percent-of-earth-is-water/

[7] Science education resource center, Carleton College, (http://serc.carleton.edu/eslabs/index.html)

[8] Hoekstra AY, Mekonnen MM. Global water scarcity: the monthly blue water footprint compared to blue water availability for the world's major river basins, Value of water research report series no. 53, UNESCO-IHE Institute for Water Education.

[9] Thu K, Chakraborty A, Kim Y-D, Myat A, Saha BB, Ng KC. Numerical simulation and performance investigation of an advanced adsorption desalination cycle. Desalination 2013;308:209–18.

[10] GWI, Desalination Markets 2007, A Global Industry Forecast (CD ROM), Global Water Intelligence, Media Analytics Ltd., Oxford, UK, 2007, www.globalwaterintel.com

[11] Dawoud MA, Al Mulla MM. Environmental impacts of seawater desalination: Arabian Gulf case study. Int J Environ Sustain 2012;1(3),22–37.

[12] Dawoud MA. The role of desalination in augmentation of water supply in GCC countries, Desalination 2005;186:187–98.

[13] Abdulrazzak MJ. Water supplies versus demand in countries of Arabian Peninsula. J Water Res PL ASCE 1995;121(3):227–34.

[14] Fath H, Sadik A, Mezher T. Present and future trend in the production and energy consumption of desalinated water in GCC countries. Int J Thermal Environ Eng 2013;5(2):155–65.

[15] Nezhad H. A report on world energy scenarios to 2050: issues and options, Decision Sciences Metropolitan State University Minneapolis, MN, Sep-2009.

[16] Al-Zubari WK. The Water-Energy Nexus in the GCC Countries Evolution and Related Policies, Sixth "Zayed Seminar" on Green Economy: Success Stories from the GCC, 8–9 May 2013, UN virtual water learning center, the Arabian regional center, Arabian Gulf University, Bahrain.

[17] Clean Water for a Healthy World, Secretariat of the Convention on Biological Diversity. 2009. Drinking Water, Biodiversity and Poverty Reduction: A Good Practice Guide, World water day March 22, 2010.

[18] Caron DA, Garneau ME, Seubert E, Howard MDA, Darjany L, Schnetzer A, Cetinic I, Filteau G, Lauri P, Jones B, Trussell S. Harmful algae and their potential impacts on desalination operations off southern California. Water Res 2010;44:385–416.

[19] Huttner KR. Overview of existing water and energy policies in the MENA region and potential policy approaches to overcome the existing barriers to desalination using renewable energies. Desalin Water Treat 2013;51:87–94.

[20] Ghaffour N, Missimer TM, Amy GL. Combined desalination, water reuse, and aquifer storage and recovery to meet water supply demands in the GCC/MENA region. Desalin Water Treat 2013;51:38–43.

[21] Langmuir I. The adsorption of gases on plane surfaces of glass, mica and platinum, J Am Chem Soc 1918;40:1361–403.

[22] Ward C, Findlay R, Rizk M. Statistical rate theory of interfacial transport. I. Theoretical development. J Chem Phys 1982;76:5599–605.

[23] Elliott JAW, Ward CA. Statistical rate theory description of beam-dosing adsorption kinetics, J Chem Phys 1997;106:5667–76.

[24] Elliott JAW, Ward CA. Statistical rate theory and the material properties controlling adsorption kinetics, on well defined surfaces. Stud Surf Sci Catal 1997;104:285–333.

[25] Elliott JAW, Ward CA. Temperature programmed desorption: a statistical rate theory approach, J Chem Phys 1997;106:5677–84.

[26] Rudzinski W, Borowiecki T, Dominko A, Panczyk T. A new quantitative interpretation of temperature-programmed desorption spectra from heterogeneous solid surfaces, based on statistical rate theory of interfacial transport: the effects of simultaneous readsorption. Langmuir 1999;15:6386–394.

[27] Rudzinski W, Lee S-L, Panczyk T, Yan C-CS. A fractal approach to adsorption on heterogeneous solids surfaces. 2. Thermodynamic analysis of experimental adsorption data. J Phys ChemB 2001;105:10857–66.

[28] Rudzinski W, Lee S-L, Yan C-CS, Panczyk T. A fractal approach to adsorption on heterogeneous solid surfaces. 1. The relationship between geometric and energetic surface heterogeneities. J Phys Chem B 2001;105:10847–56.

[29] Li A. Experimental and theoretical studies on the heat transfer enhancement of adsorbent coated heat exchangers, Doctor of Philosophy, Department of Mechanical Engineering, National University of Singapore, 2014.

[30] Wang X, Chakarborty A, Ng KC, Saha BB. How heat and mass recovery strategies impact the performance of adsorption desalination plant: theory and experiments. Heat Transfer Eng 2007;28:147–53.

[31] Saha BB, Choon NK, Chakarborty A, Thu K. Desalination System and Method, in: W.P. WO2011010205A1 (ed.), 2011.

[32] Thu K. Adsorption desalination: theory & experiments, in: Doctoral Thesis, National University of Singapore, 2010.

[33] Saha BB, Alam KCA, Akisawa A, Kashiwagi T, Ng KC, Chua HT. Two-stage non-regenerative silica gel-water adsorption refrigeration cycle, ASME Advanced Energy Systems Division (Publication) AES 2000;40:65–9.

[34] Ng KC. Recent developments in heat-driven silica gel-water adsorption chillers. Heat Transfer Eng 2003;24:1–3.

[35] Ng KC, Wang XL, Chakraborty A. in: Apparatus and Method for Desalination, 2006.

[36] Ng KC, Saha BB, Chakraborty A, Koyama S. Adsorption desalination quenches global thirst. Heat Transfer Eng 2008;29:845–8.

[37] Wu JW, Hu EJ, Biggs MJ. Thermodynamic cycles of adsorption desalination system. Appli Energy 2012;90:316–22.

[38] Wu JW, Biggs MJ, Pendleton P, Badalyan A, Hu EJ. Experimental implementation and validation of thermodynamic cycles of adsorption-based desalination. Appl Energy 2012;98:190–7.

[39] Boelman EC, Saha BB, Kashiwagi T. Experimental investigation of a silica gel-water adsorption refrigeration cycle – the influence of operating conditions on cooling output and COP. 1995;101:358–66.

[40] Khan MZI, Alam KCA, Saha BB, Akisawa A, Kashiwagi T. Study on a re-heat two-stage adsorption chiller – the influence of thermal capacitance ratio, overall thermal conductance ratio and adsorbent mass on system performance. Appl Thermal Eng 2007;27:1677–85.

[41] Khan MZI, Alam KCA, Saha BB, Akisawa A, Kashiwagi T. Performance evaluation of multi-stage, multi-bed adsorption chiller employing re-heat scheme. Renewable Energy 2008;33:88–98.

[42] Ng KC, Thu K, Saha BB, Chakraborty A. Study on a waste heat-driven adsorption cooling cum desalination cycle. Int J Refrigeration 2012;35:685–93.

[43] Thu K, Ng KC, Saha BB, Chakraborty A, Koyama S. Operational strategy of adsorption desalination systems. Int J Heat Mass Transfer 2009;52:1811–6.

[44] EI-Sharkawy I, Thu K, Ng K, Saha BB, Chakraborty A, Koyama S. Performance improvement of adsorption desalination plant: experimental investigation. Int Rev Mech Eng 2007;1:25–31.

[45] Ng KC, Wang X, Lim YS, Saha BB, Chakarborty A, Koyama S, Akisawa A, Kashiwagi T. Experimental study on performance improvement of a four-bed adsorption chiller by using heat and mass recovery. Int J Heat Mass Transfer 2006;49:3343–8.

[46] Wang X, Ng KC. Experimental investigation of an adsorption desalination plant using low-temperature waste heat. Appl Thermal Eng 2005;25:2780–9.

[47] Ng KC, Wang XL, Gao LZ, Chakraborty A, Saha BB, Koyama S. Apparatus and method for desalination, SG Patent application number 200503029-1 (2005) and WO Patent no. 121414A1 (2006).

[48] Ng KC, Thu K, Amy G, Chunggaze M, Al-Ghasham TY. An Advanced ADMED Cycle For Low-Temperature Driven Desalination, US Provisional Application No. 61/450,165.

[49] Ng KC, Thu K, Amy G, Chunggaze M, Ghasham TYA. A Regenerative Adsorption Distillation System, in: WO Patent WO2012121675A1, 2012.

[50] Shahzad MW, Ng KC, Thu K, Saha BB, Chun WG. Multi effect desalination and adsorption desalination (MEDAD): a hybrid desalination method. Appl Thermal Eng 2014;72:289–97.

[51] Thu K, Kim Y-D, Amy G, Chun WG, Ng KC. A synergetic hybridization of adsorption cycle with the multi-effect distillation (MED). Appl Thermal Eng 2014;62:245–55.

[52] Ng KC, Thu K, Oh SJ, Ang L, Shahzad MW, Ismail AB. Recent developments in thermally-driven seawater desalination: Energy efficiency improvement by hybridization of the MED and AD cycles. Desalination 2015;356:255–70.

[53] Ng KC, Thu K, Shahzad MW, Chun W. Progress of adsorption cycle and its hybrids with conventional multi-effect desalination processes. IDA J Desalin Water Reuse, DOI 10.1179/2051645214Y.0000000020.

[54] Shahzad MW, Myat A, Chun WG, Ng KC. Bubble-assisted film evaporation correlation for saline water at sub-atmospheric pressures in horizontal-tube evaporator. Appl Thermal Eng 2013;50:670–6.

[55] Lattemann S, Kennedy MD, Schippers JC, Amy G. Chapter 2; Global Desalination Situation.

[56] http://www.globalwaterintel.com/archive/13/10/general/technology-choice-still-open-yanbu-3.html

Chapter 8

Techno-Economics of Hybrid NF/FO with Thermal Desalination Plants

Abdel Nasser Mabrouk, Hassan Fath,
Mohamed Darwish and Hassan Abdulrahim

Additional information is available at the end of the chapter

http://dx.doi.org/10.5772/60207

Abstract

Hybrid desalination technology is defined as any combination of thermal and membrane processes in seawater desalination systems. So far, the two technologies have evolved rather independently with some degree of competition. Traditionally, in co-generation market applications, thermal desalination has succeeded in establishing a stronghold where large capacities are needed, energy costs are low, and seawater quality is challenging. However, in recent years, membrane systems have also succeeded in grabbing a larger share of the world seawater desalination market, mainly as a result of progress made in membrane and energy recovery technologies. Realizing the potential benefits and challenges faced by both technologies on their own, designers have been looking for ways to synergize and combine the two technologies in optimum configurations, which promise to further reduce the total cost of seawater desalination.

Several studies have been published over the past 20 years addressing the potential of integrating hybrid desalination systems. Coupling schemes worth noting for hybrid systems include RO preheating using condenser heat reject of the associated distillation unit; the use of membrane filtration upstream of MSF, MED, and RO systems; brine recirculation coupling; product blending; and the use of common intake and outfall systems. To date, commercially available hybrid desalination plants are of the simple non-integrated type. They may share common systems such as intake and outfall facilities, but otherwise they run independently at the same site. Product water from the membrane and thermal systems are usually blended to international standards on water quality. One more step ahead this chapter addresses the role of using forward osmosis (FO) or nano-filtration (NF) as a pre-treated method to the existing thermal desalination plants. The target of this hybridization is to reduce divalent ions that cause hard scale deposition at elevated temperature. The separation of divalent ion enables the increase of the desalination process temperature greater than 110°C, which consequently increases the plant performance, increases the productivity, and reduces the chemical consumption.

Integrating the NF system with new (MSF-DBM) configuration enables to increase the TBT up to 130 °C. The new NF-MSF-DBM configuration significantly reduces the unit's input thermal energy to suit the use of (the relatively expensive) solar energy as a desalination plant driver. The desalination pilot test is built to evaluate the performance of the novel de-aeration and brine mix (MSF-DBM) configuration at high top brine temperature (TBT) using NF membrane. The capacity of the desalination pilot plant is 1.0 m^3/day of water. Comparisons between the simulation and the experimental results of the pilot unit subsystems are relatively satisfactory. The newly developed NF-MSF-DBM (de-aerator and brine mix) configuration is tested at TBT = 100 °C. The gain output ratio (GOR) is calculated as 15, which is almost twice the GOR of the traditional MSF. High GOR NF-MSF-DBM system requires lower input thermal energy, which makes the integration with (the relatively expensive) RE system as a desalination plant driver, viable option.

Simulation results showed that integrating FO to existing MSF and using brine of the last stage as a draw solution at a recovery ratio of 35% reduce the Ca^+ ions in the seawater feed by 20%, which enables increasing the TBT up to 130°C safely. The simulation results show that, the production of the existing MSF plant increases by 20% as a result of working at higher temperature of TBT=130 °C. The specific operating cost (OPEX) analysis showed that up to 2.3 M$/year of chemical cost can be saved if the FO membrane is deployed to the existing MSF desalination plants in Qatar.

Keywords: Desalination, Thermal, Membrane, Hybrid, Cost

1. Introduction

1.1. Hybrid desalination concept

Hybrid technology is defined as any combination of thermal and membrane processes in seawater desalination systems. So far, the two technologies have evolved rather independently with some degree of competition. Traditionally, in co-generation market applications, thermal desalination has succeeded in establishing a stronghold where large capacities are needed, energy costs are low, and seawater quality is challenging. However, in recent years, membrane systems have also succeeded in grabbing a larger share of the world seawater desalination market, mainly as a result of the progress made in membrane and energy recovery technologies. Realizing the potential benefits and challenges faced by both technologies on their own, designers have been looking for ways to synergize and combine the two technologies in optimum configurations, which promise to further reduce the total cost of seawater desalination.

Several studies have been published over the past 20 [1-14] years addressing the potential of integrating hybrid desalination systems. Coupling schemes worth noting for hybrid systems include RO preheating using condenser heat reject of the associated distillation unit; adding thermal vapour compression to MED systems; the use of membrane filtration (NF) upstream of MSF, MED, and RO systems; brine recirculation coupling; product blending; and the use of common intake and outfall systems.

To date, commercially available hybrid plants are of the simple non-integrated type. They may share common systems such as intake and outfall facilities, but otherwise they run independently at the same site. Product water of both membrane and thermal plants are usually blended to meet the international standards water quality specifications. Examples of existing hybrid plants include Jeddah, Al-Jubail, and Yanbu in Saudi Arabia. Recently, a power and water plant has been designed and built by Doosan in Fujairah (UAE). This plant produces 500 MW of net electricity to the grid and 100,000 MIGD of fresh water, 63% of which is produced by MSF with the balance produced by RO [1].

1.2. NF's role in desalination

NF, in particular, has been advocated as a pretreatment option upstream of a thermal desalination unit [2-8]. Due to small pore size and charge at the surface of the membrane, NF is known to remove divalent ions, including a fraction of scale-producing hardness and salts, allowing in principle at least a possible increase in top brine temperature and promising improved steam economy. Studies on NF-MSF pilot tests claim scale-free operation for 1,200 hours with top brine temperatures reaching 130°C, reporting an improvement in recovery from 30% for stand-alone MSF to 70% with NF [2-8]. The integration of NF –MED system is under pilot investigation by Saline Water Desalination Research Institute (SWDRI) with Sasakura (a Japanese-based consortium) [15]. Implementation of dual-stage NF has been successful evaluated at Long Beach, California [16]. From the present analysis, NF will play a crucial role in desalination, provided that the cost of NF membrane would be decreased. Efficient removal of boron has also been reported [16]. However, the reliability and economic viability of such a design need to be confirmed, considering the higher thermal and pressure load implied by the design and the additional capital, energy, and operation and maintenance costs of membrane pre-treatment components. Adding to the uncertainty are increased risks for corrosion and the long-term reliability of such a system.

1.3. FO role in desalination

The role of using forward osmosis (FO) as a pre-treated method to the existing thermal desalination MSF/MED plants is to reduce divalent ions that cause hard-scale deposition at elevated temperature. The removal of the divalent ions, such as $CaSO_4$, from the MSF feed enables to increase the desalination process temperature greater than 110°C. Consequently the plant performance and productivity will increase. Due to the removal of the ions which cause scale deposit, the chemical additive consumption will be decreases. In the MSF process due to working at higher temperature, hard scale, such as calcium sulphate, is formed. As calcium sulphate is two orders of magnitude more soluble than calcium carbonate, the sulphate is much less likely to drop out of solution when both are present.

In the light of the recent development in the membrane filtration technologies, the cost of seawater pretreatment can be reduced if FO membranes were used with/without NF. The novel application of FO membrane for seawater filtration requires, firstly, retrofitting the FO system to the thermal desalination unit. Secondly, it also requires finding a suitable draw solution that would reduce the cost of FO pre-treatment. Fortunately, the current FO membranes exhibit high water permeability and rejection rate, which make them an ideal solution for seawater pretreatment [17]. A novel hybrid FO-thermal desalination system to remove scale deposit elements from seawater to the thermal units, is presented [17-18]. The performance of the thermal evaporator was evaluated after introducing the FO pretreatment. The scale deposition on the thermal unit was estimated by using special software to predict the precipitation on inversely soluble metal ions on the heat exchangers [17-18].

1.4. Objective

This chapter addresses the role of using FO or NF as a pre-treated method to the existing thermal desalination plants. The target of this hybridization is to reduce divalent ions that cause hard-scale deposition at elevated temperature. The separation of divalent ion enables the increase of the desalination process temperature greater than 110°C, which consequently increases plant performance and productivity, as well as reduces the chemical consumption.

2. Process description

2.1. MSF-RO hybrid

The simple type refers to co-located thermal and membrane systems that may share some common systems on site. This in turn facilitate blended to product water specifications, but otherwise are running independently. Examples include the Fujairah plant and three Saudi plants in Jeddah, Al-Jubail, and Yanbu. The Fujairah plant [1], representing a simple hybrid type, was constructed by Doosan Heavy Industries and is currently considered the largest existing hybrid type. The plant is rated at 100 MIGD, of which 63% are produced by MSF and 37% by RO. Featuring a combined cycle system, it also generates 500 MW to the grid. The thermal part of the plant includes five MSF evaporators rated at 12.5 MIGD each, with a top brine temperature of 110∘C. The membrane part includes two RO passes, using a conventional pretreatment system and energy recovery devices of the Pelton type. A specification of 200 PPM as the maximum water product salinity was met by the design.

Another option of a hybrid type is to improve the membrane performance. This type includes the integration of hybrid membrane and thermal systems, with the aim of improving membrane recovery by preheating the RO feed using heat reject from the thermal unit as shown in Figure (1). Higher temperatures are known to improve membrane flux, mainly as a result of reduced viscosity. Several investigators examined the effects of preheating in pilot tests, and about 3% of recovery improvement is reported per degree Celsius [19]. This should, however, be weighed against potential negative effects of high temperatures on membrane performance, in particular compaction damage. Membrane manufacturers have traditionally set an upper temperature limit of 40∘C for the use of their membranes, and it is not clear how close to this limit operation should be, in order to optimize life-cycle membrane performance and costs. The measure is particularly useful in winter when seawater temperatures are reduced.

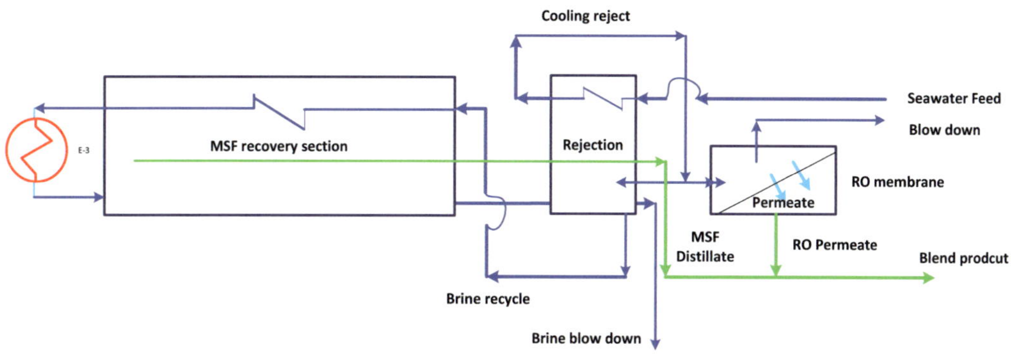

Figure 1. Process flow diagram of hybrid RO-MSF process

Preheating the Fujairah RO feed in winter is an example of integrated hybrid operation, representing type 2. A 10°C increase from 23°C to 33°C for an RO unit equivalent in capacity to the Fujairah plant would increase recovery by about 30% and, therefore, reduce feed pressure requirements. This preheating feature could then be made to good use in the winter when seawater temperatures in the Gulf drop by 15–20°C.

2.2. NF-MSF process

This type includes the integration of hybrid membrane and thermal systems with the objective of improving the gain output ratio (GOR) and steam economy of the thermal system (see Figure 2). The GOR is a function of the available temperature range and can, therefore, be improved by increasing the top brine temperature (TBT). Traditionally, the top brine temperature is limited to 110 °C for MSF and is limited to 65∘C for MED. This practice, in addition to chemical dosing and mechanical cleaning, is necessary to minimize scale deposition on heat transfer surfaces. Some investigators have advocated the use of NF membrane upstream of the thermal system as a pretreatment step to reduce scaling hardness and additionally some salt. This would, in principle, allow operation at higher temperatures, without increased scaling. SWCC investigators have tested a hybrid NF/MSF pilot unit running at a top brine temperature of 130°C for a period of 1,200 hours and reported a doubling in the recovery with no observed scale formation [10].

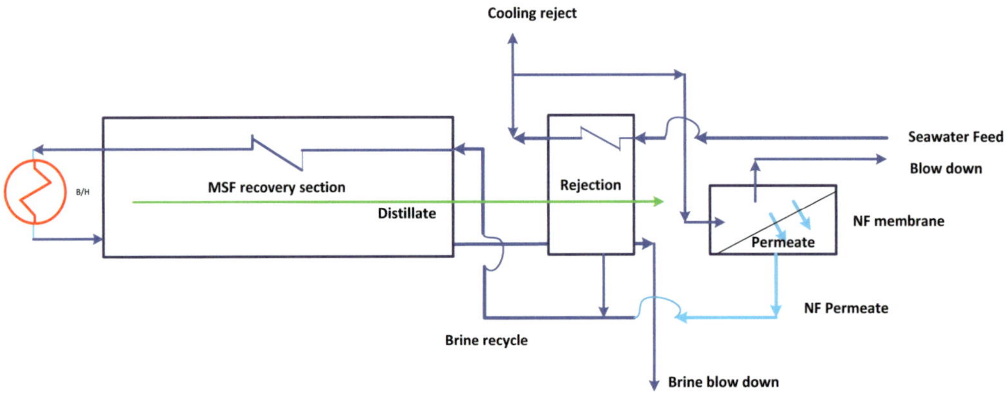

Figure 2. Process flow diagram of hybrid NF-MSF process

2.3. FO-MSF

Figure 3 shows hybrid FO-multi stage flash (FO-MSF) system for high TBT MSF. In this type of hybrid system, the brine reject from the thermal desalination process will be considered as a draw solution, while the cooling seawater exiting from the MSF heat rejection section will be used as a feed solution. Permeate water will transport across the FO membrane from the feed to the draw solution side while monovalent and multivalent ions are rejected by the FO membrane. After leaving the FO membrane, the concentrated seawater is dumped back to the sea. Simultaneously, the diluted draw solution from the FO process is circulated to the MSF recovery section. Inside the MSF plant, fresh water is extracted from the draw solution by evaporation and is condensed in the consecutive MSF chambers. The distilled water is collected at the last stage and directed to the distilled tank. The un-flashed brine through MSF stages (brine pool) is collected in the last stage of MSF evaporator at high salinity and then is directed to the FO as a draw solution.

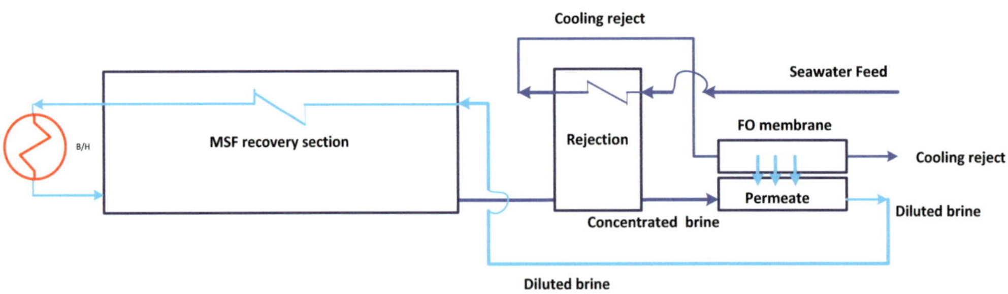

Figure 3. Process flow diagram of hybrid FO-MSF process

http://dx.doi.org/10.5772/60207

2.4. FO-MED

Figure 4 shows the hybrid FO-multi effect distillation (FO-MED) system for high TBT MED. In this type of hybrid system, the brine reject from the last effect will be considered as a draw solution stream, while the condenser cooling seawater will be used as a feed solution stream. Permeate water will transport across the FO membrane from the feed to the draw solution side while monovalent and multivalent ions are rejected by the FO membrane. After leaving the FO membrane, the concentrated seawater is dumped back to the sea water. Simultaneously, the diluted draw solution from the FO process is circulated to the MED evaporator as a makeup feed. Inside the MED evaporator, fresh water is extracted from the draw solution by evaporation and is condensed in the consecutive MED effect. The distilled water is collected at the last effect and is directed to the distilled tank. The brine (un-evaporated) through MED effect is collected in the last effect at a high salinity is directed to the FO again as a draw solution.

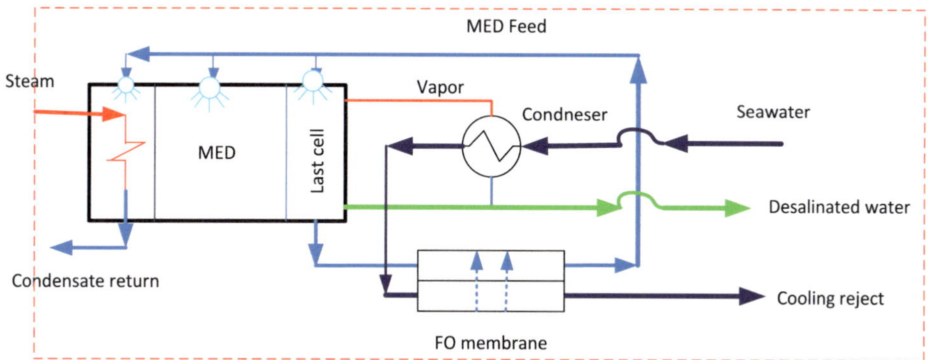

Figure 4. Process flow diagram of hybrid FO-MED process

3. Methodology

3.1. Mathematical model development of NF

Figure 5 illustrates the input and output parameters used for the mass and energy balance equations of the NF membrane [13-14].

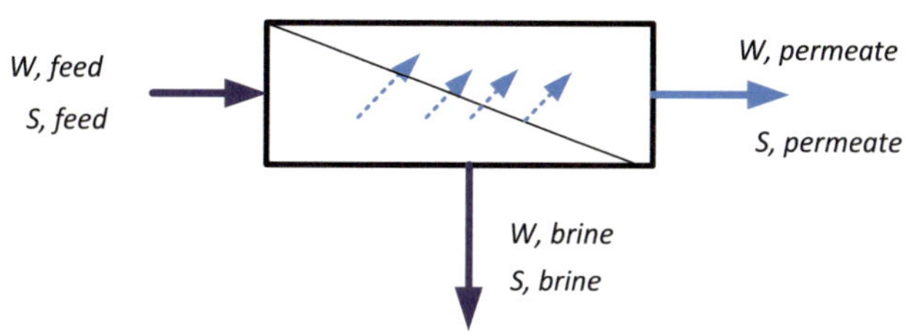

Figure 5. Schematic diagram of the NF membrane streams

Mass balance is written as follows:

$$W_{f,j} = W_{p,j} + W_{b,j} \tag{1}$$

$$S_{f,j} = S_{p,j} + S_{b,j} \tag{2}$$

The following relation defines the rate of water passage through a semipermeable membrane [14]:

$$W_{p,j} = (\Delta P_j - \sigma \Delta \pi_j) \times K_w \times A_j \times TCF \times FF \times \rho_{p,j} \tag{3}$$

$$\Delta P_j = \bar{P}_j - P_{p,j} \tag{4}$$

$$\Delta \pi_j = \bar{\pi}_j - \pi_{p,j} \tag{5}$$

$$\bar{P}_j = 0.5(P_{f,j} + P_{b,j}) \tag{6}$$

As the seawater salt concentrations ratio is almost constant, an approximation for value in kPa can be given as [13]:

$$\pi = 6.895 \times \frac{38.5 \times C_{fbNaCl} \times (T + 273)}{1000 + C_{fbNaCl}} \tag{7}$$

$$C_{fbNaCl} = 0.934348 \times C_{fb} - 0.54169 \tag{8}$$

The rate of salt flow through the membrane is defined as:

$$S_{p,j} = (C_{m,j} - C_{p,j}) \times K_s \times A_j \times TCF + (1 - \sigma) \times J_{v,j} \times \bar{C} \times K_s \times A_j \times TCF \tag{9}$$

$$J_{v,j} = (\Delta P_j - \sigma \Delta \pi_j) \times K_w \times FF \times TCF \left(m / s\right) \tag{10}$$

Where, the temperature factor correction (TCF) is calculated using the following equations [14]:

$$TCF = e^{8.859 \times \frac{T-25}{T+273}}, \text{ for } T \geq 25°\text{C} \tag{11}$$

$$TCF = e^{11.678 \times \frac{T-25}{T+273}}, \text{ for } T \geq 25°\text{C} \tag{12}$$

$$C_{p,j} = S_{p,j} \times \rho_{p,j} / (S_{p,j} + W_{p,j}) \tag{13}$$

A material balance within the mass transfer boundary layer near the membrane wall between the solute carried to the membrane by convection and the solute carried away by diffusion yields an expression that quantifies concentration polarization:

$$\varphi = \frac{C_m - C_p}{C_b - C_p} = e^{J_w/k} \tag{14}$$

The Umm-Lujj NF-RO plant [20] is considered as a case study to verify the mathematical model of the NF membrane equation (1-14), as well as to estimate the permeate constant Kw and the solute constant Ks.

This plant consists of 27 pressure vessels and six NF elements per vessel. The feed characteristic is 360 m^3/hr, the temperature is 32°C, and the salinity is 45.46 g/l. The applied feed pressure is 25 bars. The data from the Umm-Lujj plant, shown in Table (1), are used as the input data of VDS [21-25] software as shown in Figure 6.

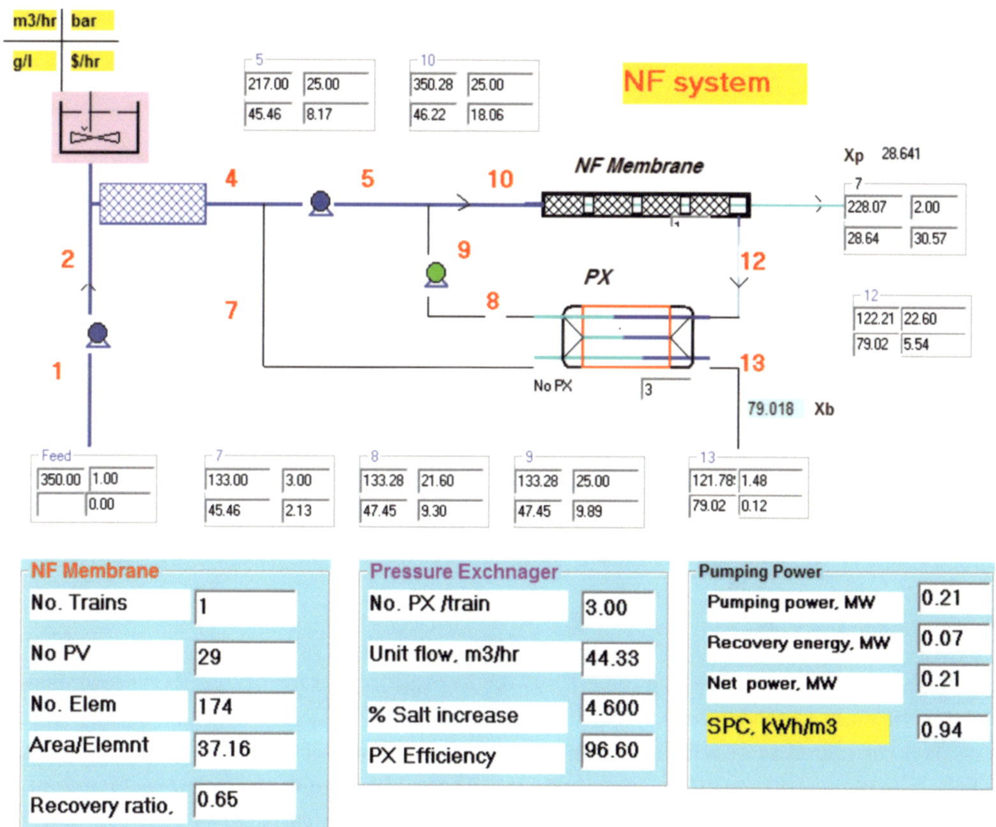

Figure 6. VDS interface of the NF system with pressure exchanger

The VDS simulates the Umm-Lujj plant of NF to estimate the permeate production and the exact value of the membrane constants Kw and Ks. After several runs, the membrane water permeability K_w of the considered NF membrane is determined as follows:

$$K_w = 5.8 \times 10^{-9} \text{ m}^3/\text{m}^2.\text{s.kPa} \quad (15)$$

The membrane salt permeability coefficient K_s is estimated as follows:

$$K_s = 9 \times 10^{-8} \quad (16)$$

Using the estimated values Kw and Ks, the VDS results are compared against the typical plant as shown in Table (1). The comparison results show a good agreement between the VDS results and the typical real plant.

Variable	VSP results	Umm-Lujj	% Error
Feed flow rate, m^3/hr, *	360	360	-
Feed salinity, TDS, g/l, *	45.46	45.46	-
Stages No.*	1	1	-
No. of pressure vessels, *	27	27	-
Feed temperature, °C, *	32	32	-
Fouling factor, *	0.95	NA	-
Feed pressure, bar, *	25	25	-
Elements No. per vessel, *	6	6	-
Permeate flow rate, m^3/hr	245	234	4.7%
Recovery ratio	0.68	0.65	4.6%
Permeate salinity, TDS, mg/l	29.11	28.26	3%

Table 1. Comparison between VDS and Umm-Lujj typical results

*Input data to the VDS

The VDS program is used to size the NF system to produce 226 m^3/hr, which represents one-third of the makeup feed, which is required for 1 MIGD MSF. The required number of pressure vessels is calculated as 29 with 174 membrane elements. The calculated system recovery ratio is 65%. The high-pressure pump is assigned by 25 bars. Three units of pressure exchangers are used to recover an electrical energy of 0.07 MW. Each unit's capacity is 44 m^3/hr, and the percentage of salt increase is only 4.6%. The net pumping power required is 0.21 MW, and the specific power consumption is 0.94 kWh/m^3.

http://dx.doi.org/10.5772/60207

3.2. Mathematical model of forward osmosis (FO)

Forward osmosis is the transport of water across a selectively permeable membrane from a region of higher water chemical potential (feed solution) to a region of lower water chemical potential (draw solution). Consequently, a less concentrated draw solution is being produced, which may be further treated to extract freshwater. Obviously, there are two key problems that must be solved to make the technology go out of the laboratory. One is continued improvement and optimization of the selectively permeable membrane, which allows passage of water but rejects most solute molecules or ions. The other is the identification of optimal osmotic agents and its corresponding recovery processes for the supply of the osmosis pressure difference, which is the driving force of the FO process. However, the FO membrane water flux is far lower than the anticipated when the membrane used is asymmetric. The primary reason for this finding is the fact that both FO are accompanied by internal concentration polarization (ICP) as shown in Figure 7.

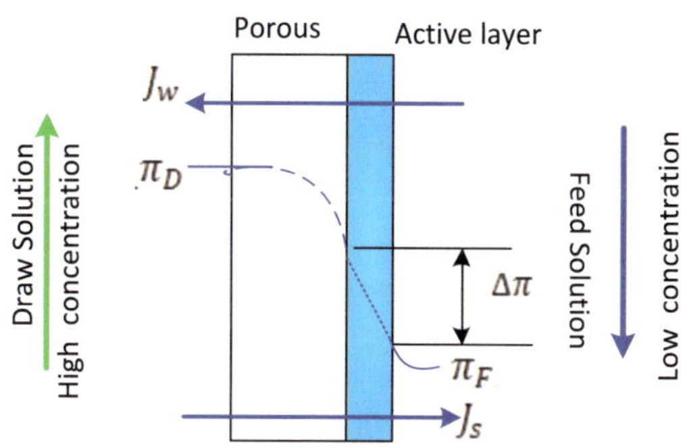

Figure 7. Concentration profile in asymmetrical FO membrane

In the work of Jung et al. [26], FO performance (permeate flux and recovery rate) of a 10 cm ×10 cm plate and frame type membrane is investigated via a numerical simulation based on the mass conservation theorem. The FO membrane orientation, flow direction of feed and draw solutions, flow rate, and solute resistivity (K) are simulated. The case of draw solution facing the active layer displays a relatively higher performance than the feed solution facing the active layer [26]. The numerical results showed that the membrane performance is much more sensitive to the physical membrane property parameter rather than the flow rate and flow direction. However, the simulation methodology does not consider fouling and reverse solute diffusion. Also, the performance of the FO in a relatively large size needs to be explored to come up with a concrete recommendation to the commercialization phase.

A simple schematic of the FO process is shown in Figure 8. The mathematical model of FO membrane is developed as shown in equations (17-26). By knowing the specifications of the FO membrane,, and the membrane area, the outlet stream can be calculated.

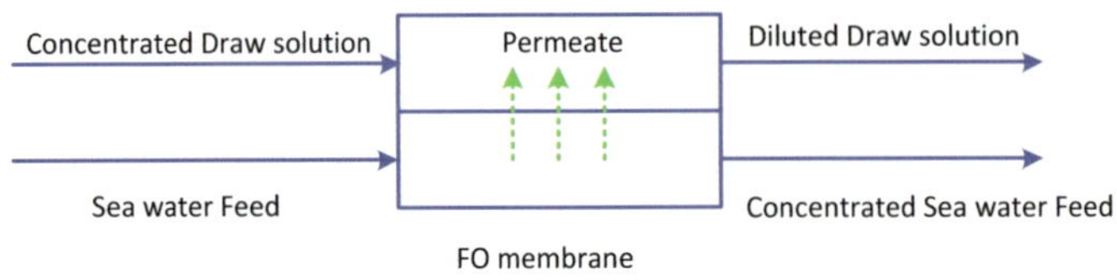

Figure 8. Schematic of the FO membrane process

$$W_{DS,outlet} - W_{DS,inlet} = W_p \tag{17}$$

$$S_{DS,out} - S_{DS,inlet} = S_p \tag{18}$$

$$W_{FS,in} - W_{FS,out} = W_p \tag{19}$$

$$S_{FS,in} - S_{FS,\ out} = S_p \tag{20}$$

$$W_p = J_w \times A \times \rho \tag{21}$$

$$S_p = J_s \times A \tag{22}$$

The general flux equation in the FO process is

$$J_w = A\left(\pi_D - \pi_F\right) \tag{23}$$

where A is the water permeability coefficient, and and denote the osmotic pressures of the draw and feed solution, respectively. The osmotic pressure can be determined by the modified Van't Hoff equation as:

$$\Delta\pi = \frac{N_{ion} R_g T \Delta C}{M_w} \tag{24}$$

where T, and indicate the ionization number of the solution, the ideal gas constant, the absolute temperature, the salt concentration difference of the solution across the membrane, and the molecular weight of the salt, respectively. For asymmetrical membrane, internal ICP occurs within the porous support as shown in Figure 7. However, when the reverse solute is considered [27], and for the case of the draw solution facing the porous layer, the water flux and solute flux are:

$$J_w = A\left(\frac{\pi_D exp\left(-\frac{J_w S}{D}\right) - \pi_F exp\left(\frac{J_w}{k}\right)}{1 + \frac{B}{J_w}\left[exp\left(\frac{J_w}{k}\right) - exp\left(-\frac{J_w S}{D}\right)\right]}\right) \tag{25}$$

$$J_s = B\left(\frac{C_D exp\left(-\frac{J_w S}{D}\right) - C_F exp\left(\frac{J_w}{k}\right)}{1 + \frac{B}{J_w}\left[exp\left(\frac{J_w}{k}\right) - exp\left(-\frac{J_w S}{D}\right)\right]}\right) \tag{26}$$

where B is the salt permeability coefficient.

The water flux and reverse solute equations are difficult to solve analytically because these equations are dependent on water flux and solute passage including in concentration polarization terms. Thus, using a program to solve these equations will be presented.

The FO is still facing many challenges such as internal concentration polarization, which requires breakthrough in the molecular design of high-performance FO membrane. On the other hand, development of draw solution with low cost and low energy consumption required for recovery is urgently needed. To mitigate ICP, the FO membranes must have characteristics of high permeability and hydrophobicity with a small structure parameter, while the preferred draw solutes must have diffusion coefficient, reasonable molecular size, and low viscosity [52].

The VDS Software interface is developed to consider a case study as shown in Figure 9. As shown in Figure 9, the specified feed solution of (1680 t/h and 45 g/l) is pumped to the FO membrane against a draw solution (NaCl) (1200 t/h and 90 g/l). Three elements per vessel are arranged. Two pressure vessels are placed in parallel. The VDS simulated the case, and the results are presented in the same interface as shown in Figure 9. The diluted flow rate is calculated as 1744 t/h and 64.34 g/l, while the concentrated feed solution is calculated as 1135.8 t/h and 66.46 g/l. Based on these calculations, the system recovery ratio is calculated as 45%, and the specific power consumption is calculated as 0.55 kWh/m^3. So far, the effect of concentration polarization is not considered yet.

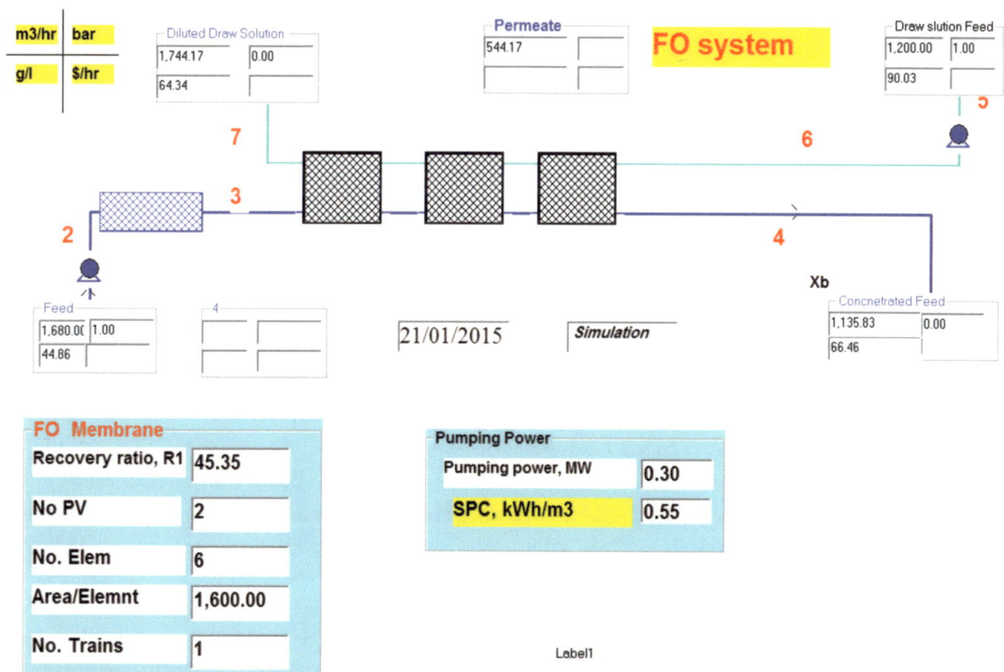

Figure 9. Interface of VDS software for FO process

4. Technoeconomic analysis

4.1. NF-MSF

The first MSF of 0.5 MGD per unit evaporator was built in 1957 in Kuwait using the once through MSF-OT configuration by the Westinghouse Company [28]. The design was modified according to the recommendation of the client, the Ministry of Electricity and Water, in Kuwait and of engineers for reliable operation. For some time, the market was dominated by the once through (MSF-OT) due its simplicity and high thermodynamic efficiency. However, due to high oxygen and CO_2 gas liberation in addition to large amount of feed water to be pre-treated, the market was forced to shift to the brine recirculation configuration (MSF-BR). The first 19-stage 1 MGD MSF-BR plant was built by Weir Company in 1959 in Kuwait [28]. The developed specifications led to more reliable, easy-to-operate-and-maintain, and longer life units. Now, the MSF evaporator production capacity was increased dramatically through the years to reach 20 MGD in UAE, and designs of 25-30 MIGD are available. The disadvantage of the MSF-BR system is the higher brine concentration, which increases the potential for having scale deposits on the heat transfer surfaces and for the liquid boiling point elevation (BPE), thus penalizing the coefficient of heat transfer and the available condensing temperature difference, respectively.

Increasing the MSF unit production (for both new designs and operating units) can be carried out either by: i) increasing the re-circulating brine flow rate, or ii) increasing the flashing range. Increasing the re-circulating brine flow rate is limited, however, by the available pump capacity and the chamber load (flashing brine flow velocity). Increasing the flashing range (TBT–BBT) can be carried out by increasing the top brine temperature (TBT), with hard-scale solution, or reducing the bottom brine temperature (BBT), with lower heat sink temperature (naturally in fall/winter/spring or utilizing deep intake or cooling towers). Increasing TBT is the parameter addressed in this paper.

At high TBT, scale deposits of high seawater brine concentration present a real problem in MSF plants, as they directly affect the heat transfer rates on the heating surface. The main scale-forming constituents are calcium (Ca^{++}), magnesium (Mg^{++}), bicarbonate ($HCO3^{-}$), and sulphate ($SO4^{--}$) ions. On heating, bicarbonate decomposes into carbonate $CO3^{-}$, which reacts with Ca^{++} forming calcium carbonate (CaCO3) that precipitates on the heat transfer surface (if saturation limits are exceeded). At high temperature, magnesium hydroxide (MgOH) will also be formed. At higher temperature of >120°C, non-alkaline calcium sulphate (CaSO4) precipitates if saturation limits are also exceeded, due to inverse solubility. Formation of alkaline scale (CaCO3 and MgOH) can be controlled by lowering the pH (acid additives) or by anti-scalant. Non-alkaline (hard) scale (such as CaSO4) is only controlled nowadays by limiting the TBT below 120°C.

Scale deposits have a direct influence on the thermal units' performance and water cost. Sulphate scales are a result of the direct crystallization of anhydrite (CaSO4), hemihydrate (CaSO4 0.5H2O), or gypsum (CaSO4 2H2O) from seawater once their solubility limits are exceeded as shown Fig. 10 [29]. Most of the deposited calcium sulphate found in seawater

http://dx.doi.org/10.5772/60207

desalination plants is in the form of hemi-hydrate. The sulphate minerals are insoluble in common chemicals, and their development inside a distiller should be avoided by all means.

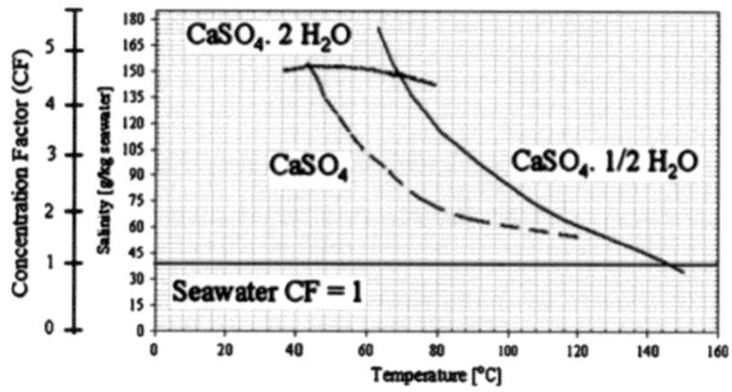

Figure 10. Phase diagram of CaSO4

Increasing the TBT with hard-scale solution can be carried out by: i) introducing high-temperature anti-scalant, and ii) reducing hard-scale ions to avoid it from reaching the saturation conditions. The first is not yet available through the use of nano-filtration (NF) membrane system for make-up feed water pretreatment.

The application of NF in seawater desalination has gained significant attention in the desalination industry due to the selective removal of divalent ions. The SWCC R&D team [30-32] carried out extensive experiments on an MSF test pilot unit with NF as the pretreatment. NF pressure was 24 bars, and its recovery ratio ranged from 60% to 65%. The total concentration of the sulphate and calcium ions of the brine recycle was at a TBT of 130°C, and the makeup entirely formed from NF permeate was below the solubility limits. This result indicated the possibility of operating the MSF plant safely and without any scaling problem at TBT equal to or higher than 130°C. However, many questions on the adding of capital cost which might result in saving in operational cost still need clear answers.

The NF was originally applied to reject electrolytes and obtain ultra-pure water with high volume flux at low operating pressure, as most membranes have either positive or negative charge due to their compositions [33]. The NF membrane possesses a molecular weight cut-off of about hundreds to a few thousands, which is intermediate between reverse osmosis (RO) membranes and ultra-filtration (UF). The pore radii and fixed charge density of practical membranes were evaluated from permeation experiments of different neutral solutes of sodium chloride. The pore radii of these NF membranes were estimated to range from 0.4 to 0.8 nm [33].

The flexible and powerful tool "Visual Design and Simulation program (VDS)" is used to perform process and techno-economical calculations. VDS was developed for the design and simulation of different types and configurations of the desalination processes [21-25]. Typical desalination processes are simulated to show the wide scope and high capability of the developed package. The description of the VDS software and discussions on how to access and handle the package are presented in [21-25]. In this work, the scope of the VDS program will be extended to develop and build up an NF system and a new MSF configuration model. The NF system's mathematical model will be verified using typical NF-RO plant data.

Table (2) shows the CAPEX cost analysis of the NF system, which produces 226 m^3/hr. The direct costs of the purchased equipment (membrane section, filters, pumps, valves, and piping) are included. The indirect costs of the building structure, engineering, and project development are also included. The intake cost is not included and is assumed to be included in the MSF CAPEX cost. The levelized cost is calculated (based on the 7% interest rate and 15 year life span) as 0.0775 $/$m^3$ of the NF permeate as shown in column three of Table (2).

Items	Cost, $	Specific cost, $/$m^3$
Direct cost		
PV, pass 1	26,100.00	0.001510285
Element, NF	87,000.00	0.005034285
Pumps	118,026.20	0.006829626
PX/turbine	115,359.61	0.006675323
Piping and valves	188,939.88	0.010933071
Filters	249,015.74	0.014409381
Others, building, start-up	258,803.85	0.014975773
Subtotal	1,043,245.28	0.060367744
Indirect cost		
Engineering design	131,455.62	0.007606724
Financial	164,319.52	0.009508405
Sub-total	295,775.14	0.017115129
TCI	1,339,020.42	0.077482874

Table 2. CAPEX cost analysis of the NF system [14]

Table (3) shows the operational cost of the NF system, which includes labor, O&M, NF membrane replacement, electricity, and chemicals. The analysis showed that the cost of electricity represents the biggest chunk of the total OPEX, and the specific operational cost is 0.0566 $/$m^3$ of the NF permeate. From both Table (2) and Table (3), the calculated unit permeate cost is 0.134 $/$m^3$.

Figure 11 shows the interface of the existing 5,000 m^3/day MSF-BR desalination plant at TBT=110°C [14]. The evaporator consists of 20 stages – 17 for the heat recovery section and 3 stages for the heat rejection section. The extracted steam from the power side is directed to the brine heater as a heat source. Sea water flows through the tubes of the heat rejection section condensers as a coolant. Part of this coolant outlet is used as a make-up, and the remaining coolant is rejected back to the sea. The make-up is directed to the de-aerator, and pretreatment chemicals are added, then mixed with a portion of the last-stage brine. The circulation pump circulates the diluted mixed brine to the condensers of the heat recovery section. The tube

http://dx.doi.org/10.5772/60207

Items	Cost, $/year	Cost, $/m³
Labor	21,909.27	0.011546899
Maintenance	16,431.95	0.008660173
Insurance	6,572.78	0.003464069
Membrane replacement	1,305.00	0.000687777
Electricity	34,740.04	0.018309132
Chemicals	26,401.62	0.013914513
Total	107,360.66	0.056582563

Table 3. OPEX cost analysis of the NF system [14]

materials used in this plant are CuNi 90/10 for the brine heater and heat recovery section and CuNi 70/30 for the heat rejection section. The evaporator length is 29 m, the width is 7 m, and the height is 2.5 m. The design conditions are 27°C for seawater, and the brine velocity inside the tube is 2 m/s. The working pressure of the de-aerator is 0.055 bars, which is lower than the make-up saturation temperature of 38°C.

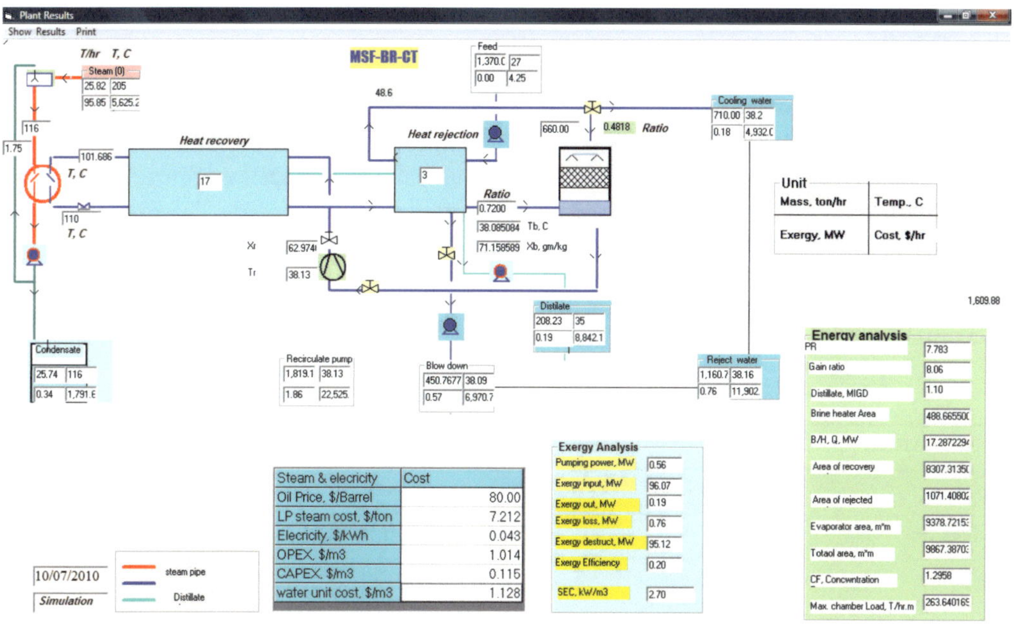

Figure 11. Interface of the existing MSF-BR desalination plant at TBT=110 C [14]

Figure 12 shows the interface the MSF-BR with the NF system, which allows increasing the TBT to 130°C. The NF system treats one-third of the make-up. The feed of the NF system is

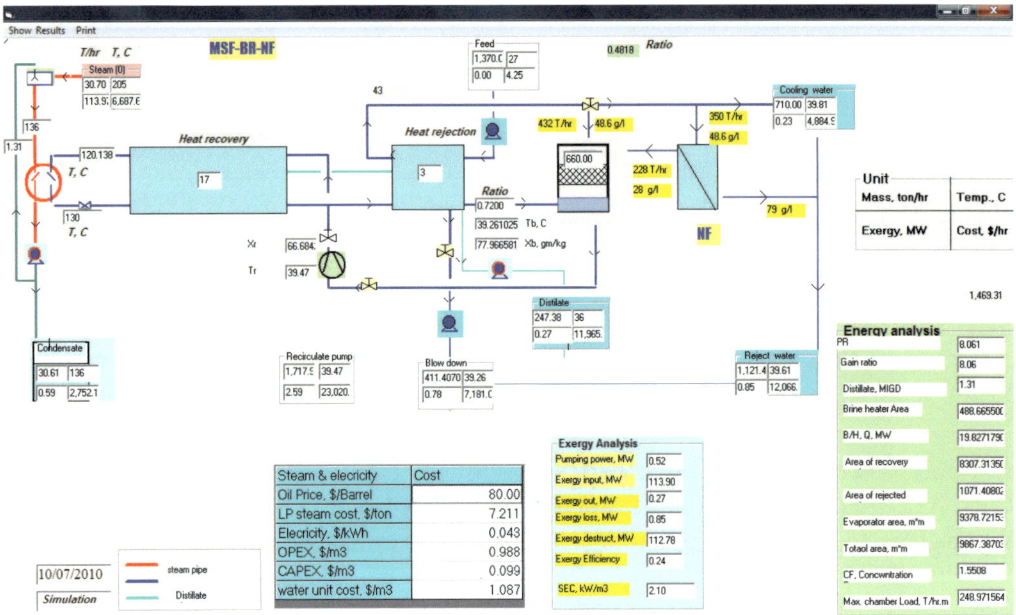

Figure 12. Interface of the NF-MSF-BR desalination plant [14]

extracted from the cooling reject stream (48 g/l) as shown in Figure (5). The NF permeate is mixed with the remaining make-up and directed to the de-aerator. The mixed make-up of low salinity of 43 g/l (15 % les) flows to the last stage of the heat rejection section. Due to the increase in the TBT from 110°C to 130°C, the distillate production increases by 19%. There is no increase of the GOR, as the heating steam increased also by 19%.

Table (4) shows that the CAPEX of the NF-MSF-BR system is 65.5% higher than that for the conventional MSF. Table (5) shows that the operating cost of the NF-MSF-BR system also increased by 22.4% higher than that for the conventional MSF.

CAPEX, 1 MIGD	Conventional	NF-MSF	% diff
Item	COST, US$	COST, US$	
Evaporator	1,040,551.34	1,040,551.344	-
Pumps	306,223.35	306,223.350	-
Pipes, valves, I&C	302666.77	302,666.770	-
Intake	394560	394,560.000	-
NF system	-	1,339,020.42	
Total	2,044,001.46	3,383,021.88	65.5%

Table 4. CAPEX analysis of MSF and NF-MSF-BR [14]

http://dx.doi.org/10.5772/60207

Items	MSF-BR	MSF-BR-NF	% diff
LP steam cost	186.23	221.4	18.89%
Electricity	23.8	23.8	-
Chemicals	0.48	0.48	-
Total	210.92	244.21	15.78%
NF (OPEX)		13.61753678	
Total	210.92	257.8275368	22.24%

Table 5. OPEX analysis of MSF and NF-MSF-BR [14]

Table (6) shows that the levelized CAPEX cost of MSF-BR at TBT=130°C is 16% lower than that for the conventional MSF at TBT=110°C. This is due to the increase in distillate production of 19%. Also, due to the increase in the productivity, the specific OPEX reduced by 2.5%. However, due to adding the NF system, the levelized OPEX of NF-MSF at TBT=130°C is 2.65% higher, while the specific CAPEX of NF-MSF is 28.7% higher than that for the conventional MSF.

The unit product cost of NF-MSF is 5.4% higher than that of the MSF plant. The analysis of these CAPEX and OPEX results shows that the OPEX cost has significant effect on the total unit water cost. This can be concluded that adding the NF system to an existing MSF plant (just to increase the production) is not enough to reduce the unit product cost.

	MSF-BR	MSF-BR (TBT=130°C)	% diff
Interest rate	0.07	0.07	
Life span	20.00	20.00	
Amortization factor	0.09	0.09	
Annual investment	192,468.70	192,468.70	
Hourly production	208.07	247.21	18.81%
Hourly investment	24.41	24.41	
Specific CAPEX, MSF	0.12	0.10	-15.83%
Specific OPEX, MSF	1.01	0.99	-2.50%
NF			
NF, specific OPEX		0.05	
NF, specific CAPEX		0.05	
MSF-NF, specific CAPEX	0.12	0.15	28.68%
MSF-NF, specific OPEX	1.01	1.04	2.65%
Total water unit cost	1.13	1.19	5.35%

Table 6. Levelized cost of MSF and NF-MSF-BR [14]

A modified MSF-DM configuration has been proposed as shown in Figure 13. In this MSF-DM configuration, the heat rejection section is removed, and the bottom part of the de-aerator is utilized as a mixer where part of the last stage brine is mixed with de-aerated make-up. The new configuration is half-way between brine recirculation MSF-BR and once through MSF-OT and will benefit from both techniques and overcome the limitation encountered through operation. The GOR of the MSF-DM configuration at TBT=110°C could be as high as 12.

The MSF-DM design configuration is targeting high MSF GOR to be adopted in solar energy applications (high GOR is also needed, as the cost of energy is increasing). As the capital cost in solar energy systems is expensive, it will be cost-effective to develop high-performance MSF to reduce the CAPEX of the solar energy systems. A high-performance MSF system requires a combination of more evaporating stages, and more heat transfer surface area sequentially increases the MSF CAPEX. The increase in the MSF CAPEX could be balanced by reducing the MSF OPEX, and accordingly, CAPEX reduction of the solar energy system will be the main contribution to the developed system.

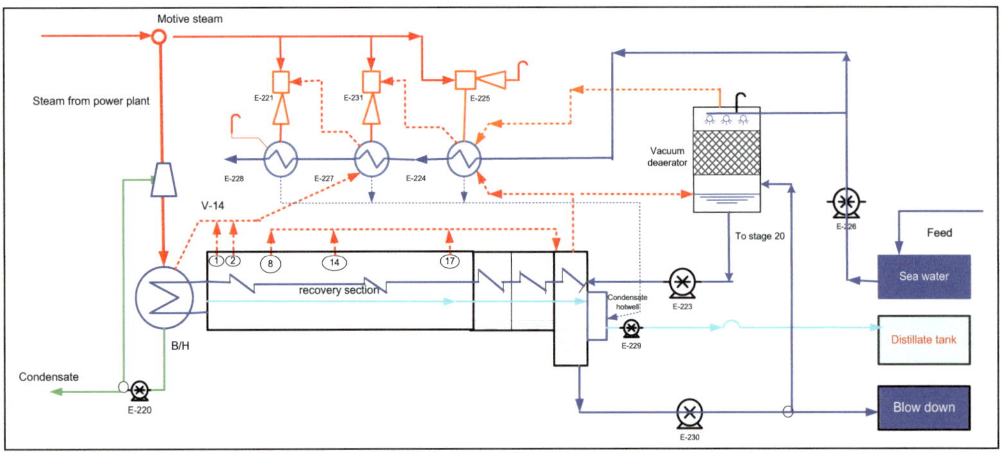

Figure 13. The interface of the new MSF-DM for desalination plant [14]

Figure 14 shows the configuration of NF with the newly developed de-aeration brine mix NF-MSF-DM system to reduce the operational cost (OPEX). NF enables increasing the TBT to 130°C, while the MSF-DM enables increasing the GOR. As shown in Figure 14, the stage number of MSF-DM increased to 35, which is 75% higher than that of the conventional MSF-BR. The 61.5% of the last-stage brine is mixed with the de-aerated make-up flow of 675 m^3/hr. The make-up is diluted from 48 g/l to 43 g/l using NF system permeate of TDS=28 g/l. This mixed is directed to the MSF condensers at 32.8°C, which is 15% lower than that for the conventional MSF (38°C). This lower temperature of coolant enhances the heat transfer process (condensation). However, the reducing cooling water reduces the LMTD across condenser compared with that of the conventional one. This explains why the heating surface area of MSF-DM was increased by 72%. One feature of increasing the heat transfer area of the heat

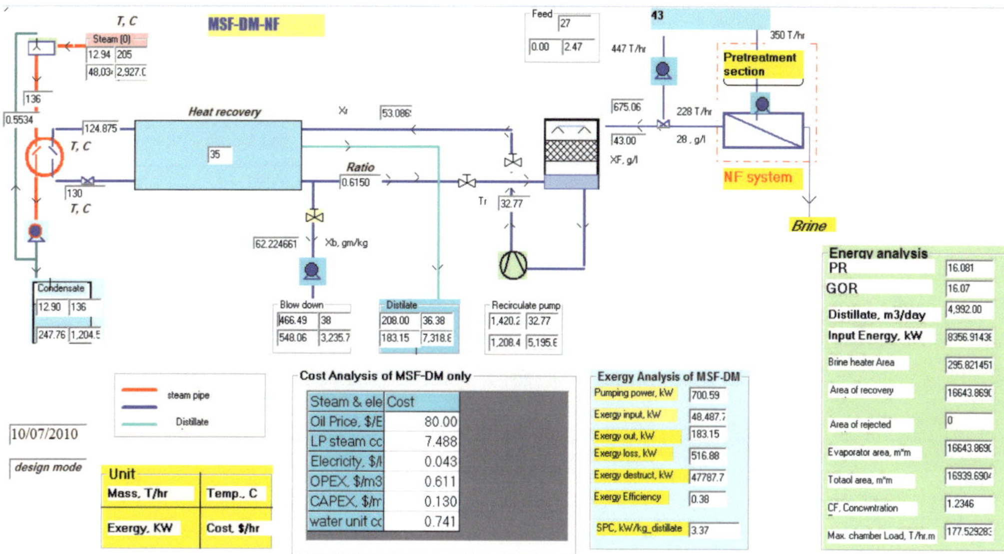

Figure 14. NF-MSF-DM configuration

recovery section is reducing the temperature difference across the brine heater, which sequentially increases the brine heater surface area. Increasing the heat transfer area of the heat recovery section increases the recovered energy, thus minimizing the external source of heating. Reducing the source of heating (steam) for fixed capacity will increase the GOR. The process calculations show that the GOR is 100% higher than that of MSF-BR (see Table (7)). This means that steam consumption is reduced by 100 %.

Table (7) shows the process calculation of MSF-DM at TBT=130°C compared with the conventional MSF at TBT=110°C. The GOR of MSF-DM-NF is twice of the conventional MSF; however, the heat transfer area is increased by 72%. Table (7) and Figure 14 show that the intake seawater of MSF-DM is 42% lower than that of MSF-BR. This, in turn, would reduce the seawater supply pump capacity, as well as the intake civil work. One feature of MSF-DM, the make is the same value of the conventional, which leads to having the same chemical cost of treatment and same manufacturing cost of de-aeration.

Table (7) shows that the specific power consumption of MSF-DM is 27% higher than that of MSF-BR. This is because of the increase of the friction loss due to the increase of the stage number by 75%. The evaporator length is increased by 142% in the case of MSF-DM; the evaporator width is decreased by 2%, while height is increased by 7% as shown in Table (7).

The purchased equipment cost (PEC) of these components is estimated based on recent market prices. In cases when data about the real installation cost of the desalination plant are scarce, the PEC of the individual components could be calculated based on cost relations. These relations of estimating the capital and operating costs of the components, such as pumps, valves, piping, and instrumentations are presented in [14].

	MSF-BR	NF-MSF-DM	% diff
Ton/hr	208	208	0%
TBT, C	110	130	18 %
Sea water flow rate, m^3/hr	1370	797	-42 %
Make-up, m^3/hr	660	675	2%
Sea salinity, g/l	48.60	48.6	0%
Recycle ratio	0.72	0.615	-15%
Recycle salinity, g/l	62.90	60	-5%
Blow down salinity, g/l	70.00	70	0%
No. of stages	20.00	35	75%
Heat transfer area, m^2	9,868	16,940	72%
Tube length, m	7.45	8	7%
Tube diameter, m	19.05	17	-11%
GOR	8	16	100%
Velocity, m/s	1.98	1.91	-4%
SPC, kWh/m^3	2.70	3.42	27%
Evaporator length, m	29.30	70.9	142%
Evaporator height	2.50	2.44	-2%
Evaporator width, m	7.45	8	7%

Table 7. Process calculation of MSF and MSF-DM [14]

A detailed cost breakdown is shown in Table (8). The evaporator (shell and tubes, de-aerator) cost of MSF-DM is 47% higher due to the increase in the heat surface area by 75% as shown in Table (7). Evaporator manufacturing cost, including the labor cost, of MSF-DM is 52% higher than that of MSF-BR. The costs of pumps, piping, valves, and I&C control of MSF-DM are lower than that of the conventional system due to the removal of the heat rejection section. The cost analysis shows that the intake construction cost of MSF-DM is 42% lower than that of the conventional one due to lower seawater flow rate. So the increase of MSF-DM evaporator cost is partially compensated by the cost reduction in auxiliaries and intake cost. The total capital cost (CAPEX) of the proposed configuration, MSF-DM, is 6% higher than that of the conventional MSF-BR. However, the total CAPEX cost of the NF-MSF-DM system is 71% higher than that of the conventional MSF. The main increase in CAPEX is contributed to the additional NF system.

Table (9) shows the OPEX items for both the conventional MSF-BR and the MSF-DM configurations. The cost of the steam and electricity is calculated based on an average of 80 $/barrel oil price and the recent purchase cost of power generation cycle [14]. The cost of the low-pressure steam is directed to the desalination plant, and the steam utilized for power generation is allocated based on exergy analysis [14]. Using levelization method through 20 years and 7%, the specific cost of low pressure steam is calculated as 7.5 $/$m^3$ of steam, and the cost of the generated electricity is 0.043 $/kWh.

Items	MSF-BR (TBT=110)	NF-MSF-DM (TBT=130)	% diff
Evaporator	1,066,100.45	1,570,420	47%
Pumps	306,223.35	246,014	-20%
Piping, valves, I&C	302,666.77	179,202	-41%
Intake	394,560.00	194,416	-51%
Total	2,069,550.57	2,190,052	6%
NF	-	1,339,020.42	100%
Total	2,069,550.57	3,529,072.713	71%

Table 8. CAPEX analysis of MSF and NF-MSF-DM configurations [14]

	MSF-BR	NF-MSF-DM	% diff
LP steam cost	186.23	96.9	-48%
Electricity	23.8	30.23	27%
Chemicals	0.48	0.49	2%
Total, MSF	210.92	127.63	-39%
NF		12.9	
Total, MSF-NF	210.92	140.53	-33%

Table 9. OPEX analysis of MSF and MSF-DM-NF [14]

The OPEX cost analysis (Table (9)) shows that the cost for NF-MSF-DM is 31% lower than that for the conventional MSF. The reduction in OPEX contributed to the reduction of the heating steam cost due to higher GOR. The levelized cost showed that the unit product cost of NF-MSF-DM is 21% lower than that of the conventional MSF as shown in Table (10).

As shown in Table (9), the low-pressure steam cost for the MSF-DM configuration is 48% lower than that for the conventional MSF-BR, and the steam consumption of MSF-DM is 100% lower than that consumed by the conventional MSF-BR. This is mainly due to the different steam cost invoked from the power side, as the heat steam temperature is higher in MSF-DM (TBT=130). The electricity cost of the MSF-DM is 27% higher than that of the conventional MSF-BR due to higher pumping power of the same order. The chemical cost is only 2% higher than that for the conventional MSF-BR. This is mainly due to the increase of make-up to be treated. The total number of OPEX items in the proposed configuration, MSF-DM, is 33% lower than that in the conventional MSF-BR, mainly due to low amount of steam consumption.

The annual investment cost (fixed capital cost depreciation rate (FCCDR) per year) of each component in the desalination plant is calculated according to the following relation:

$$\text{Annual Investment } = \text{CAPEX } \times \frac{i \times (1+i)^n}{(1+i)^n - 1} \tag{27}$$

Using an interest rate, i, of 7% and number of amortization years, n, of 20 years: then, the operation and maintenance cost is calculated by multiplying the equipment purchase cost by a factor of the equipment cost index. The hourly cost ($/hr) of the desalination plant is calculated as follows:

$$hourly-CAPEX = \frac{\text{Total annual investment}}{365\times 24\times 0.9} \tag{28}$$

Similarly, the hourly OPEX is calculated as follows:

$$hourly-OPEX = \text{LP steam +Electricity+Chemiclas} \tag{29}$$

Then the unit product cost of the desalted water is calculated as follows:

$$\text{Unit product cost, \$/m}^3 = \frac{hourly\_CAPEX + hourly\_OPEX}{hourly\_\text{Pr}oduct} \tag{30}$$

The levelized cost of capital purchased components and operating invested (chemicals, steam, electricity, O&M) to produce water is calculated as shown in Table (10). The specific OPEX of the MSF-DM is 34% lower than that of the conventional MSF-BR. The specific CAPEX of the MSF-DM is 12% higher than that of the conventional MAF-BR. However, the sum of the total cost invested using the MSF-DM is 34% lower than that of the conventional MSF-BR. Due to adding of the NF system, the specific OPEX of NF-MSF-DM is 20% lower that of the conventional one, while the specific CAPEX increased by 84% as shown Table (10). The total unit product cost of NF-MSF-DM is 9.5% lower than that of the conventional MSF-BR.

Levelization cost, $/m^3	MSF-BR	NF-MSF-DM	% diff
Interest rate	0.07	0.07	-
Life span	20	20	-
Amortization factor	0.094392926	0.094392926	-
Annual investment	192468.7044	215680.4949	12.06%
Hourly production	208.07	208	-0.03%
Hourly investment	24.41257032	27.35673451	12.06%
Specific CAPEX	0.117328641	0.131522762	12.10%
Specific OPEX	1.014038462	0.613605769	-39.49%
Total	1.131367102	0.745128531	-34.14%
NF			
Specific OPEX, NF	-	0.062023194	-
Specific CAPEX, NF	-	0.08493315	-
Specific CAPEX, NF-MSF-DM	0.117328641	0.216455912	84.49%
Specific CAPEX, NF-MSF-DM	1.014038462	0.807151725	-20.40%
	1.131367102	1.023607638	-9.52%

Table 10. Levelized cost of MSF and NF-MSF-DM configurations [14]

4.2. Hybrid FO-MSF

In this section, technical approach to consider the impact of TBT and varying FO recovery on the process performance is presented. The VDS software [18] will be used as a powerful simulation tool. In this program, a reference MSF plant of 16.2 MIGD working at TBT=111°C is simulated. The performance ratio, distillate production, concentration and flow rates, and temperatures of all streams are calculated. The software is adapted and developed to consider the FO membrane. The hybrid MSF-FO is simulated at fixed brine recycle flow rate and brine concentration (draw solution) and by varying the FO recovery ratio with the TBT. For a fixed performance ratio, the distillate of MSF (D) and the required heat transfer surface area (A) are calculated at different operating conditions. For comparison, the specific heat transfer area (SA) is calculated as:

$$SA = \frac{A}{D}\,\mathrm{m}^2/\mathrm{MIGD} \tag{31}$$

For the same seawater and draw solution flow rate across the FO membrane, the permeate (D_m) and the membrane area (A_m) are calculated at different recovery ratios. For comparison, the specific area of FO membrane (SA) is calculated as follows:

$$SA = \frac{A_m}{D_m}\,\mathrm{m}^2/\mathrm{MIGD} \tag{32}$$

At certain FO recovery ratio, the reduction in the Ca+ ions in the MSF feed is calculated and compared to the reference MSF process, which was operated without the FO process. The potential of $CaSO_4$ scale formation in the MSF feed after dilution is estimated at different TBTs (115-135°C) using the Skillman index.

Using VDS, all process stream characteristics are determined (mass, temperature, pressure, entropy, and rated cost), and the heat transfer surface area (number of tubes), evaporator size, internal dimensions, and pumps are sized. So, a detailed CAPEX analysis is performed and estimated. The VDS software calculates the heating steam consumption rate and the consumed chemicals (anti scales, anti-foam, and chlorination), as well as the pumping power (OPEX items). The price of electricity and heating steam is estimated and calculated as illustrated in [18]. Then the final tariff of water unit cost is obtained.

Figure 15 shows the reduction in Ca^+ ions in the feed of MSF desalination system at different recovery ratios of the FO membrane system. The reduction of Ca ions increases as the FO recovery ratio increases. At 40% recovery ratio, the reduction in Ca ions is calculated as 20%.

Figure 16 shows that the FO membrane water flux decreases as a result of the increase in the FO recovery ratio. The membrane flux decreases at higher recovery ratio due to dilution of the draw solution, which decreases the osmotic pressure driving force.

Figure 17 shows that the specific membrane area increases as the recovery ratio increases; this is due to lower water flux per unit area at higher recovery ratio. The higher value of specific membrane area is reflected in higher capital cost.

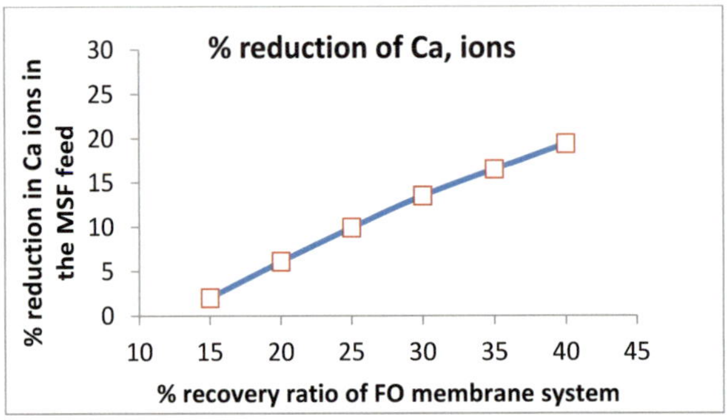

Figure 15. reduction in Ca^{+} ions in MSF feed after FO dilution

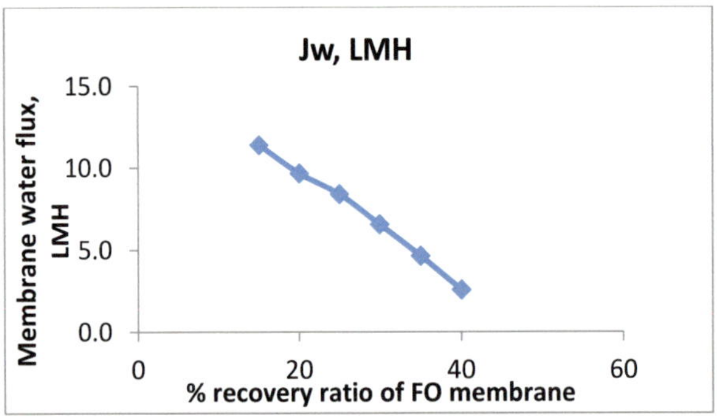

Figure 16. FO membrane flux at FO membrane recovery ratio variation

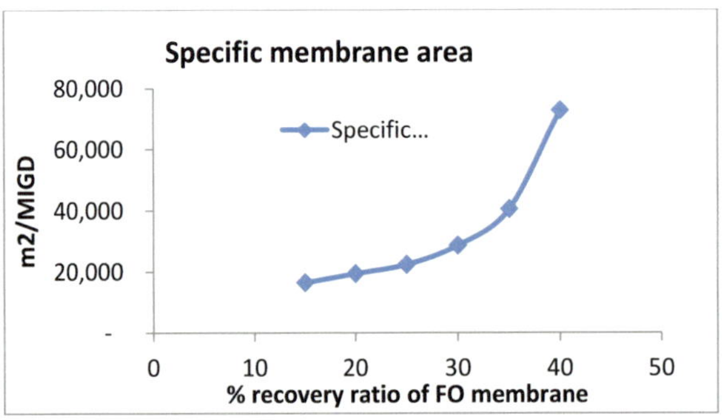

Figure 17. Membrane area variation with FO recovery ratio variation

Figure 18 shows the Skillman index (SI) at different TBTs and the variation of FO recovery ratio. The SI of calcium sulphate solubility in case of the reference MSF plant without FO operating at TBT=111°C is calculated as 1.33. As the calculated SI of traditional MSF is greater than 1, the precipitate of calcium sulphate occurs. However, in practical MSF plant anti-scalant is used to disperse the crystalized scale. The SI=1.33 is considered as the reference for comparison; the value above 1.33 indicates scale formation, while the lower value indicates safe operation. As shown in Figure 5, the SI at different TBTs decreases as the FO recovery ratio increases. This is due to the increase in the removal of divalent ions. Figure 5 shows that the Skillman index increases as TBT increases. The MSF can operate at TBT=135 safely without scale problems at an FO recovery ratio of 40%. MSF at TBT=130°C can operate safely at a recovery ratio of 35%. Also, the MSF at TBT=125°C can operate safely at a recovery ratio of 30%. The MSF can operate safely at TBT=120°C and FO recovery ratio of 25%, while the MSF can operate safely at TBT=115°C and FO recovery ratio of 20%.

For the same performance ratio of MSF (PR=9), the reduction in the specific heat transfer surface area of MSF is calculated at different TBTs and different FO recovery ratios as shown in Figure 6. This figure shows that the reduction in specific heat transfer of MSF increases as the TBT increases. The increase of TBT resulting in increase of the logarithmic mean temperature difference between hot and cold streams of MSF, in turn, reduces the heat transfer area. Figure 6 shows that the reduction in SA slightly increases with the increase of the FO recovery ratio.

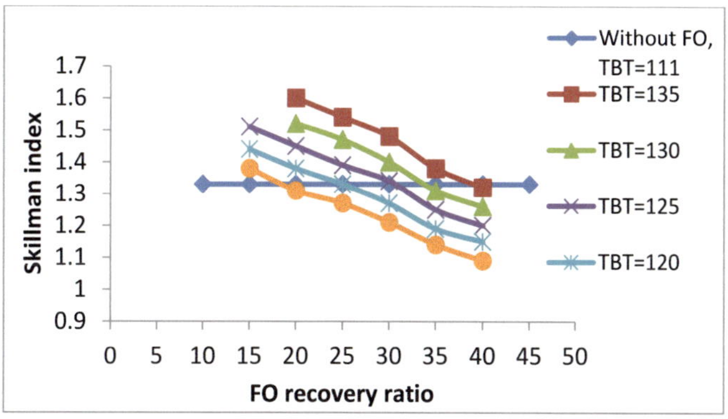

Figure 18. Skillman index at different MSF TBTs and different FO recovery ratios

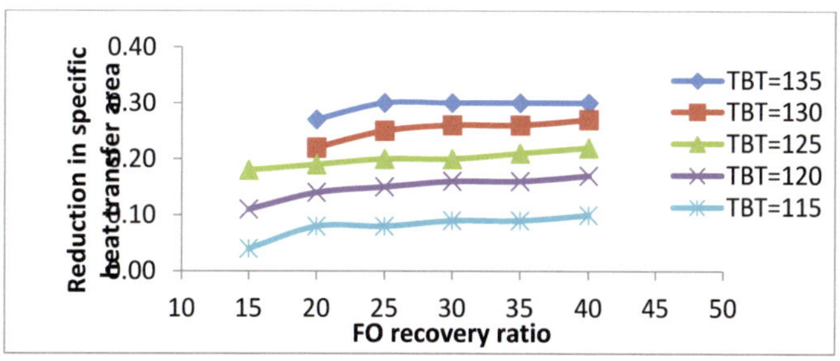

Figure 19. Reduction in the heat transfer area (CAPEX) at different FO recover ratios and TBTs

For the existing MSF plant, at TBT=130°C, the production will increase up to 30% as shown in Figure 19. The question is whether or not the existing material can withstand the 135°C temperature.

Figures 17, 18, and 19 indicate that it is beneficial to work at higher TBT to reduce the heat transfer area or to increase the production of the existing plant of MSF (CAPEX/OPEX reduction); however, this requires higher recovery ratio of the FO membrane system, which requires higher membrane area (CAPEX increase). So, an economical evaluation and compromise to reach the trade-off point is still required.

The existing capacity of water desalination plants in Qatar is approximately 1.5 Mm^3/day. MSF represent the main technology in Qatar. The make-up of seawater feed is chemically treated (anti-scalant, anti-foulant, and sodium sulphate) before being introduced to the heat recovery section. The amount of make-up flow rate is 3 times of the water production capacity, which is 4.5 Mm^3/day. As shown in Table (11), the chemical cost is 2.35 M$/year.

Chemical	Dosing rate, ppm	Consumption, kg/day	$/year
Anti-scalant	2.5	2.81E+03	2.28E+06
Anti-foam	0.1	1.35E+01	3.08E+04
Sodium sulfate	0.5	2.25E+02	4.11E+04
Total			2.35E+06

Table 11. Chemical cost analysis of thermal desalination plant

Thus, it can be concluded that the integrated FO as a pretreatment unit to the seawater feed to the existing MSF desalination plant in Qatar is technically visible in terms of production capacity increase and chemical consumption decrease. However, cost analysis is required to balance the OPEX reduction with the addition caped of the FO membrane unit. Integrating FO to the existing MSF and using the brine of the last stage as a draw solution at a recovery ratio of 35% reduce the Ca^+ ions in the seawater feed by 20%, which enables increasing the TBT up to 130°C safely. The simulation results show that at TBT=130°C, the production of the existing MSF plant increases by 20%. The OPEX analysis showed that an amount of 2.3 M$/year of chemical cost can be saved if the FO is deployed to the existing MSF plant in Qatar. The trade-off point between the additional CAPEX of the FO membrane system and the savings in OPEX will be considered under different operating condition in the present work.

5. Experimental study of hybrid NF-MSF

The process design and simulation for the test pilot is developed to prepare specifications of different components. Some units are manufactured by an Egyptian contractor, while others are purchased from vendors. The site is prepared where civil work and foundation are constructed. The test pilot components are installed and assembled, and finally, individual commissioning for each component is carried out. The site is located at the "Wadi El-Natroun" remote area, which is almost 150 km south-west of Alexandria city (Egypt). The site belongs to Alexandria University.

http://dx.doi.org/10.5772/60207

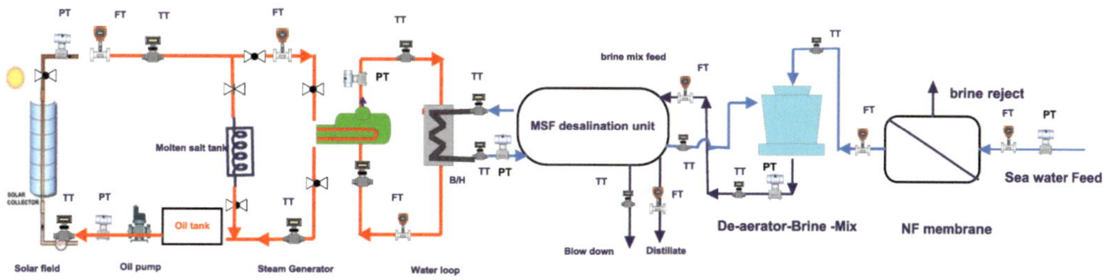

Figure 20. Basic flow diagram of CSP-NF-MSF-DBM pilot plant [13]

Figure 20 shows the pilot test of the solar energy and desalination units. The concentrated solar parabolic trough with thermal energy storage facility provides the necessary heating to generate the required steam of the MSF desalination unit. The system is also equipped with a backup boiler for steam compensation. Solar PV and wind turbine (not present in Figure 20) are installed and run separately. However, in this phase, diesel engine is used to provide the pump electricity until the match and synchronization between the PV and the wind turbine are finalized.

5.1. MSF desalination unit

The MSF pilot unit consists of 28 flashing chambers with 28 connected condensers as shown in Figures 21.a and 21.b. The stages are arranged in double deck to reduce the foot print. There are four sets; each set consists of 7 stages. MSF chambers are equipped with glass windows for monitoring of the flashing process. The shell material of MSF is fabricated from 2 mm-thick stainless steel 316L. The flash chamber is 0.5 m in length and 0.5 m in width, while the height is 1.0 m. The condenser tube is 8 mm in diameter, 6 m long, and made of stainless steel (0.7 mm thick). The number of tubes is 2 per condenser, which are arranged in multi-pass inside a 0.5 m shell length. The unit is manufactured in Egypt and assembled at the project site.

The orifice opening area is controlled using gate valve, which is located between the flash chambers. The inter-stage valves controls the inter-stage flow rates to guarantee the brine flashing at each stage. The splash plate is designed just above of the inlet opening to reduce the carry-over. A demister is placed near the vapor outlet vapor pipes to reject the brine carry-over before going to the condenser. The shells are insulated to minimize energy losses. In addition to the brine heater, different supporting systems are added including vacuum system and chemical injection systems. The vacuum system has control valves at each stage to adjust the venting rate of non-condensable gases (NCG) and the stage pressure.

The MSF is the main subsystem where distillation is produced using the flashing process. Different instrumentations are installed to measure and record the temperatures, pressures, and flow rates as shown in Figure 20. In the heating section, steam input and output temperatures, in addition to pressure and flow, are measured using proper transducers. All chambers are equipped with temperature and pressure indicators. The first and last chambers are equipped with temperature transmitter (TT) and pressure transmitter (PT), and the two

a- Double deck MSF desalination test pilot unit

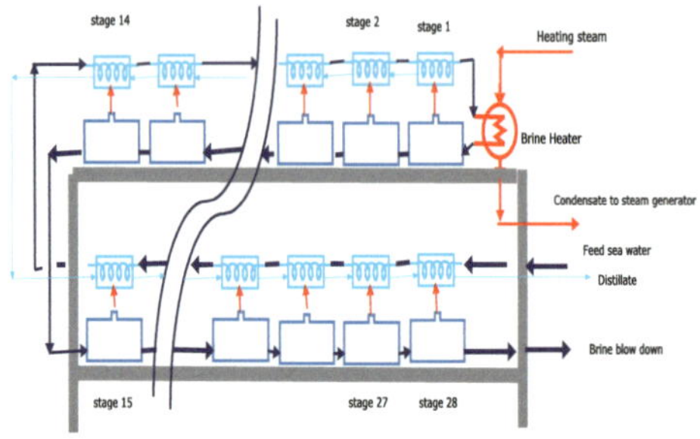

b- Process flow diagram of Double deck MSF

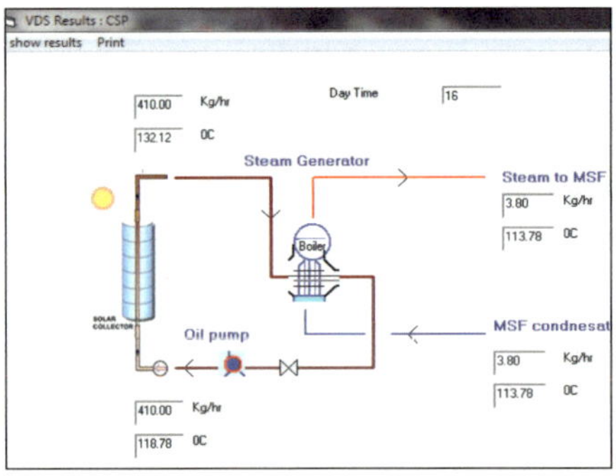

a- VDS simulation results of CSP[13]

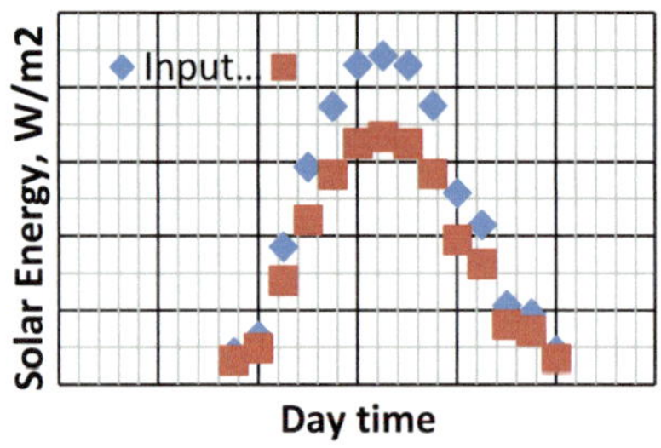

b- Solar and absorbed energy [13]

Figure 21. MSF desalination unit with double deck [13]

additional movable PT and TT are supplied to be inserted in the chambers of the operator choice. Input seawater flow and output brine and distilled water flow rates are measured using flow transmitters.

5.2. Concentrated solar power (CSP) system

Four modules of solar concentrator (parabolic trough) are purchased and assembled in series at the site of the project as shown in Figure 22. Each module is 3.6 m in length and 1.524 in width. The collector area per module is 5.6 $m^{2,}$ while the collector reflective area is 5 m^2. The assembled collector length becomes 14.6 m, while the total area is 22.4 m^2. The receiver absorptivity is 0.92, the mirror reflectivity is 0.91, while the receiver emittance is 0.23. The black-coated pipes are 1.0 inch in diameter and placed in 2.0-inch-diameter glass pipes to minimize convection losses. The concentrators have a tracking system and were placed east-west and facing south.

The CSP system contains a steam generator to supply the MSF brine heater with the required heating steam. Thermal oil is circulated through the collecting pipes, gains the solar thermal energy, and flows through the steam generator and energy storage tank. The steam generator

Figure 22. Four modules of the concentrated solar collector in series [13]

consists of shell and tube and has a separate vapor header. The shell diameter is 10 inches and is 2 m in length. The hot oil passes through tubes, while the water flows through the shell. The tube length is 4 m, and the diameter is 6 mm; the number of tubes is 24, which are arranged in two passes. The CSP system is instrumented with temperature transmitters (TT), flow meters (FT), and pressure transmitters (PT), as shown in Figure 20, to monitor the temperatures, flow rates, and pressure in both steam and oil loops.

5.3. NF pretreatment

Figure 23.a shows the P&I diagram of the NF system. The system consists of dual media filter, cartridge filter, high-pressure pumps, chemical injection pumps, and nano-filtration (NF) membrane. One dual media filter vessel is installed with a specified feed flow rate of 1.5 ton/hr and 3.5 ton/hr for back wash. The vessel contains sand, gravel, and anthracite. The cartridge filter of 5 micron is installed after high pressure pump and just before the membrane section. The membrane section consists of 4 pressure vessels running in parallel; each vessel contains one membrane element of NF270 4040 type. The whole NF system, except feed, permeate, and brine tanks, is placed inside one container with its control panel, as shown in Figure 23.b. For water salinity, samples are collected periodically to measure the conductivity using a mobile conductivity meter.

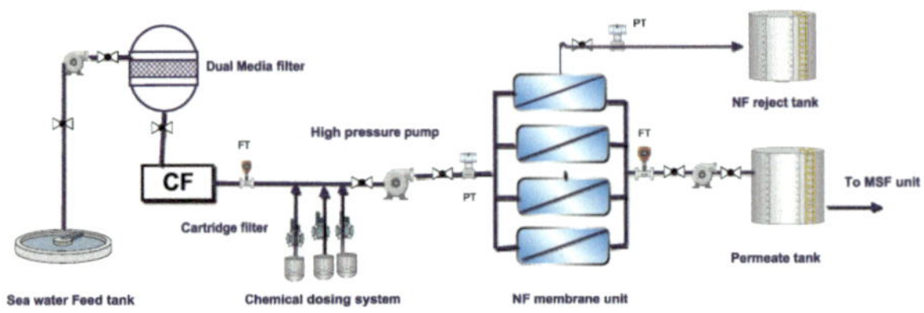

a- Process Flow Diagram of NF system

b- Photo of NF room

Figure 23. Nano-filtration (NF) system [13]

5.4. NF test performance

The NF system testing is carried out using the site brackish water (TDS=2000 ppm). A mathematical model of the NF membrane is developed and verified against typical operating NF unit data using the VDS software developed by the authors [13-15]. The VDS simulation results of the NF system were derived at different feed pressures of 8 and 10 bars and compared with experimental results as shown in Table (12). The NF performance was carried out and assessed by the recovery ratio and salt rejection. The recovery ratio (permeate/feed) increases as the feed pressure increases. The salt rejection (1- (permeate salinity/feed salinity)) is calculated as shown in Table (12). The salt rejection decreases as the feed pressure increases due to the increase in permeate salinity. The measured recovery ratio is slightly lower than the simulation results, although the salt rejection determined in the experiment is lower than that of the simulation. The differences between the measured values of permeate flow, salinity, recovery ratio, salt rejection, and simulation results are within the acceptable range.

	VDS	Exp	% diff	VDS simulation	Exp	% diff
Feed pressure, bar		8		10		
Feed flow rate, ton/hr		1.325		1.375		
Feed salinity, ppm		2000		2000		
Permeate flow rate, ton/hr	0.67	0.6	-10%	0.85	0.76	-11%
Permeate salinity, ppm	648	600	-7%	761	650	-15%
Brine flow rate, ton/hr	0.65	0.725	12%	0.52	0.6	15%
Brine salinity, ppm	3395	3158	-7%	4027	3760	-7%
Recovery ratio, %	51	45	-12%	62	55	-11%
Salt rejection, %	67.6	70	4%	62	68	10%

Table 12. Typical NF experimental results compared with the VDS results [13]

5.5. Concentrated solar power (CSP) test performance

The CPS system, including the solar collector and steam generator, is simulated using the VDS program. The mass and heat balance equations of the solar collector, steam generator, and pumps are developed. The oil and water thermo-physical property correlations at different temperatures are considered in the program. The characteristic surface of collector reflectivity, receiver emission and absorptivity, and glass tube material transmittance are specified in the VDS program. The specifications of the solar collector and steam generator are defined and fed to the program. The measured weather conditions (solar intensity, ambient temperature, and wind velocity) at each hour are fed to the program. The duration time starts at 7:00 AM and ends at 8:00 PM with 1 hour step.

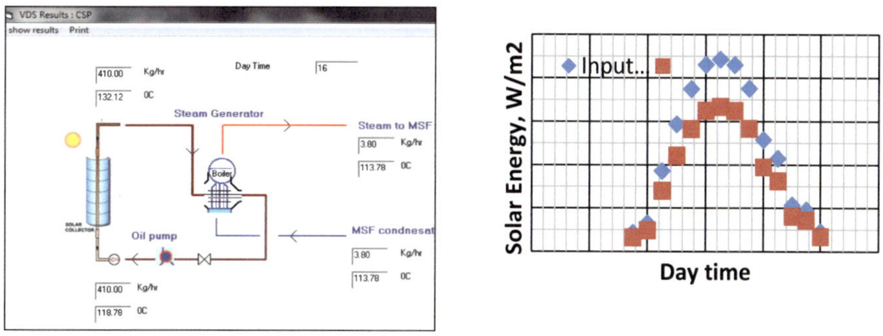

a- VDS simulation results of CSP[13]

b- Solar and absorbed energy [13]

Figure 24. ADD CAPTION

Figure 24.a shows the interface of the VDS program results at 4:00 PM. The oil mass flow rate and the temperature at inlet and outlet for both the solar collector and steam generator are presented. The collected energy is transferred to the steam generator to generate 3.8 kg/hr of saturated steam at 113.8°C. The solar intensity (I) and calculated absorbed energy by the receiver are shown in Figure 24.b. The difference is noticeable at mid-day time.

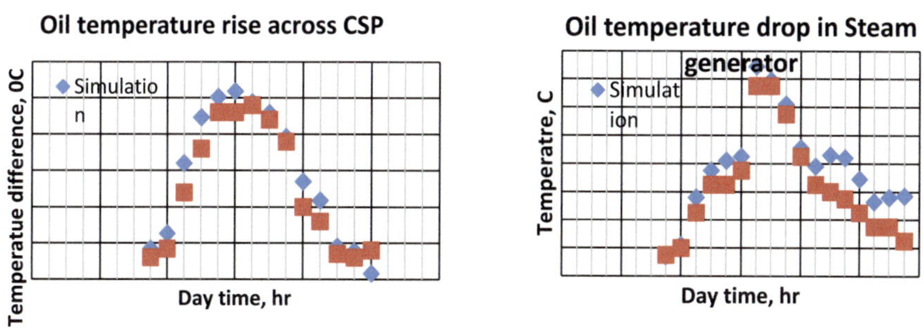

a- temperature rise in CSP

b- Temperature drop in steam generator

Figure 25. Oil temperature variation through CSP trough and steam generator [13]

Figure 25.a shows comparison between the VDS simulation and experimental results of oil temperature rise through the solar collector during day time. The oil temperature difference increases as the solar intensity increases, while the maximum difference at mid-day reached 25°C. Figure 25.b shows a comparison between the simulation and experimental results of the oil temperature drop in the steam generator unit. The maximum heat transfer occurs during mid-day, and the maximum temperature drop is 14°C. It is similarly noticed that at day time, the temperature drop in steam generator unit is less than the temperature rise in the solar collector. This means that part of the gained energy in the collector is absorbed in the steam generator, and the remaining is maintained with the outlet oil stream from the steam generator and comes back the collector. This explains the increase of oil temperature at the concentrated solar collector inlet at day time.

The CSP system average efficiency () is calculated as the average useful gained power/average solar input power:

$$\eta_{CSP} = \frac{\dot{m} C_p \Big|_{oil} (T_{o,oil} - T_{i,oil})}{I \times A_{CSP}} \tag{33}$$

Figure 26 shows the simulation and experimental results of the collector efficiency at day time. The collector efficiency decreases during day time due to the increase in the average oil temperature, which increases the energy loss to the ambient. The experimental collector efficiency shows relatively low value than that of the simulation collector efficiency due to: i) inaccurate tracking system that could not follow the sun movements accurately, and ii) the inefficient concentrated tube location in the CSP focus and possible convection loss.

Figure 27 shows comparison between the simulated and the measured generated steam temperatures. The water inlet and steam exit valve remain closed while the oil valves are open to allow energy transfer from oil to heat the enclosed water in the boiler. The water feed and steam valves are opened when the water temperature reaches 77°C. The generated steam temperature increases as the solar intensity increases, and the maximum temperature reached is 110°C at mid-day time.

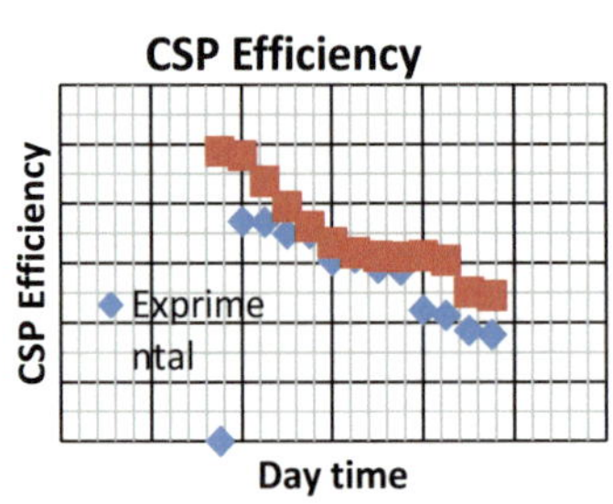

Figure 26. ADD CAPTION

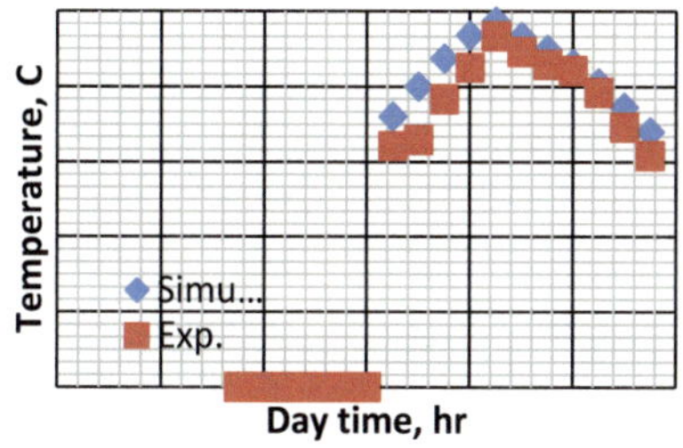

Figure 27. ADD CAPTION

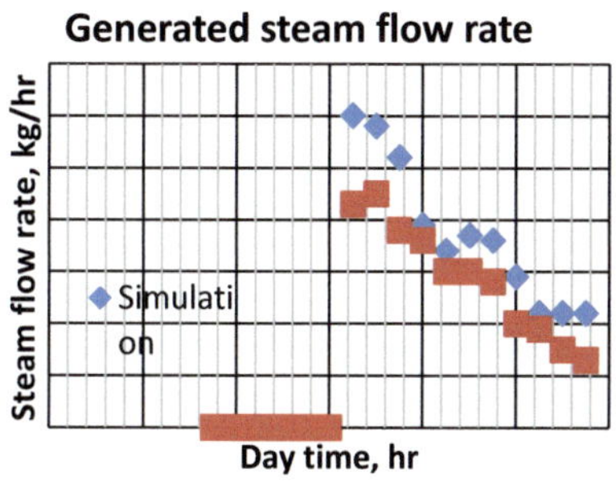

Figure 28. Generated steam flow rate [13]

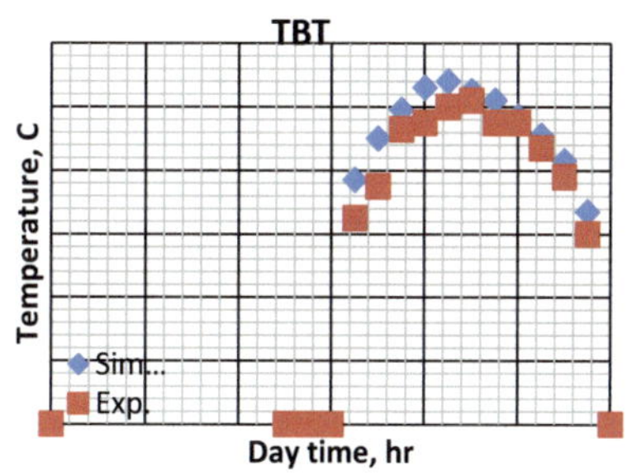

Figure 29. TBT variation [13]

The steam valve is opened at 1:00 PM at a steam flow rate of 4.3 kg/hr. The generated steam is directed to the MSF desalination unit as a heating source. The condensate steam in the brine heater of MSF is fed back to the steam generator. The amount of generated steam flow rate decreased linearly, as shown in Figure 28, due to the decrease in the solar collector efficiency. The measured generated steam flow rate shows lower values than the simulation results due to the thermal losses encountered through insufficient insulation of steam generator and throughout the connection pipe between the CSP outlet and steam generator. As shown in Figure 28, the operation of the steam generation extends up to 11:00 PM due to the heat storage in the CSP system.

Figure 29 shows the simulation and experimental values of TBT variation during day time. As the CSP steam condenses in the brine heater of MSF, the TBT rises up due to the gained energy of the latent heat. Under the ambient and operating condition in June 2012, the midday TBT reaches up to 100°C while the CSP steam condenses at 106°C, that is, 6°C temperature difference is maintained.

5.6. New MSF with de-aerator and brine mix (DBM)

The permeate water of the NF system is directed to the de-aeration and brine mix tower, where the feed is sprayed for oxygen removal. The deaerated water is mixed with parts of the brine blow down, then it is pumped to the MSF condensers. The brine mix feed absorbs latent heat energy in condensers before passing though the brine heater where the brine reaches its top temperature (TBT). The brine is then directed to the first flash chamber where flashing process occurs and vapor releases. The released vapor condenses to form product water. The flashing process occurs in the successive stages until the last stage is reached, where the un-flashed brine exits as brine blown down. The condensate of all stages is collected and directed to the water product tank. The brine level is adjusted above the interconnecting pipes (inter-stage gates) to guarantee the sealing of the flash chambers.

Under the same feed saline water flow rate (NF permeate) of 370 kg/hr and feed temperature of 27°C and controlling the brine mix ratio at 20-70% of the MSF brine blow down, the distillate water is measured and recorded at different TBTs, as shown in Figure 30, as compared with design calculated values. The pressure of saline water before the first chamber is controlled and fixed at an absolute value of 1.5 bars (above saturation conditions) by partially closing the valve. Also, the orifices among chambers are controlled by partially closing the valve between the two successive chambers. The in-tube water velocity is controlled at 2 m/s.

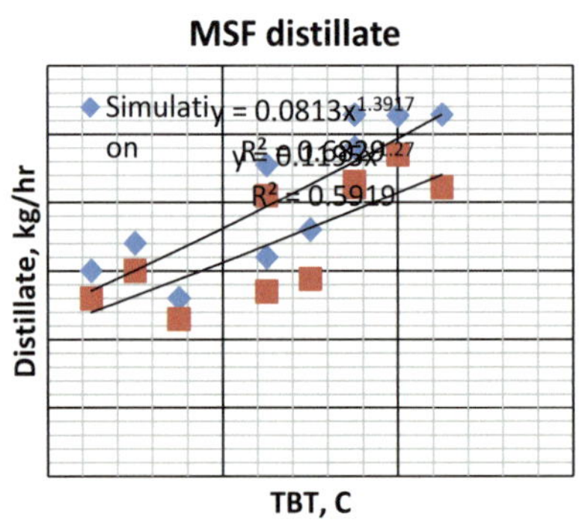

Figure 30. MSF distillate productivity with TBT variation [13]

The rates of both the design and the measured distilled water increase as a result of the TBT increases as shown in Figure 30. The amount of distillate is lower than the expected; this may be due to the partial loss of flashed vapor through vacuum system and the irreversibility of the flashing process that occurs within the orifices and weirs.

Figure 31 shows the design and the experimental GOR of the MSF variation with TBT. This is defined as the ratio between the distillate flow rate and the heating steam consumption,. The

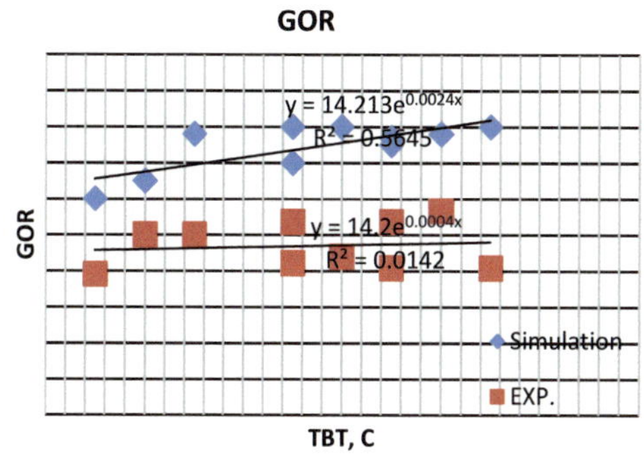

Figure 31. The GOR of MSF system variation with TBT [13]

average value of the unit design GOR is 17, which is almost twice of the conventional MSF GOR. The average measured GOR is 15 as shown in Figure 31. The small difference between the measured and designed values of GOR is due to the lower distillate productivity under fixed amount of heating steam flow rate.

The MSF specific power consumption (SPC) is defined as the ratio between the pumping power consumption (kW) and the rate of water distillate (m^3/hr),.

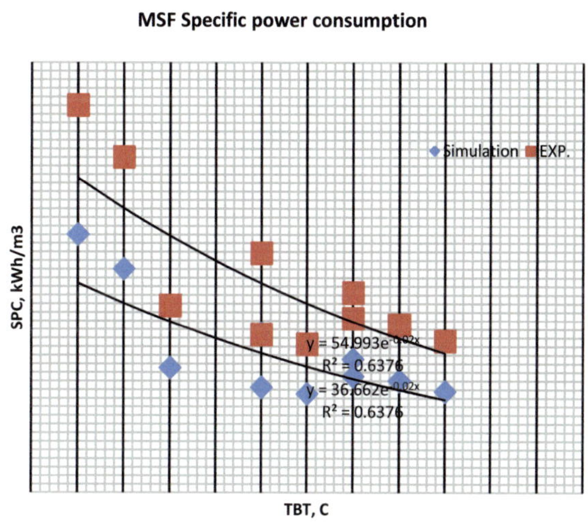

Figure 32. Specific power consumption (SPC) of MSF with TBT variation [13]

Figure 32 shows that the SPC decreases as the TBT increases mainly due to the increases in water productivity. The experimental SPC is calculated based on the measured distillate flow rate and the rated power consumption. The experimental SPC is higher, however, than the design value mainly due to lower experimental distillate for the same saline water feed and may be due to the pressure drop in piping and valves that were not considered properly in the design stage. The SPC of the test pilot unit is relatively higher than that of the commercial value of large scale MSF desalination plant due to the very small test pilot unit productivity.

6. Conclusion

To date, commercially available hybrid desalination plants are of the simple non-integrated type. They may share common systems such as intake and outfall facilities, but otherwise, they run independently at the same site. Product water from the membrane and thermal systems are usually blended to international standards on water quality. One more step ahead this chapter addresses the role of using FO or NF as a pre-treated method to the existing thermal desalination plants. The target of this hybridization is to reduce divalent ions that cause hard-scale deposition at elevated temperatures. The separation of divalent ion enables increasing the desalination process temperature greater than 110°C, which consequently increases the plant performance, increases the productivity, and reduces chemical consumption.

Integrating the NF system with new (MSF-DM) configuration at TBT=130°C, the gain output ratio could be as high as 16, which is double of that for the conventional MSF-BR. The new NF-MSFDM configuration significantly reduces the unit's input thermal energy to suit the use of (the relatively expensive) solar energy as a desalination plant driver.

Simulation results showed that integrating FO to the existing MSF and using the brine of the last stage as a draw solution at a recovery ratio of 35% reduce the Ca^{+} ions in the seawater feed by 20%, which enables increasing the TBT up to 130°C safely. The simulation results show that at TBT=130°C, the production of the existing MSF plant increases by 20%. The OPEX analysis showed that an amount of 2.3 M$/year of chemical cost can be saved if the FO is deployed to the existing MSF plant in Qatar.

The desalination pilot test is built to evaluate the performance of the novel de-aeration and brine mix (MSF-DBM) configuration at high TBT using the NF membrane. The capacity of the desalination pilot plant is 1.0 m^3/day of water. Comparison between the simulation and the experimental results of the pilot unit subsystems is relatively satisfactory. The newly developed NF-MSF-DBM (de-aerator and brine mix) configuration is tested at TBT=100°C, and the GOR is calculated as 15, which is almost twice of the traditional MSF under the same operating conditions. Using the new high-performance NF-MSF-DBM and the unit's input thermal energy, which make the integration with (the relatively expensive) RE as a desalination plant driver, is a viable option.

Author details

Abdel Nasser Mabrouk[1,2*], Hassan Fath[3], Mohamed Darwish[1] and Hassan Abdulrahim[1]

*Address all correspondence to: aaboukhlewa@qf.org.qa abdelnaser.mabrouk@suezuniv.edu.eg

1 Qatar Environmental & Energy Research Institute, Qatar

2 Suez University, Egypt

3 American University, Sharjah, UAE

References

[1] K.P. Jong. Application of hybrid technology to the largest desalination plant, Fujairah, UAE, IDA World Congress, Bahamas (2003).

[2] H. Ludwig. Hybrid systems in seawater desalination, Desalination, 164 (2004) 1.

[3] Ata M. Hassan. Process for desalination of saline water especially water, having increased product yield and quality, Patent No. US 6,508,936 B1.

[4] Hassan Ata. Fully integrated NF-thermal seawater desalination, process and equipment, US Patents No 2006/0157410 A1, July 20, 2006.

[5] M. Al Sofi, Atta Hassan, G. Mustafa, A. Dalvi, N. Kithar. Nanofiltration as means of achieving higher TBT of >120 C in MSF, Desalination 118 (1998) 123–129.

[6] Atta Hassan, A. Al Sofi, A. Al-Amoudi, A. Jamaluddin, A. Farooque, A. Rowaili, N. Dalvi, G. M Kithar, I. Al-Tisan. A new approach to membrane and thermal seawater desalination processes using nanofilteration membranes (Part I), Desalination 118 (1998) 35–51.

[7] L. Awerbuch. Water desalination process using ion selective membranes, US Patent, No. 2005/0011832 A1, January 20, 2005.

[8] O. Hamed, A. Hassan, K Al-Shail, K. Bamardouf, S. Al-Sulami, Ali Hamza, M. Farooque, A. Al-Rubaian. Higher TBT, more product, less scaling, Desalin, Water Reuse 13/3, IDA, November/December (2003).

[9] S. May. Hybrid desalination systems, Middle East Desalination Research Center (MEDRC) Report (2000).

[10] O. Hamed. Overview of hybrid desalination systems, Desalination, 186 (2005) 207.

[11] Al-Mulla. Integrating hybrid systems with existing thermal desalination plants, Desalination, 174 (2005).

[12] P.K. Eriksson. Evaluation of nanofiltration as pretreatment to reverse osmosis in seawater desalination, IDA World Congress, Singapore (2005).

[13] Abdel Nasser Mabrouk and Hassan El-banna Fath. Experimental study of high performance hybrid NF-MSF desalination pilot test unit driven by renewable energy, Desalination and Water Treatment, 2103, DOI:10.1080/19443994.2013.773860.

[14] Abdel Nasser Mabrouk and Hassan El-banna Fath. Techno-economic analysis of hybrid high performance MSF desalination plant with NF membrane. Desalination and Water Treatment, 1944-3994/1944-3986@2012, http://dx.doi.org/10.1080/19443994.2012.714893.

[15] O. A. Hamed. Development of trihybrid NF/RO/MED desalination systems, Gulf water conference, Riyadh, 2007.

[16] S. Adham, R.C. Cheng, D.X. Vuong, K.L. Wattier. Long Beach's dual-stage NF beats single stage SWRO, Desalination and Water Reuse, 13 (2003), 18.

[17] Ali Altaee, Abdel Nasser Mabrouk, Karim Borouni. A novel forward osmosis membrane pretreatment of seawater for thermal desalination processes. Desalination, 326, 19–26, October (2013). http://dx.doi.org/10/1016/j.desal.2013.07.008. (ISSN: 00119164).

[18] Ali Altaee, Abdel Nasser Mabrouk, Karim Borouni. Forward osmosis pretreatment of seawater to thermal desalination: High temperature FO-MSF/MED Hybrid System. Desalination, 339, (2014), 18-25. http://ex.doi/10.1016/j.desal.2014.02.006. (ISSN: 00119164).

[19] E. El-Sayed, M. Abdeljawad, S. Ebrahim, A. Al Saffar. Performance evaluation of two membrane configurations in a MSF/RO hybrid system, Desalination, 128 (2000) 231.

[20] Atta Hassan, M. Al Sofi, A. Al-Amoudi, A. Jamaluddin, A. Farooque, A. Rowaili, A. Dalvi, N. Kithar, G. Mustafa, I Al-Tisan. A new approach to membrane and thermal seawater desalination processes using nanofilteration membranes (Part I), Desalination, 118, 35–51 (1998).

[21] A. A. Mabrouk, A. S. Nafey, H. E. S. Fath. Thermoeconomic analysis of some existing desalination processes, Desalination, 205 (2007), 354–373. DOI: 10.1016/j.desal. 2006.02.059, (ISSN: 00119164).

[22] A. A. Mabrouk, A. S. Nafey, H. E. S. Fath. Analysis of a new design of multi stage flash-mechanical vapor compression (MSF-MVC) desalination process, Desalination 204 (2007), 482–500. DOI: 10.1016/j.desal.2006.02.046, (ISSN: 00119164).

[23] A. S. Nafey, H. E. S. Fath, A. A. Mabrouk. Thermoeconomic investigation of multi effect evaporation (MEE) and hybrid multi effect evaporation-multi stage flash (MEE-MSF) systems, Desalination, 201 (2006), 241–254. DOI: 10.1016/j.desal.2005.09.044, (ISSN: 00119164).

[24] A. S. Nafey, H. E. S. Fath, A. A. Mabrouk. Exergy and thermoeconomic evaluation of MSF process using a new visual package, Desalination, 201 (2006), 224 –240. DOI: 10.1016/j.desal.2005.09.043, (ISSN: 00119164).

[25] A. S. Nafey, H. E. S. Fath, A. A. Mabrouk. A new visual package for design and simulation of desalination processes, Desalination, 194 (2006), 281–296. DOI: 10.1016/ j.desal.2005.09.032, (ISSN: 00119164).

[26] Da Hee Jung, Jijung Lee, Do Yeon Kim, Young Geun Lee, Minkyu Park, Sangho Lee, Dae Ryook Yang, Joon Ha Kim. Simulation of forward osmosis membrane process: Effect of membrane orientation and flow direction of feed and draw solutions, Desalination, 277 (2011), 83–91.

[27] Tiraferri A., N. Yip, A. Straub, S. Castrillon. A method for the simultaneous determination of transport and structural parameters of forward osmosis membranes, J. of Membrane Science, 444 (2013), 523–538.

[28] M.A. Darwish, M.M. El-Refaee, M. Abdel-Jawad. Developments in the multi-stage flash desalting system, Desalination, 100 (1995), 35–64.

[29] Aiman E. Al-Rawajfeh, Hassan E.S. Fath, A.A. Mabrouk. Integrated salts precipitation and nano-filtration as pretreatment of multistage flash desalination system, Heat Transf. Eng., 33 (3) (2011), 272–279.

[30] Ata Hassan. Fully integrated NF-thermal seawater desalination process and equipment, US Patent No. 2006/0157410 A1, July 20, 2006.

[31] M. Al Sofi, Atta Hassan, G. Mustafa, A. Dalvi, N. Kithar. Nanofiltration as means of achieving higher TBT of >120°C in MSF, Desalination, 118 (1998), 123–129.

[32] Atta Hassan, A. Al Sofi, A. Al-Amoudi, A. Jamaluddin, A. Farooque, A. Rowaili, N. Dalvi, G. M. Kithar, I. Al- Tisan. A new approach to membrane and thermal seawater desalination processes using nanofilteration membranes (Part I), Desalination, 118 (1998), 35–51.

[33] X. Wang, T. Tsuru, M. Togoh, S. Nakao, S. Kimura. Evaluation of pore structure and electrical properties of nano filtration membranes. J. Chem. Eng. Jpn, (1995), 186–192.

Chapter 9

The Expanded Electrodeionization Method for Sewage Reclamation

Yansheng Li, Zhigang Liu, Ying Wang and Yunze Hui

Additional information is available at the end of the chapter

http://dx.doi.org/10.5772/60981

Abstract

The aim of the chapter is to introduce a new technology using the flow-divided electrodes for sewage reclamation based on redox and concentrating ions. This creative system can be used directly for sewage treatment without the need for costly ion-exchange membranes and other chemical reagents. Under experimental conditions, the removal percentage of TDS (total dissolved salts) is 83 %and the removal percentage of COD (chemical oxygen demand) is 92 %.

Keywords: Flow-divided electrodes, expanded electrodeionization, sewage reclamation, ion-exchange

1. Introduction

In general, urban sewage discharge accounts for 75_85% of the consumption and the industrial sewage discharge accounts for 80_90% of its consumption. Sewage, that is actually utilized in a biological or chemical way and disappears in the form of vaporization loss is less than 25%. Because of the low rate of sewage treatment in China, the majority of the surface water resources and the underground water resources are polluted to different extents, resulting in extra pressure, caused by the shortage of water resource, on the sound and sustainable development of economy and society. Sewage reclamation is not only favorable for enterprises to increase income and decrease expenditure as well as reduce consumption and improve efficiency, but is also an inevitable option for enhancing the supply capacity of urban water.

It is more important to improve the revitalization environment of aquatic ecosystems, benefiting not only the present generation, but also future generations. Advanced sewage treatment means that, in order to cause urban sewage or industrial wastewater, which has been subjected to routine treatment, to reach a certain standard of recycled water and to be reused for production or living, treatment units are added in order to further remove the pollutants that cannot be removed by routine treatment, such as heavy metals, COD (chemical oxygen demand), TN (total nitrogen), TP (total phosphorus), TDS (total dissolved salt) and the like.

The electrochemical process has the prominent characteristics of wide adaptability, good cleaning property and easy control in removing the pollutants in sewage water, and its course is mainly mass transfer control. Firstly, it is necessary to eliminate the ohmic resistance between electrodes with a supporting electrolyte such as sodium chloride. However, the wide application of this method is limited by water recycling. Secondly, the reaction products (especially gases such as hydrogen from cathode and oxygen and carbon dioxide from anode as shown in Figs.1and 2) inhibit the reactants at the reaction zone and increase the ohmic resistance markedly. Thirdly, the scaling on cathode accumulates over the reaction zone and decreases current efficiency.

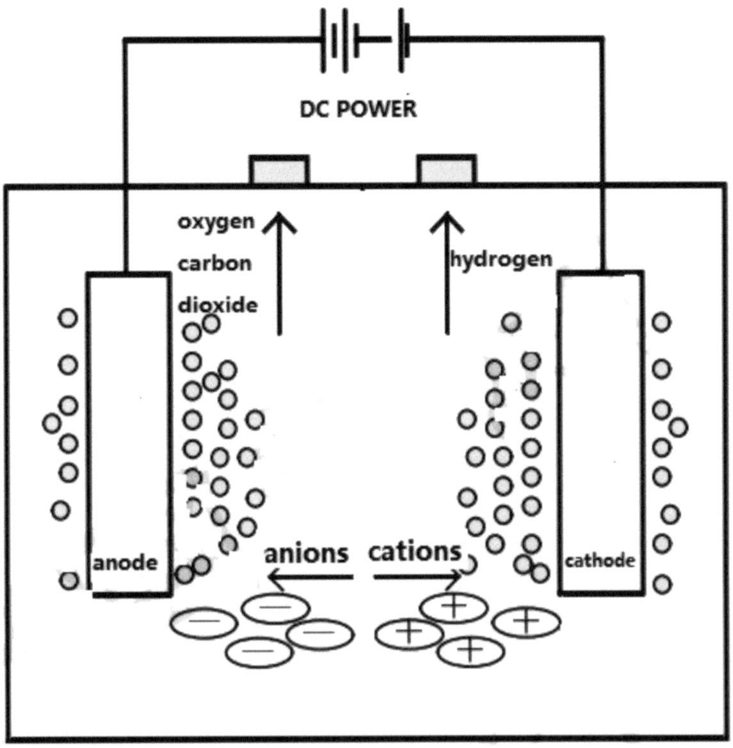

Figure 1. A schematic illustration of a sewage electrolysis system

http://dx.doi.org/10.5772/60981

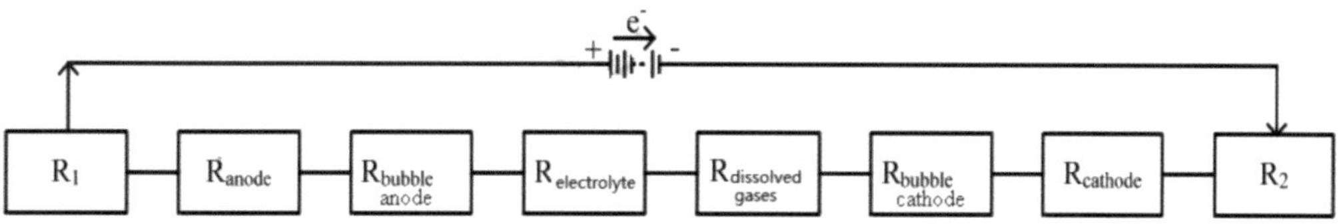

Figure 2. Electrical circuit analogy of resistances in a sewage electrolysis system

As shown in Fig.2, electrochemical resistance consists of the electrical resistance (R_1 R_2) reaction resistance (R_{anode} $R_{cathode}$) and transfer resistance ($R_{anode\ bubble}$ $R_{electrolyte}$ $R_{dissolved\ gases}$ $R_{cathode\ bubble}$). Transfer resistance in a sewage electrolysis system is negatively related to the reaction resistance. Decreasing energy consumption of the electrochemical process is a challenge.

Combination of ion exchange and electrochemical technology research from the beginning that the ion exchange membrane i.e. the electrodialysis technology. Ion exchangers and selective ion exchange membranes having ion conduction capability are simultaneously used in, electrodeionization (EDI) and to overcome the defects that electrolyte addition causes in the process of electrolytic reaction and chemical regeneration in the process of ion exchange. However, the particularity of hydrogen ion and hydroxyl ions (migration behavior in the electric field and selectivity of hydrogen ion to strongly acidic cation exchange resin and hydroxyl ions to strongly basic anion exchange resin) limits the efficiency of the electrical energy. Polarization caused by the verticality of membrane surface and flow field to electric field, as well as their impacts still exist as key factors that restrict the improvement of operation efficiency and the widening of applicable scope, their practical applications still require improvement.

The aim of this chapter is to provide a new technology using flow-divided electrodes for sewage reclamation based on redox and concentrating ions. This creative system can be used directly for sewage treatment without use of costly ion-exchange membranes and other chemical reagents. Under experimental conditions, the removal percentage of TDS (total dissolved salts) is 83 %and the removal percentage of COD (chemical oxygen demand) is 92 %.

2. Technological process

Fig 3 is a schematic diagram of the super advanced sewage treatment process. We proposed a novel EEDI system by introducing porous flow-divided electrodes (FDE) as the electroosmotic unit. During operation process, the flow of wastewater and gases from electrodes is controlled so as to penetrate into the internal part of FDE, which we call as the flow-divided model (Fig 3). The flow field is parallel to the electric by means of the porous FDE,thus avoiding mutual interference between the cathode reaction and the anode reaction, and reducing the ohmic resistance and mass transfer resistance ($R_{anode\ bubble}$ $R_{cathode\ bubble}$), and realizing separation of the anolyte (containing oxygen and carbon dioxide) and catholyte (containing hydrogen) without any diaphragm(Fig 4).

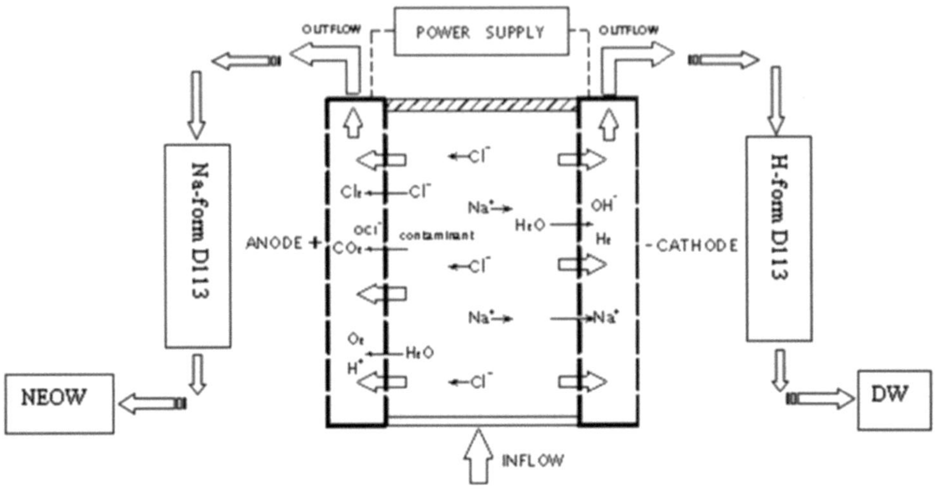

Figure 3. Schematic diagram of the Super Advanced Sewage Treatment process

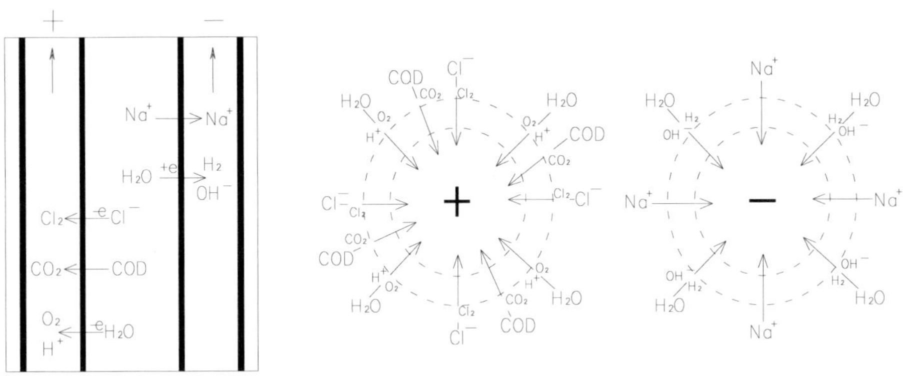

Figure 4. Mechanisms of ion migration to the electroosmotic section

Furthermore, polarity of the electrodes can be flexibly switched, thus eliminating cathode scaling during extended operation. The EEDI utilizes electrochemical redox and ion-exchange simultaneously to produce the concentrated electrolyzed oxidizing water (EOW), desalted water (DW) and hydrogen as by-product.

Since the anodes and cathodes in the EEDI are completely uniform and no ion-exchange membranes are used, it can be operated by switching anodes and cathodes flexibly for eliminating the scaling on the surface of cathodes. The strongly basic anion exchange resin grains placed among the anode and cathode act as supporting electrolyte, enapabling the treatment of wastewater with low salt. The concentrated and neutralized anolyte containing active chlorine is effective for disinfection and contaminant removal.

http://dx.doi.org/10.5772/60981

3. Materials and methods

3.1. Titanic porous FDE and electroosmotic unit

The electroosmotic unit was constructed with three hexagonal anodes and cathodes (area 0.0113 m^2) within a insulated polymethyl methacrylate shell(Fig 5). The distance of between anode and cathode is 25 mm and they are packed with strong basic anion exchange resin grains (201×7, properties shown in Table 2). The electrolyzed oxidizing water (containing gases) and electrolyzed reducing water were pressed to flow out of the anode and cathode, respectively (see Fig 3).

	Dimension (mm)	Porous dimension (μm)	Porosity factor (%)	Penetrability of gas (L/cm²minPa)	Compression resistance (Mpa)	Most pressure difference (Mpa)
value	Φ12×100	30-50	28-50	0.02-20	0.5-1.5	0.6

Table 1. Properties of the porous FDE

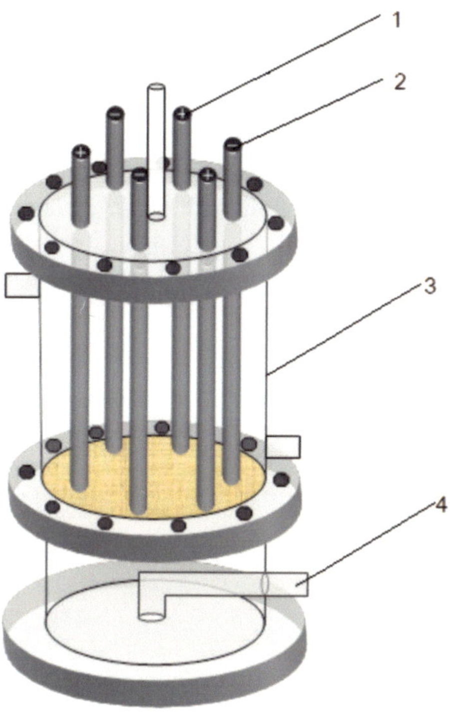

Figure 5. a structural schematic diagram of the electro-osmotic unit 1—anode, 2—cathode, 3—organic glass shell, 4—sewage inlet.

Group	Exchange capacity (mmol/ml)	Diameter (mm)	Water content (%)	Density (g/ml)	Expansibility (%)	Upper limit temperature (°C)	pH range
$-N^{+}(CH_3)_3$	≥1.4	0.6_1.4	42_48	1.06_1.11	Cl → OH:18_22	100	1_14

Table 2. Properties of 201×7type strong basicity anion exchange resin

3.2. Ion exchange treatment and regeneration

The electrolyzed oxidizing water out of anode was pumped into Na-form D113 column (regeneration), whose properties are shown in Table 3. The electrolyzed reducing water out of cathode was pumped into H-form D113 column (treatment).

Group	Exchange capacity (mmol/ml)	Diameter (mm)	Water content (%)	Density (g/ml)	Expansibility (%)	Upper limit temperature (°C)	pH range
$-COO^{-}$	≥10	0.35_0.55	45_52	1.15_1.20	(H → Ca) 20_30	100	5_14

Table 3. Properties of D113 weak acid cation exchanger

3.3. Sewage treatment

The flux of the simulated wastewater (properties shown in Table 4) is 7.5 L/h. The voltage of the electroosmotic unit is controlled at 25 V. The ratio of electrolyzed oxidizing water and electrolyzed reducing water is 1. The COD was determined by bichromate method. Concentration of Cl^{-} was determined by titration with silver nitrate solution. A power supply (LW10J10) was used to maintain constant DC voltage. Conductivities were measured by a conductometer (DDS-IIA). The pH of water was measured by an acidimeter (DELTA320). Temperature was held at room temperature during all experiments.

3.4. Cooling water treatment

The flux of the cooling water (properties shown in Table 5) is 3.0L/h. The voltage of the electroosmotic unit is controlled at 15 V. The ratio of electrolyzed oxidizing water and electrolyzed reducing water is 0.5. Apparent current density was between 88 and 90 Am^{-2}.

4. Results and discussion

4.1. Treatment of the simulated wastewater

The properties of water samples, including the feed, electrolyzed oxidizing water, electrolyzed reducing water and the desalted water are shown in Table 4. The removal percentage of TDS and the removal percentage of COD for the desalted water are 83 % and 92 %, respectively.

This result demonstrates that the novel flow-divided EEDI can remove contaminants effectively without chemical reagents.

	Cl^- concentration (mg/L)	pH	Conductance(μs/cm)	COD(mg/L)
Simulated wastewater	689	7.28	2100	426
Electrolyzed oxidizing water	2899.10	2.24	---	
Electrolyzed reducing water	120.00	12.36	4500	
Desalted water		4.16	350	32.6

Table 4. Properties of the feed and the electrolyzed oxidizing and electrolyzed reducing water

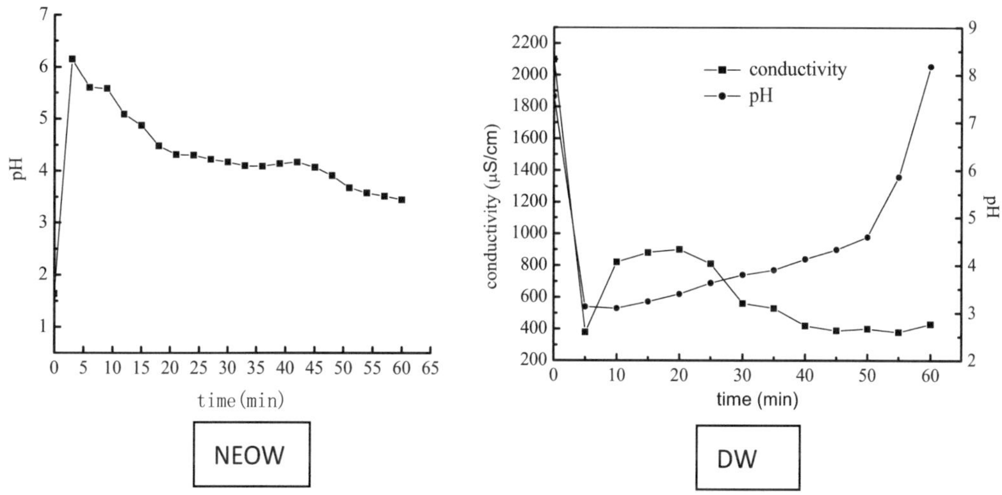

Figure 6. The pH of NEOW and DW versus time

Fig 6 shows pH variation for EOW and DW from flow-divided EEDI versus time. It can be seen that pH of EOW gradually decreased, while pH of DW gradually increased with time. The change of pH can periodically provide information about the transforming degree of the ion exchanger for the polarity reversal operation.

4.2. Mechanism of ion migration and COD removal

The strong base anion exchange resins placed between anode and cathode play an important role in gathering anions, in adjusting ion migration and acting as the supporting electrolyte. There are three routes for anion migration: (1) ions migrate alternatively through solution or ion-exchange resins, (2) ions migrate through ion-exchange resin grains and (3) ions migrate in solution. However, there is only one route for cation migration, that is, through solution (Fig 7). Furthermore, the diffusion coefficient for ion migration in ion-exchange resins is much higher than that in solution. As a result, the amount of anions in anode affinity is much higher than the cations in cathode affinity. Chlorine ions thus gathered in anode around and formed concentrated chlorine water. The relevant reactions are shown as follows:

$$2\,Cl \quad 2e \rightarrow Cl_2 \tag{1}$$

$$H_2O \quad Cl_2 \rightarrow H^{+} + Cl \ + HClO \tag{2}$$

$$H_2O \quad 2e \rightarrow H^{+} \quad 1/2\,O_2 \tag{3}$$

$$\text{Organic substances} \quad ne \rightarrow \text{intermediate} + CO_2 \tag{4}$$

The organic substances (or HCO_3^-) around the anode are oxidized directly or indirectly into carbon dioxide or intermediates. The hydroxide ions are concentrated and passed through the filter cylinder cathode. Hydrogen formed by water electrolysis penetrates into the inner part of FDE by the force of electric fields and flow fields. In this case, concentration and ohm resistance can be reduced remarkably. Hydrogen evolution from water electrolysis is shown as follows:

$$2H_2O \quad Na^{+} \quad 2e \rightarrow H_2 \quad Na^{+} + 2OH \tag{5}$$

$$H^{+} \quad 2e \rightarrow H_2 \tag{6}$$

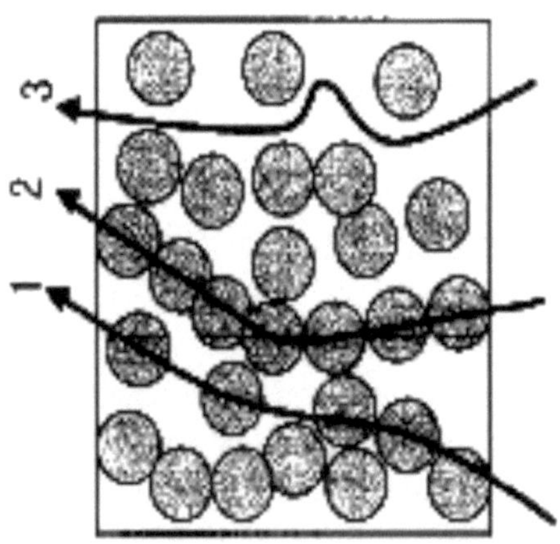

Figure 7. Schematic representation of the three paths that the current can take

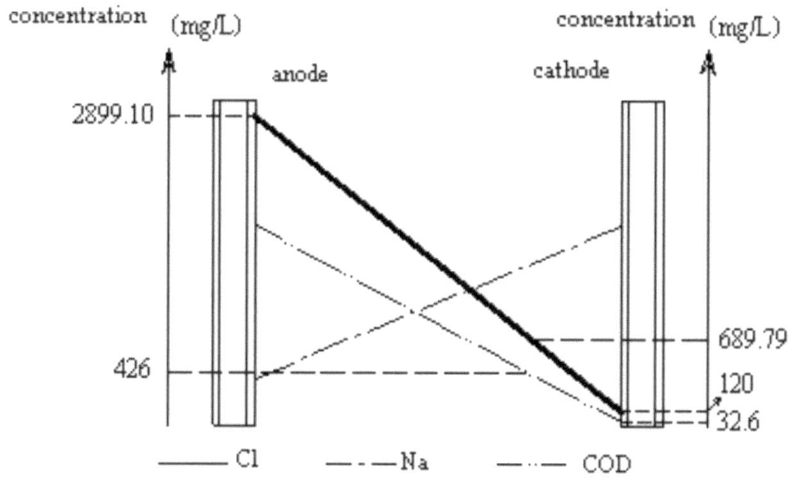

Figure 8. Distribution of the species between anode and cathode

4.3. Reaction of ion exchange

The neutralization reaction between the electrolyzed reducing water and the H-form D113 resin is as follows:

$$\text{R-COOH} \quad Na^{+} \quad OH \rightarrow \text{R-COONa} \quad H_2O \tag{7}$$

Na^+ Ions are adsorbed on ion-exchange resin, thereby desalting the feeds. The Na-form D113 ion-exchange resin was regenerated into H-form D113 with electrolyzed oxidizing water for reuse. The reaction is shown as follows:

$$\text{R-COONa} + H^{+} \rightarrow \text{R-COOH} + Na^{+} \tag{8}$$

Since the cathodes and anodes in EEDI are completely uniform and no ion-exchange membrance are used in this system, this flow-divided EEDI can be operated by switching anodes and cathodes flexibly, eliminating the scaling on the surface of cathodes.

4.4. Mass balance

The mass balance of this novel flow-divided EEDI was calculated according to the ratio of electrolyzed oxidizing water and electrolyzed reducing water as shown in Fig 5. As mentioned in Section 3.2, NaCl was concentrated in anode around up to 83 %(mass ratio) with the help of strong base ion-exchange resin. Organic contaminants were oxidized in anode with COD removal percentage of 92 %.

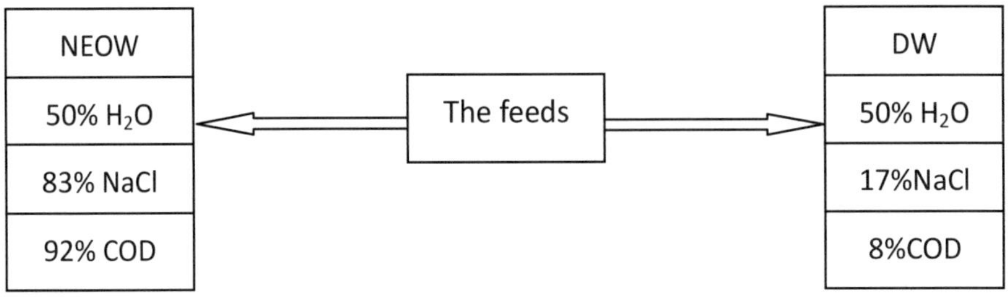

Figure 9. Schematic representation of mass balance

4.5. Treatment of the cooling water

4.5.1. Material equilibrium of the cooling water treament

The properties of the cooling water and the desalted water are shown in Table 5. The removal percentage of the hardness and the removal percentage of Cl^- for the desalted water are 98 %and 74 %, respectively. The removal percentage of dissolved oxygen is 36 %.

	Conductance (μs/cm)	DO (mg/L)	pH	Cl^- (mg/L)	Hardness ($CaCO_3$mg/L)
Cooling water	2100	7.5	7.55	479.85	750
Desalted water (DW)	580	4.6	6.57	189.94	30

Table 5. Properties of the feed and the desalted water

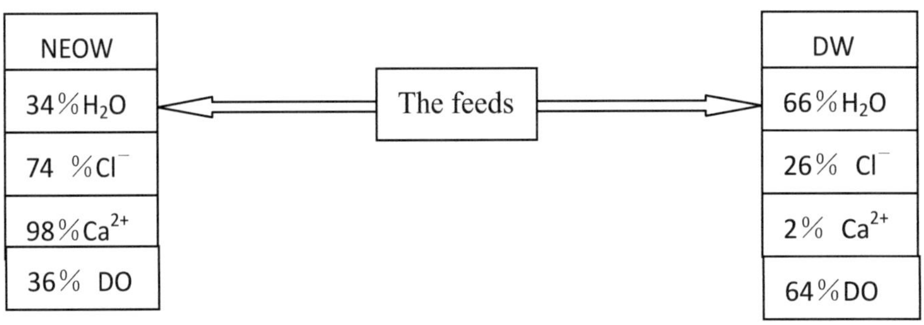

Figure 10. Mass balance of the cooling water treatment

4.5.2. Mechanism of DO and hardness ions removal

The water passes through the cathode and the following reactions take place:

$$O_2 + 4H^+ + 4e^- \rightarrow 2H_2O \quad (9)$$

The above reaction decreases the dissolved oxygen content of the cooling water and contributes to corrosion control in the circulating system.

The electrolyzed reducing water with the higher pH goes into the series type hydrogen D113. Neutralization and exchange reaction is shown as follows:

$$\text{R-COOH} \quad \text{Na} \quad \text{OH} \rightarrow \text{R-COONa} \quad H_2O \quad (10)$$

$$\text{2R-COONa} \quad \text{Ca}\left(\text{Mg}\right)^{2+} \rightarrow (\text{R-COO})_2\text{Ca}\left(\text{Mg}\right) \quad 2Na^+ \quad (11)$$

4.5.3. The sterilization effect

Fig 11 shows bacterial concentration in the feed and DW samples. It can be observed that there are groups of bacteria in the feed sample. When water passes through the electrode, the bacteria are destroyed by directly touching the electrodes. As a result, no bacteria exist in the culture dish. The DW does not come in contact with the active chlorine, but the bacteria are destroyed by cathode and electric field.

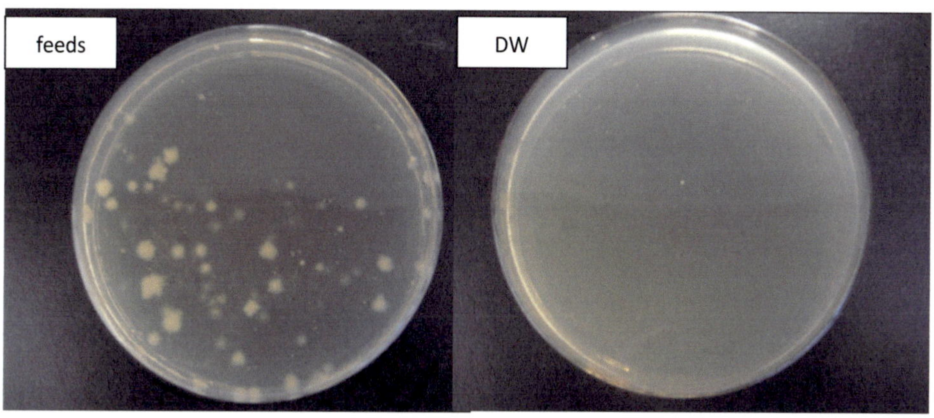

Figure 11. Comparison of bacterial concentration in the feed and the DW

5. Conclusion and expectation

The electro-osmosis ion exchange method presented here has wide adaptability in the aspect of the super advanced purification of sewage, no chemical agent is needed and no secondary pollution is caused.

The electro-osmosis unit does not require the use of membranes, the cathode/anode-equivalent micropore titanium filter electrode array and the designadopted,such that the flow field is parallel to the electric field, avoids mutual interference between the cathode reaction and the anode reaction and reduces current reduction and mass transfer hindrance, which are caused by gas coverage generated in the process of electrolysis. Furthermore, the elimination of scaling of electrodes through electrode reverse operation and the regeneration and continuous operation of subsequent ion exchange resin are facilitated.

Filling resin between the electrodes results in an electrolyte supporting effect, so as to avoid the limitation on the electrical conductivity of influent. Electrolyte addition is not required when sewage with low electrical conductivity is treated.

The invention provides a new way for pollution control and resource recycling, has become an extremely promising technology as the recycling of H_2 is turned into another point for economic growth and will bring wide prospect to ion exchange application technology in the field of sewage reclamation.

A wide space is provided for the application and development of the electro-osmosis ion exchange method due to diversified commercial ion exchangers.

Author details

Yansheng Li[1*], Zhigang Liu[1], Ying Wang[1] and Yunze Hui[2]

*Address all correspondence to: hjx@djtu.edu.cn

1 College of Environmental and Chemical Engineering, Dalian Jiaotong University and Liaoning Key Laboratory for Environmental science and Technology, Dalian, China

2 Newcastle University, Australia

References

[1] Mark A. Shannon, Paul W. Bohn, Menachem Elimelech, et al. Science and technology for water purification in the coming decades Nature, 2008, 452 (20) :301-310.

[2] Konstantinos Dermentzisa, Konstantinos Ouzounis Continuous capacitive deionization–electrodialysis reversal through electrostatic shielding for desalination and deionization of water Electrochimica Acta 53 (2008) 7123–7130

[3] ANDREW TURNER.Electrochemical ion exchange[J].Membane Technology, 1996, 75: 6-9.

[4] Yansheng Li, Yongbin Li, Zhigang Liu, et al. A novel electrochemical ion exchange system and its application in water treatment[J] Journal Of Environmental Sciences 2011,23Supplement:14-17

[5] K. H. Yeon, J. H. Seong, S. Rengaraj, and S. H. Moon Electrochemical Characterization of Ion-Exchange Resin Beds and Removal of Cobalt by Electrodeionization for High Purity Water Production Separation Science And Technology 2003, 38 (2) : 443–462.

[6] LI Yan-sheng, ZHOU Han, LI Yong-bin, JIANG Lin Research on ultra deep treatment technology of circulating & cooling Water Journal Of Harbin Institute Of Technology 2012,44 (Supplement2) :53-56

Section 4

Special Industrial Application

Chapter 10

Cr(VI) Adsorption Onto Biomass Waste Material-Derived Activated Carbon

Dilek Duranoğlu and Ulker Beker

Additional information is available at the end of the chapter

http://dx.doi.org/10.5772/60206

Abstract

Activated carbons derived from biomass waste material can be favorable compared to conventional activated carbons with respect to their adsorptive properties, low cost, and renewable sources. Literature about Cr(VI) adsorption onto biomass waste material based activated carbons has been reviewed. Activated carbon derived from peach stone was used for Cr(VI) adsorption from aqueous media. The effects of process parameters like pH, temperature, concentration, and adsorbent amount on Cr(VI) removal were examined with aspect of equilibrium and kinetics.

Peach stone based activated carbon showed high Cr(VI) adsorption capacity. Cr(VI) adsorption was strongly pH-dependent and superior at low pH. Freundlich model well expressed Cr(VI) adsorption. Adsorption kinetics followed the pseudo second-order kinetic model which consists of chemisorption. Possible adsorption mechanism was suggested in three steps: (i) Cr(VI) adsorption onto positively charged groups, (ii) reduction to Cr(III), (iii) adsorption of Cr(III) while a little amount of it releases to the aqueous phase.

Keywords: Cr(VI), activated carbon, biomass, peach stone, adsorption

1. Introduction

Water pollution by chromium is due to both natural and anthropogenic sources like leather, textile, steel producing, magnetic tapes, wood protection, paper, chemical manufacturing, metal plating, battery, and pigment industries. According to the EPA (2000) [1], chromium is the second common metal which exists in contaminated fields. It exists in different oxidation states ranging from –2 to +6. However, the most frequently encountered chromium species are trivalent and hexavalent forms since the +2, +4, and +5 forms are unstable and are easily converted to trivalent form.

Cr(III), which is the most thermodynamically stable chromium form, is relatively innocuous due to the poor biological cell permeability and is an essential trace element for living organisms. According to Eh–pH diagram (Figure 1), chromium hydroxyl species exist as $CrOH^{2+}$, $Cr(OH)_2^+$, $Cr(OH)_3$, and $Cr(OH)_4^-$ at pH > 3.5 while Cr(III) predominates at pH < 3.0. The second stable chromium form, Cr(VI), is generally produced by anthropogenic activities. Cr(VI) is a strong oxidant and reacts readily with any oxidizable organic material, hence converts to trivalent form. In the absence of organic matter, especially under aerobic conditions, it can be stable for long time [2]. It exists primarily as salts of chromic acid (H_2CrO_4), hydrogen chromate ion ($HCrO_4^-$), and chromate ion (CrO_4^{2-}), depending on pH (Figure 2). Predominant Cr(VI) species in the solution is $HCrO_4^-$ at pH between 1.0 and 6.5 while CrO_4^{2-} ions are dominant above pH 6.5 [3]. Dichromate anions ($Cr_2O_7^{2-}$) can be found only at high concentrations about more than 1 g/L.

Chromium is considered the fourteenth most noxious heavy metal [4]. Cr(VI) compounds are more toxic and carcinogenic even at low concentrations and over a wide range of pH values. High concentrations of Cr(VI) cause water quality problem in natural water sources, and health problem included lung cancer as well as kidney, liver, and gastric damage [5, 6] for human. Contact with chromium can also cause skin irritation [7]. Because of its toxicity and carcinogenic nature, Cr(VI) compounds must be removed before discharging to the environment. The permissible limit of Cr(VI) for discharge into inland surface water is 0.1 mg/L [8, 9]. However, American Environmental Protection Agency (EPA) limits Cr(VI) concentration in potable water as 0.05 mg/L [1].

The most common methods of Cr(VI) removal from aqueous system are chemical precipitation, ion exchange, membrane processes, electrodialysis, and adsorption [7, 10–13]. Among them, chemical precipitation is the most commonly used method. However, chemical precipitation yields large quantities of sludge, which is difficult to dispose directly or treat further. Moreover, chemical precipitation method is not applicable for low concentrations.

However, adsorption by activated carbon is considered to be an effective process for chromium removal due to high specific surface area, pore volume, and specific surface functionalities of activated carbon; application in this field is not common. The limited selectivity and high cost are the main obstacles for using commercial activated carbon in order to remove chromium species from aqueous media. Many researchers have studied on chromium adsorption onto activated carbon during the past ten years. A significant amount of them have used several biomass waste materials as activated carbon source. Activated carbons derived from biomass waste material can be favorable compared to conventional activated carbons with respect to their adsorptive properties, low cost, and renewable sources; in addition, evaluation of this biomass waste materials as activated carbon precursor prevents solid waste pollution problem.

http://dx.doi.org/10.5772/60206

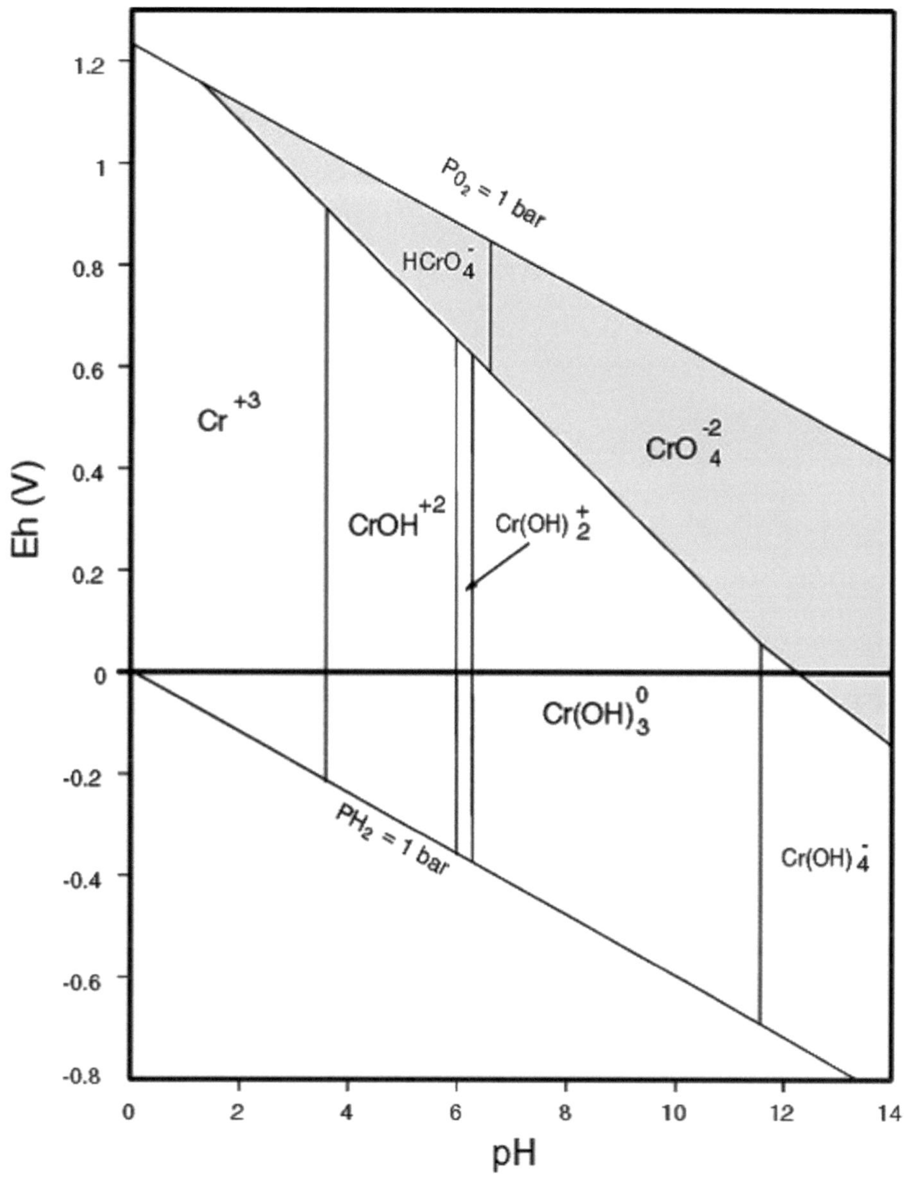

Figure 1. Eh–pH diagram for chromium [1]

In this chapter, the studies about Cr(VI) adsorption onto biomass based activated carbons have been reviewed and Cr(VI) adsorption onto activated carbon derived from peach stone has been studied. The effects of process parameters like pH, temperature, concentration, and adsorbent amount on chromium removal have been examined. Experimental results have been discussed in aspect of equilibrium and kinetics. In addition, possible chromium adsorption mechanisms have been discussed.

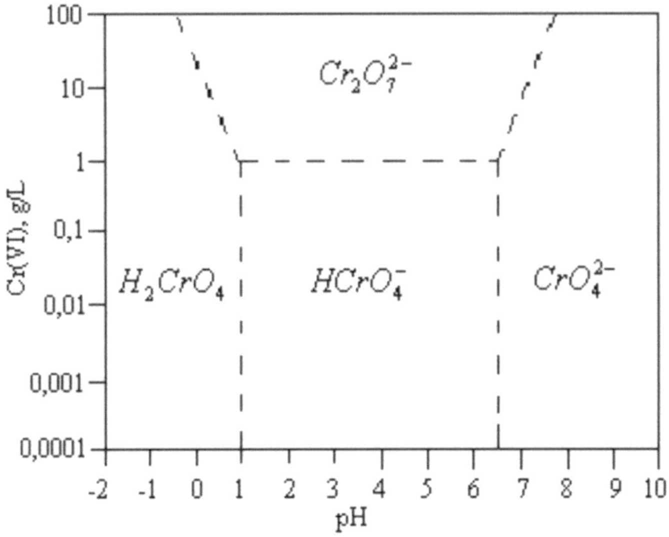

Figure 2. Speciation diagram of Cr(VI) [3]

2. Cr(VI) adsorption onto biomass-derived activated carbons

Activated carbons derived from wood, coal, and coconut shells are only available on the market although the activated carbons derived from biomass waste materials can be favorable with respect to their adsorptive properties, low cost, and renewable sources. The biomass waste materials are usually used as an energy source, burned in the field or spilled into the environment in a noncontrolled way. Evaluation of biomass waste materials as activated carbon precursors also aids to prevent solid waste pollution problem. The availability and the abundance of biomass waste material are two critical points in order to decide which biomass waste material will be used. Various biomass waste materials have been used to develop low cost activated carbons that have relatively high adsorption capacity.

Groundnut was treated with concentrated H_2SO_4 which is then carbonized at 150–155°C during 24 h (GNH) [14]. A part of produced carbon was impregnated with silver nitrate (SIGNH). Cr(VI) adsorption capacity of both SIGNH and GNH were 11.4 and 7.0 mg/g, respectively. Maximum Cr(VI) adsorption was at pH 3. Freundlich model represented the experimental data for both samples.

Researchers produced steam activated carbon from olive bagasse in reference [15] in order to remove Cr(VI) from aqueous solution. Produced microporous activated carbon had 718 m^2/g of BET surface area and 0.3699 cm^3/g of total pore volume. Cr(VI) adsorption capacity of activated carbon was dependent upon pH and maximum removal was obtained at pH 2 with 48.5 mg/g Cr(VI) adsorption capacity. High adsorption capacity at low pH was explained with neutralizing negatively charged adsorbent surface, thereby reducing the hindrance to the diffusion of chromate ions. In addition, the authors reported that Cr(VI) could be reduced to trivalent form at highly acidic medium in the presence of activated carbon, but there was no reduction in their case. They only proved this with the pH increment from 2 to 4 during the experiment, and Cr(III) do not precipitate at this pH. On the other hand, one should keep in mind that Cr(III) can adsorb by carbon surface which oxidized as a result of reduction reaction, and oxidized carbon surface is more attractive for Cr(III) cations than the previous [16]. In order to verify this, Cr(III) concentration in the solution and in the solid phase should be determined. Equilibrium time for Cr(VI) removal was 10 h [15]. Adsorption of Cr(VI) on the olive bagasse based activated carbon followed pseudo second-order kinetic model. Langmuir model fitted better than the other models (Freundlich, Dubinin–Radushkevich (D–R), Temkin and Frumkin equations). Adsorption is chemical in nature and the process is endothermic ($\Delta H° = 39.45$ kJ/mol). Adsorption energy was in the range of 18.99–21.11 kJ/mol, which is bigger than the physical adsorption. Overall, it can be concluded from the results of thermodynamics and energy calculations that main mechanisms of Cr(VI) adsorption onto olive bagasse based activated carbon are chemisorption or reduction coupled adsorption.

Researchers used almond shell based activated carbon, which was activated with H_2SO_4 at 200°C during 24 h, for Cr(VI) adsorption [17]. The activated carbon has 412 m^2/g BET surface area and pH_{PZC} of carbon is 5.26. Cr(VI) adsorption was favored at pH 1 at 293 K with 144 mg/g capacity value. The authors explained the high adsorption capacity at low pH with the electrostatic attraction between the positively charged groups of carbon surface and $HCrO_4^-$ anions. Experimental data fitted the Freundlich model well. Thermodynamic studies indicated the spontaneous nature of the adsorption process was endothermic. The rise of adsorption with temperature was interpreted with enlarging the pore size to some extent by the authors.

Date palm seed was treated with concentrated H_2SO_4 for 5 h, then boiled for 20 h in a fume hood [18]. Maximum adsorption was noticed at pH 2 due to the electrostatic attraction between the positively charged groups of carbon surface and $HCrO_4^-$ anions. Various theoretical adsorption models (Langmuir, Freundlich, Temkin, Redlich–Peterson, Koble–Corrigan, and Dubinin–Radushkevich) were applied and all models represented the data reasonably well except Dubinin–Radushkevich equation. Adsorption equilibrium time was 180 min. Kinetic data obeyed the pseudo first-order kinetic model. Moreover, Elovich kinetic model had also very high correlation coefficient values.

Oil palm fiber was treated with concentrated H_2SO_4 at 150°C for 24 h in a muffle furnace [19]. Treated oil palm fiber was used for Cr(VI) removal. Maximum Cr(VI) adsorption capacity was 40 mg/g at 200 mg/L of initial Cr(VI) concentration, 5 g/L of adsorbent dosage, pH 1.5, and 28°C of temperature. Cr(VI) removal sharply reduced when the initial pH of the solution was increased to higher values. The authors explained this as same in reference [15] that low pH causes neutralizing of adsorbent surface charge by H^+ ions in the solution, hence Cr(VI) anions can diffuse to the surface easily. Equilibrium time was 6 h and Cr(VI) adsorption kinetics represented well by pseudo second-order kinetic model. The authors interpreted this compliance with chemical adsorption or chemisorption between the adsorbent and the adsorbate. Although Cr(VI) adsorption could be represented by both Freundlich ($R^2 = 0.8775$) and Langmuir ($R^2 = 0.8715$) isotherms, the Freundlich model showed better fit to the experimental isotherm data. The authors also reported that desorption of Cr(VI) from treated palm fiber was very difficult.

Sugarcane husk and eucalyptus wood were pyrolized at 700°C for 2 h, then the char carbonized at 900°C under CO_2 flow during 4 h after treating with HNO_3 in reference [20]. For chemical activation, sugarcane husk and eucalyptus wood were activated at 900°C under nitrogen for 2 h after impregnation with H_3PO_4. The produced activated carbons have 260–367 m^2/g BET surface area and 0.865–1.234 cm^3/g total pore volume. Activated carbons have higher acidic groups and pore volume comparing to the literature. Comparing sugarcane husk and eucalyptus wood for Cr(VI) adsorption, sugarcane husk based one was more effective than the other for both physical and chemical treatment. Activated carbons obtained by the chemical method show a higher Cr(VI) adsorption capacity and the highest capacity (28 mg/g) was obtained at pH = 5 with chemically activated sugarcane husk based carbon.

Activated carbon was prepared from tamarind wood with zinc chloride in order to adsorb Cr(VI) from aqueous solution [21]. Micro porous activated carbon was obtained with 1322 m^2/g BET surface area and 1.042 cm^3/g pore volume. The authors observed that Cr(VI) adsorption percent decreased with pH in the range of pH 1–6. After pH 6.5, no more adsorption took place at all. Maximum adsorption capacity of about 28 mg/g was obtained at pH 1. It is very well known that pH is the most important parameter which affects both adsorbent and adsorbate ionic state. Researchers reported the adsorption mechanism with electrostatic attraction between negatively charged chromate/dichromate species and positively charged surface functional groups on the carbon surface at acidic media [21]. Adsorption rate was reduced above pH 3.0. According to the authors, physical adsorption has taken an important role, and exchange mechanism might have reduced after pH 3.0. The pseudo second-order kinetic model fitted the data better than pseudo first-order kinetic model. Webber Morris and Boyd kinetic models have been applied and it was found that the film diffusion is the rate control step for chromium adsorption. The Cr(VI) adsorption isotherm followed both linearized Freundlich and Langmuir model. Cr(VI) adsorption onto activated carbon produced was endothermic in nature. However, there was no explanation about the magnitude of adsorption enthalpy ($\Delta H°$ = 34.54–53.53 kJ/mol).

Researchers carbonized rice husk and saw dust at 150°C in hot air oven after treated with H_2SO_4 in [22]. The specific surface area of the produced RHC was 1.12 m^2/g and SDC was 1.16 m^2/g. Cr(VI) adsorption was favored at pH 2 and the maximum adsorption capacity was found as 47.8 mg/g and 53.1 mg/g at 250 mg/L initial Cr(VI) and 4 g/L adsorbent concentration for RHC and SDC, respectively. Similar to the others, the authors interpreted the pH effect with the electrostatic attraction between the highly protonated carbon surface and the $HCrO_4^-$ anions. The authors reported that Langmuir, D–R models, and Temkin isotherm fitted experimental data well. However, when looking at the linearized plots of both models, only Langmuir and Freundlich model showed the straight line with the experimental data. Adsorption energy values were calculated by using D–R model as lesser than 8 kJ/mol, which indicated the physical adsorption. As a result of dynamic investigation, it was obtained that Cr(VI) adsorption onto RHC and SDC has increased with time and remained constant after 120 min. Adsorption kinetics followed the pseudo second-order kinetic model very well

http://dx.doi.org/10.5772/60206

according to the correlation coefficients and the differences between the calculated and experimental adsorption capacities. The authors desorbed 55.48% and 53.03% of Cr(VI) by KI solution and 65.04% and 68.84% by HCl solution from RHC and SDC, respectively.

Activated carbon was prepared by burning the rice straw in the air in order to use as an adsorbent for Cr(VI) removal [23]. Cr(VI) was completely removed from solution in 12 h at pH 1 and the removal capacity was 20 mg/g, increasing pH resulted in decreasing the removal rate and capacity. Cr(III) was detected in the solution and as bounded to the carbon surface at pH 1–3. XPS results showed that Cr bound to carbon at any studied pH (pH 1–5) was predominately in its trivalent form. According to their mechanism, Cr(VI) anions first adsorbed on the carbon surface and adsorbed Cr(VI) reduced to Cr(III), consequently. The Cr(III) ions are either in the solution or bound to carbon surface.

Researchers in reference [24] prepared activated carbon (THAC) from Trapa natans husk, water chestnut, by using H_3PO_4 impregnation. In addition, Fe was loaded onto activated carbon (THAC-Fe). BET surface area and pore volume of THAC and THAC-Fe adsorbents were 782.86 m^2/g, 920.13 m^2/g, and 0.498 cm^3/g, 0.618 cm^3/g, respectively. pH_{PZC} values were 6.20 and 3.77 for THAC and THAC-Fe, respectively. Both produced sorbents were used for Cr(VI) adsorption from aqueous solution. Cr(VI) removal efficiency decreased dramatically from 100% to 2.1% as the initial solution pH increased from 3 to 11. The authors explained this with the electrostatic attraction between protonated carbon surface at $pH < pH_{PZC}$ and Cr(VI) anions as same as the most workers. It was reported that iron significantly enhanced the Cr(VI) adsorption capacity of the activated carbon. This was explained with larger surface area and pore volume, more acidic functional groups of THAC-Fe. Furthermore, the iron oxides have high affinity and selectivity toward Cr(VI) oxyanions. Pseudo second-order model well represented the kinetic data. It was noticed that both external mass transfer and intra-particle diffusion controlled the Cr(VI) adsorption rate. Tempkin and Freundlich models gave the best correlations for experimental equilibrium data. The maximum adsorption capacity was 11.83 mg/g. The authors interpreted the positive value of $\Delta H°$ with the endothermic nature of the adsorption. However, there was no explanation about the magnitude of adsorption enthalpy ($\Delta H° = 70.844$ kJ/mol). Very little desorption of Cr(VI) was interpreted with chemisorption.

Syzygium jambolanum nut, an agricultural waste, was chemically activated with ammonium per sulfate in the presence of sulphuric acid at 140–160°C and then subjected to thermal activation at 800–850°C in CO_2 atmosphere [25]. BET surface area and bulk density of produced activated carbon was 660 m^2/g and 0.63 g/mL, respectively. Maximum Cr(VI) removal was obtained at pH 2. High Cr(VI) removal in highly acidic media was interpreted with the electrostatic attraction of Cr(VI) in the anionic form and protonated adsorbent surface. Kinetic studies indicated that 4-hour-equilibration time was sufficient for maximum Cr(VI) removal and the removal process followed pseudo second-order kinetic model. It was also noticed that the film diffusion process was the rate-determining step. Freundlich adsorption model better represented the experimental data than Langmuir and Tempkin models. Eighty-five percent of Cr(VI) was eluted with 10% H_2O_2 and 1 M NaOH. Negative $\Delta G°$ value commented the feasibility and the spontaneous nature of the adsorption process.

Palm shell based steam activated carbon was modified with polyethyleneimine in [26]. Maximum 20 mg/g of Cr(VI) adsorption capacity was obtained with polyethyleneimine impregnated palm shell based activated carbon at 200 mg/L of initial concentration and pH 3–4. It was informed that adsorption capacity increased with pH up to pH 4, then it reduced by further pH increment. The adsorption took place between the anionic Cr(VI) species and polyethyleneimine, which is protonated at pH < 10. Cr(VI) adsorption was interpreted with the interaction between the Cr(VI) species and the protonated amine groups of polyethyleneimine. Freundlich model described the Cr(VI) adsorption onto modified activated carbon very well.

Algae bloom residue was converted to activated carbon by applying H_3PO_4 impregnation at 400°C [27]. The Cr(VI) adsorption capacity was superior at pH 1 due to the electrostatic attraction between protonated surface functional groups on carbon and negatively charged chromium species. Adsorption capacity was sharply declined by increasing pH since the deprotonation of surface functional groups. Maximum Cr(VI) adsorption capacity was 155.52 mg/g at 200 mg/L initial concentration, at pH 1. Both Langmuir and Freundlich models represented the data with high correlation coefficient (R^2=0.991, 0.994, respectively), indicating both monolayer and heterogeneous surface conditions may exist. The authors indicated the physical adsorption is the major process since calculated mean free adsorption energy was 3.28 kJ/mol, which was lesser than 16 kJ/mol. Tempkin constant related to the heat of Cr(VI) adsorption onto activated carbon was estimated at 18.58 kJ/mol, indicating weak interaction between Cr(VI) ions and the carbon surface. Kinetic studies showed that 300 min of equilibration time was sufficient for maximum Cr(VI) removal and the removal process followed both pseudo first-order and pseudo second-order kinetic models according to the high correlation coefficients. However, the pseudo second-order model fits better the experimental data. Overall, the authors informed that there was more than one mechanism involved in the actual process.

Sugarcane bagasse-derived activated carbon was developed with chemical $ZnCl_2$ activation [28]. Central composite design method was applied to optimize the experimental conditions like adsorbent dosage, initial Cr(VI) concentration, temperature, and pH. Maximum Cr(VI) adsorption capacity was about 9.8 mg/g (87.01 %) at optimum conditions like 40°C, 77.5mg/L initial Cr(VI) concentration and pH 8.58. Converse to the similar works, the authors found high pH for optimum Cr(VI) removal with produced activated carbon. The authors did not give any explanation about this situation.

Highly microporous activated carbon was developed (S_{BET}: 1462 m^2/g V_{tot}: 0.638 cm^3/g) from apricot stone by using H_3PO_4 impregnation at 400°C [29]. The produced activated carbon was used for Cr(VI) adsorption from aqueous solution in batch system. Box-Behnken design method was used to optimize the process parameters. The maximum Cr(VI) adsorption capacity (234.28 mg/g) was found at pH 2 and the capacity decreased with increasing pH. The authors explained high adsorption capacity at low pH with the electrostatic attraction between the protonated carboxyl groups on carbon surface and the negatively charged hydrogen chromate ions. The carboxyl groups are protonated ($AC\text{-}COOH_2^+$) at low pH. Consecutive formation of neutral ($AC(COOH)^0$) and ionized ($AC\text{-}COO^-$) form with the rising pH leads to a decreased electrostatic attraction between these groups and anionic Cr(VI) species. Moreover, further increase in pH causes a dual competition between negatively charged Cr(VI) species and hydroxyl ions. Predictive mathematical model indicated that the optimum process conditions were at pH 2 and at 60°C with the initial Cr(VI) concentration of 60 mg/L. Pseudo second-order model described the kinetic data very well. The positive sign of enthalpy change confirmed the endothermic nature of adsorption process. Cr(VI) adsorption mechanism was interpreted as complexation or chemisorption with the respect of high enthalpy value (54.05 kJ/mol) and fitting of pseudo second-order kinetic model.

Researchers in [30] carbonized Eichhornia crassipes root at 600°C after impregnating with concentrated H_2SO_4 at 150–165°C for 24 h. The adsorption capacity of activated carbon was 36.34 mg/g, at optimum conditions of pH 4.5 and temperature of 25°C. The experimental adsorption data were fitted to Langmuir isotherm model. Adsorption equilibrium was reached after 60 min. Pseudo second-order model is more likely to predict adsorption kinetic. The authors interpreted this result with the chemical adsorption being the rate-controlling step. The rise in Cr(VI) adsorption with the increased temperature indicated that the adsorption process is endothermic in nature. $\Delta H°$ values were 11.9 kJ/mol and 12.5 kJ/mol for different initial concentrations. Mean free adsorption energy was found to be 17.36 kJ/mol. This is due to different chemical processes accompanying the ion exchange process [30]. The maximum Cr(VI) removal capacity in column studies was 0.8 mg/g.

KOH activated carbon was developed from peanut shell [31]. During the activation process half of produced carbon was oxidized by air. The authors investigated Cr(VI) adsorption onto these unoxidized and oxidized carbons with respect of equilibrium, kinetics, and thermodynamics. BET surface area and total pore volume of oxidized and unoxidized samples are 95.51 m^2/g, 0.35 cm^3/g, and 88.85 m^2/g, 0.33 cm^3/g, respectively. Maximum adsorption capacity (about 8.4 mg/g for oxidized and 7 mg/g for unoxidized carbon) was obtained at pH 2 and a sharp decrease in the adsorption capacity occurred between pH 2 and 7, thereafter the capacity became negligible. The authors explained the high Cr(VI) adsorption with electrostatic force of attraction at low pH. It is reported that adsorption was fast in 7 h and then became slow up to 20 h, finally equilibrium occurred almost at the 24 h for both oxidized and unoxidized carbon. Chromium adsorption onto oxidized and unoxidized activated carbons followed both pseudo first-order and pseudo second-order kinetic models according to high R^2 values. However, if one can look up the experimental data and models, it is obvious that pseudo second-order model represents the results much better than the first-order one. The Cr(VI) adsorption isotherms of both carbons fitted the Langmuir model better than Freundlich model in the temperature range studied. The authors calculated enthalpy value as positive, which showed the endothermic nature of Cr(VI) adsorption. Overall, oxidized carbon had higher Cr(VI) adsorption capacity than unoxidized carbon related with high surface area, carbon content, and micropore volume of oxidized carbon.

The used biomass waste materials are tamarind wood, peanut shell, rice husk, saw dust, sugarcane bagasse, olive bagasse, almond shell, groundnut, date palm seed, sugarcane husk, eucalyptus wood, Eichhornia crassipes roots, bael fruit shell, rice straw, oil palm fiber, Trapa natans (a water chestnut), jambolanum nut, palm shell, algae bloom residue and apricot stone in the reviewed articles. As well as the nature of precursor, activation process is also very effective on activated carbon properties. Various activation methods from physical activation like steam, CO_2, and air to chemical activation like KOH, $ZnCl_2$, H_2SO_4, and H_3PO_4 activation have been used in order to produce activated carbon. Amongst them, H_3PO_4 activation, which was the most common used, distinctively yielded with the highest surface area activated carbons. Activated carbon produced by zinc chloride activation also has high surface area. Although high surface area and high pore volume are effective for Cr(VI) adsorption, linear relationship between the high surface area and Cr(VI) adsorption capacity was not observed amongst the reviewed articles.

High surface area and pore volume are not only critical parameters but also surface chemistry of activated carbon is very important during the adsorption of charged ions like heavy metals. The possible mechanisms that could be occurred during the Cr(VI) adsorption onto activated carbon are physical adsorption, ion exchange, electrostatic attraction, chemical adsorption, and reduction to Cr(III). Surface functional groups have very important role for the all mentioned occurrences except physical adsorption by van der Waals forces. High surface functionalities may be resulted in high chromium adsorption. In case of heavy metal adsorption, surface functional groups on activated carbon surface should be defined by different analytical methods like FTIR and Boehm's titration [32]. In addition, potentiometric titration method, which was described in [33], could be used for characterizing the surface functionalities of activated carbon.

Surface functional groups can be scanned by FTIR, comprehensively, while Boehm's titration gives quantitative amount of acidic surface functionalities. The relative concentrations of acidic surface functional groups in carbons can be determined by Boehm's titration with sodium hydrogen carbonate, sodium carbonate, and sodium hydroxide [32, 34]. There are assumptions that all carboxylic, lactonic, and phenolic groups can be reacted with NaOH; Na_2CO_3 consumed by both carboxylic and lactonic groups; and $NaHCO_3$ can neutreulize only carboxylic groups.

Surface basicity of carbon is not understood yet. Delocalized π electrons on the edges of graphene sheets could be responsible for this basicity. However, it was declared in reference [34] that the basicity of these π electrons is relatively weak and some groups like quinine, diketone, chromene, and γ-pyrones on the carbon surface may cause these basic sites [35]. Titration with NaOH gives the amounts of total basic groups on carbon surface after neutralizing by HCl.

The shape of proton-binding curve, which is obtained by potentiometric titration, also indicates the presence of surface functional groups on the activated carbon surface. Carboxylic groups dissociate in the pH interval 3–6 and phenolic groups dissociate above this pH; consequently, one can say that the sharp falls of the proton binding curves on the corresponded pH range give the proton dissociations of these mentioned groups [36]. It is assumed that the anion and cation exchanges are in equilibrium at the crossover point with the pH axis on the proton binding. Carbon surface is positively charged at pH below this point, which is called point of zero charge (pH_{PZC}).

Cr(VI) adsorption was strongly pH-dependent. pH of the solution not only affects the Cr(VI) species but also the activated carbon surface. On raising pH, the possible reactions that occurred on the activated carbon surface are expressed as follows:

$$\text{AC-OH}_2^+ \leftrightarrow \text{AC}(\text{OH})^0 + \text{H}^+ \tag{1}$$

$$\text{AC}(\text{OH})^0 \leftrightarrow \text{AC-O}^- + \text{H}^+ \tag{2}$$

These surface hydroxyl groups are called protonated, neutral, and ionized, respectively. The activated carbon surface is nondissociated until pH_{PZC} and show great interest to the anions like $HCrO_4^-$ because of the electrostatic attraction, especially at low pH. After pH_{PZC} value, proton dissociation from the surface occurs; consequently, deprotonated surface repulses the anionic chromium species but attracts to the cations. Although it is very important to determine pH_{PZC} of activated carbon in order to characterize carbon surface during the pH variation, consequently affinity of carbon surface to the Cr(VI) species, only a few of the reviewed studies [17, 24] reported the pH_{PZC} of the produced activated carbon.

Overall, the following points are essentially effective on Cr(VI) adsorption onto activated carbon derived from biomass waste material:

1. Structural properties of activated carbon like surface area and pore volume
2. Surface chemistry of activated carbon [type and amount of surface functional groups, pH of point of zero charge (pH_{PZC})]
3. pH of the solution
4. Initial chromium concentration
5. Adsorbent dosage
6. Temperature
7. Contact time

These matters will be discussed in detail in the following section by using our data about Cr(VI) adsorption onto peach stone, one of the biomass waste material, based activated carbon.

3. Application of peach stone based activated carbon for Cr(VI) removal

Peach stone based activated carbon was used for Cr(VI) removal from aqueous solution. Effects of process parameters like pH, carbon dosage, contact time, and temperature on Cr(VI) adsorption will be discussed within this section. Furthermore, Cr(VI) adsorption kinetics and thermodynamics will be investigated and Cr(VI) adsorption mechanisms will be tried to illuminate.

3.1. Peach stone based activated carbon

Peach stone was converted to activated carbon by using one-step steam activation at 800°C. Detailed production conditions of peach stone based activated carbon were reported previously [37]. The some characteristics of peach stone based activated carbon are given in Table 1. N_2 adsorption measurements were performed using a Quantachrome Autosorb-1-C surface analyzer. Elemental analysis of the carbon was carried by VarioEL III CHNS elemental analyzer. The Density Functional Theory (DFT) model was applied to N_2 adsorption–desorption data to determine pore volume and pore size distribution, which is given in Figure 3.

BET surface area (m^2/g)	608
Pore volume (cm^3/g)[a]	0.341
Micropore volume (cm^3/g)[a]	0.331
Carboxylic groups (meq/g)[b]	< detection limit
Lactonic groups (meq/g)[b]	0.062
Phenolic groups (meq/g)[b]	0.254
Total Acidic groups (meq/g)[b]	0.316
Basic groups (meq/g)[b]	0.319
Point of Zero Charge (pH_{PZC})[c]	8.0
C (%)[d]	93.18
H (%)[d]	0.940
O (%)[d]	5.670
N (%)[d]	0.215

[a]Density functional theory model, [b]Boehm method, [c]obtained from potentiometric titration data, [d]elemental analysis

Table 1. Characteristics of peach stone based activated carbon [37]

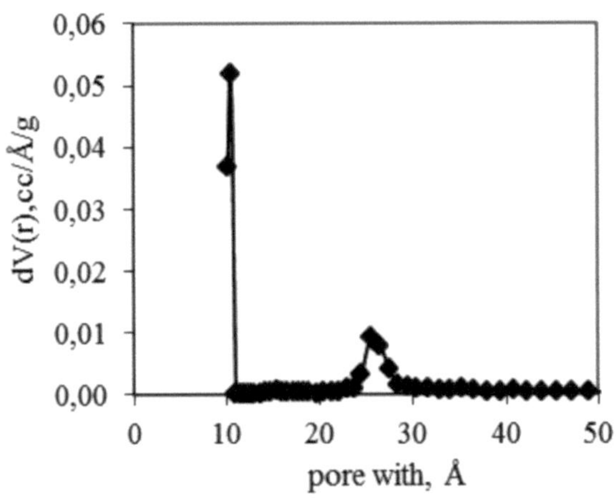

Figure 3. Pore size distribution of activated carbon

http://dx.doi.org/10.5772/60206

It is clear from Figure 3 that distribution intensity is quite high at lower than 10Å, which shows microporous structure of produced carbon. There are also an amount of pores centered between 24Å and 28Å (Figure 3). As can be seen from Table 1, approximately 97% of pore volume consists of micropores. In addition, the amounts of acidic and basic groups on activated carbon surface are almost the same.

3.2. pH effect on Cr(VI) adsorption

Experimental Cr(VI) adsorption isotherms obtained at different pH values are shown in Figure 4. All isotherm curves show typical Type 1 shape, according to IUPAC classification at pH 2, 4, and 6. Type 1 isotherms are generally given by microporous adsorbents in case of physical adsorption. Furthermore, monolayer adsorption, hence, chemical activation could be attributed for this behavior.

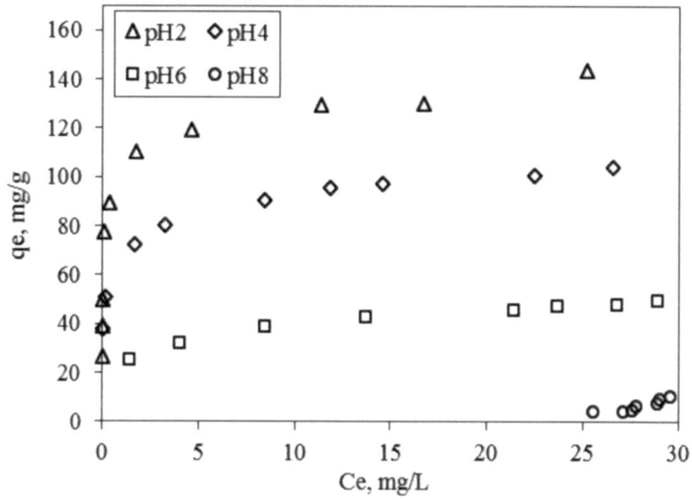

Figure 4. Experimental Cr(VI) adsorption isotherms at various pH (initial CrVI) concentration: 30 mg/L; solution volume: 50 mL; activated carbon dosage: 2–60 mg)

Cr(VI) adsorption was strongly pH-dependent and superior at low pH (Figure 4) as observed by the most researchers [14, 15, 17–19, 21–25, 27, 29, 31]. This can be evaluated by examining the adsorbate and the adsorbent surface. Cr(VI) species at different pH are shown in the following equations:

$$H_2CrO_4 \leftrightarrow H^+ + HCrO_4^- \, pK_a = 0.8 \tag{3}$$

$$HCrO_4^- \leftrightarrow H^+ + CrO_4^{2-} \, pK_a = 6.5 \tag{4}$$

$$2HCrO_4^- \leftrightarrow Cr_2O_7^{2-} + H_2O pK_a = 1.52 \quad (5)$$

All functional groups on the carbon surface are nondissociated at studied pH (2, 4, 6, 7), due to the pH_{PZC} of carbon about 8.0. These groups are protonated throughout the $[H]^+$-rich region, i.e., at low pH (Eqs. 1–2) There is an electrostatic attraction between protonated carbon surface and negatively charged hydrogen chromate ions (Eq. 3) at low pH. Maximum Cr(VI) adsorption capacity was obtained at pH 2, which was about 140 mg/g and it was decreased with increasing pH and was equal to approximately 100 mg/g, 48 mg/g, and 12 mg/g at pH 4, 6, 8, respectively (Figure 4). Activated carbon surface becomes neutral ($AC(OH)^0$) due to the proton releasing with increasing pH. Consequently, rising pH was resulted in lowering Cr(VI) adsorption capacity. Although almost all chromium was removed with 20 mg of activated carbon at pH 2, only 56% of chromium was removed at pH 6 and this value went through the 7% when studied at pH 8 (Figure 5).

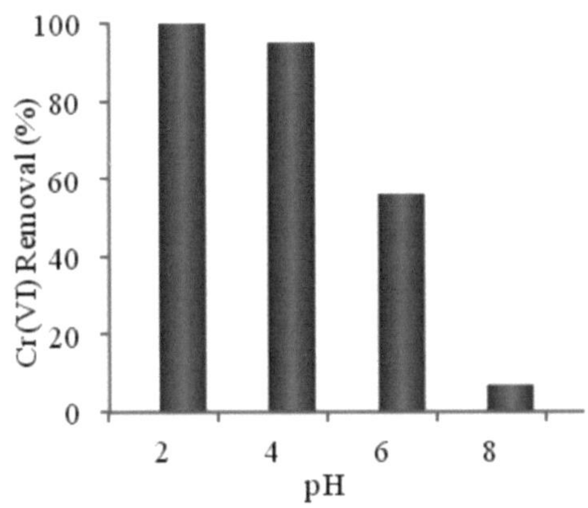

Figure 5. Cr(VI) removal percentage at different pH (initial Cr(VI) concentration: 30 mg/L; solution volume: 50 mL; activated carbon dosage: 20 mg)

As can be seen from Figure 4, chromium adsorption isotherm shapes are different and capacities are very low at studied pH 8. This can be interpreted by the competitiveness between chromate ions (CrO_4^{2-}) and hydroxyl ions; besides, one chromate ion needs two active sites because of two minus charges of it. It should also be kept in mind that proton dissociation of carbon surface starts at pH 8.0, which is the pH_{PZC} of carbon; consequently, deprotonated surface repulses the chromate anions.

3.3. Effect of pH on Cr(VI) reduction

Hydrogen chromate can reduce to Cr(III) at acidic medium [38] as demonstrated in Eqs. 6–8.

$$HCrO_4^- + \text{Organic matter} + H^+ \rightarrow Cr^{3+} + H_2O \quad (6)$$

$$HCrO_4^- + 7H^+ + 3e^- \rightarrow Cr^{3+} + 4H_2O \quad (7)$$

$$HCrO_4^- + 3C + H^+ \rightarrow Cr^{3+} + H_2O + 3/2\ CO_2 \quad (8)$$

Reduction of Cr(VI) was followed with the difference between the concentrations of total chromium and Cr(VI) in the solutions after treatment with activated carbon (figure was not given.). Cr(III) ions were observed only at pH 2 not at higher pH. It was also noticed that there were no big differences at low adsorbent dosages. This could be explained with low dissolved organic carbons at low adsorbent dosages. There were no differences between the total Cr and Cr(VI) concentrations at pH 4, 6, and 8. There are two possibilities: (i) there is not enough protons to reduce Cr(VI) to Cr(III) at these pH or (ii) the reduction reactions are occurred and then reduced Cr(III) ions are adsorbed on the carbon surface easily due to the low proton concentration, which compete with positively charged Cr(III) ions. Likewise, Cr(VI) could be reduced to Cr(III) in the presence of organic compounds (Eq. 6) even at neutral pH [39]. Similarly, it was suggested in [23] that Cr(VI) adsorption is mainly through a reduction to Cr(III) on activated carbon.

3.4. Effect of temperature on Cr(VI) adsorption

Increasing temperature from 22°C to 32°C caused a decrease of Cr(VI) adsorption and then capacity increased with further increment of temperature to 45°C (Figure 6). There could be several processes like ion exchange, physical adsorption, chemisorption, and reduction, which have different effects by temperature, during the Cr(VI) adsorption.

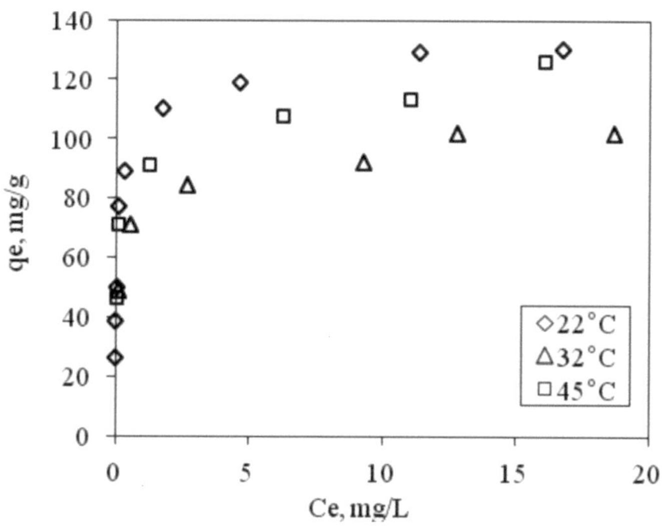

Figure 6. Experimental Cr(VI) adsorption isotherms at various temperature (initial CrVI) concentration: 30 mg/L; solution volume: 50 mL; activated carbon dosage: 2-60mg, pH:2)

It is known that ion exchange and physical adsorption capacity decreases with temperature while chemisorption and Cr(VI) reduction reaction are endothermic in nature. Cr(VI) adsorption trend by temperature could be interpreted with the combining effect of the mentioned mechanisms, that means, while ion exchange or/and physical adsorption was the predominant mechanism in the range of 22–32°C, chemisorption or/and reduction process was more effective at elevated temperature. It should also be considered that adsorption capacity could increase due to the hydrophobic effects. Elevated temperature is resulted in a decrease of hydration of $HCrO_4^-$ molecules, hence less hydrated ions could bond to the relatively hydrophobic carbon surface easily.

3.5. Effect of initial Cr(VI) concentration

Initial ion concentration is one of the process parameters that affected the adsorption capacity. Cr(VI) adsorption capacities increased with increasing initial Cr(VI) concentrations for all studied pH (Table 2).

pH		C_i: 5 mg/L	C_i: 30 mg/L	C_i: 60 mg/L
	2	62.9	89.1	111.7
q_e, (mg/g)	4	42.6	72.2	86.7
	6	19.9*	25.3	35.0

(*) obtained at 90% removal (due to could not be reached to 95% removal)

Table 2. Cr(VI) adsorption capacities at different initial Cr(VI) concentrations (capacities obtained at 95% Cr(VI) removal efficiency)

Increasing Cr(VI) adsorption capacity by initial concentration could be explained that high initial ion concentration provides an important driving force in order to overcome all mass transfer resistances between the adsorbate molecules and the adsorbent surface [40].

3.6. Adsorption isotherms

Adsorption isotherm is a unique way in order to evaluate an adsorption process as a unit operation. It not only indicates the adsorption capacity under studied conditions but also gives information about the mechanism and favorability of adsorption [41]. Langmuir and Freundlich adsorption isotherm models, which express the nonlinear relationship between the adsorbed ion on the adsorbent phase and the ion in the solution phase, are the most frequently used in the literature. These two-parameter models are simple and give a good description of experimental behavior in a large range of operating conditions [41].

The Langmuir theory, the first adsorption isotherm model developed in 1918, is on the basis of kinetic view point, and considers assumptions mentioned below:

1. Adsorption takes place in monolayer. Each active site adsorbs only one molecule.
2. There is no interaction between adsorbed molecules.
3. Adsorption enthalpy of all adsorbate molecules is equal [42].

Langmuir model is expressed by the following equation:

$$q_e = \frac{Q \cdot b \cdot C_e}{1 + b \cdot C_e} \tag{9}$$

where q_e is the amount of ions adsorbed per unit mass of adsorbent (mg/g), C_e is the equilibrium concentration of the adsorbate in the bulk solution (mg/L), Q and b are the Langmuir constants related to maximum adsorption capacity and the free energy of adsorption, respectively. Values of Q and b can be graphically determined from the linear form of the Langmuir model (Eq. 10).

$$\frac{1}{q_e} = \frac{1}{Q \cdot b \cdot C_e} + \frac{1}{Q} \tag{10}$$

Langmuir isotherm can be expressed in terms of a dimensionless separation factor, R_L, which describes the type of isotherm;

$$R_L = \frac{1}{1 + b \cdot C_0} \tag{11}$$

where C_0 is the initial concentration of Cr(VI). The magnitude of R_L determines the feasibility of adsorption process. If $R_L > 1$, adsorption is unfavorable; if $R_L = 1$, adsorption is linear; if $R_L < 1$, adsorption is favorable; and if $R_L = 0$, adsorption is irreversible [43].

The Freundlich model is one of the earliest known empirical equations based on heterogeneous surface and adsorption heat. It is expressed by the following equation:

$$q_e = K_F \cdot C_e^{1/n} \tag{12}$$

where K_F and n are indicators of adsorption capacity and adsorption intensity, respectively.

The constants k and n of the Freundlich model are respectively obtained from the intercept and the slope of the linear plot of Freundlich model (Eq. 13).

$$\ln q_e = \ln K_F + (1/n)\ln C_e \tag{13}$$

The conformity between the experimental data and the theoretical data was investigated using correlation coefficients (R^2) and error analysis. Error (%) values were calculated from Eq. 14;

$$Error(\%) = \frac{\sum_{1}^{N} \left| 100 \cdot \frac{(q_e - q_{er})}{q_e} \right|}{N} \tag{14}$$

where q_e (mg/g) and q_{et} (mg/g) are experimental and theoretical equilibrium adsorption capacities, respectively and N is the total data number.

Langmuir and Freundlich adsorption isotherm model parameters, evaluated from the linear plots, are presented in Tables 3 and 4, respectively. According to the correlation coefficients and error values, Freundlich model expressed Cr(VI) adsorption onto activated carbon better than Langmuir model.

T (°C)	C_i (mg/L)	pH	*Q* (mg/g)	*b* (L/mg)	R^2	Error (%)	R_L
22	5	2	100.219	14.422	0.936	8.599	0.01301
22	5	4	73.392	11.435	0.928	4.652	0.01623
22	5	6	34.978	3.109	0.917	4.851	0.05723
22	5	8	–2.371	–0.145	0.806	10.534	4.22181
22	30	2	112.167	97.453	0.843	13.880	0.00033
22	30	4	89.106	17.331	0.862	10.233	0.00183
22	30	6	47.960	0.743	0.939	5.067	0.04079
22	30	8	–1.041	–0.031	0.885	9.473	36.3553
22	60	2	123.436	84.386	0.865	14.781	0.00019
22	60	4	99.618	9.599	0.908	7.459	0.00163
22	60	6	53.373	35.983	0.759	16.076	0.00046
22	60	8	–2.624	–0.014	0.878	7.393	8.00448
32	30	2	92.302	13.705	0.926	7.110	0.00254
45	30	2	101.079	103.184	0.883	12.512	0.00034

Table 3. Langmuir adsorption isotherm constants

T (°C)	C_i *(mg/L)*	pH	K_F *(mg/g)*	*n*	R^2	Error (%)
22	5	2	89.203	4.566	0.981	3.277
22	5	4	61.780	6.024	0.992	1.371
22	5	6	24.302	4.386	0.991	1.722
22	5	8	0.022	0.283	0.736	11.609
22	30	2	97.597	8.122	0.979	4.160
22	30	4	65.655	6.793	0.986	2.997
22	30	6	23.772	4.578	0.995	1.159
22	30	8	$6.14.10^{-11}$	0.131	0.884	11.982
22	60	2	105.026	8.849	0.942	7.247
22	60	4	60.663	5.814	0.964	9.886
22	60	6	37.629	7.874	0.951	5.859
22	60	8	$1.52.10^{-7}$	0.224	0.899	8.817
32	30	2	71.769	7.675	0.970	4.086
45	30	2	88.146	8.288	0.979	4.005

Table 4. Freundlich adsorption isotherm constants

Although a significant amount of researchers [17, 19, 24–26] indicated the same result, there are also many workers [15, 30, 31] who reported that Cr(VI) adsorption onto activated carbon fitted the Langmuir model. Furthermore, there are some studies [18, 21, 22, 27] which reported that both Freundlich and Langmuir models described Cr(VI) adsorption onto activated carbon very well.

Dimensionless separation factors, R_L, show that the adsorption onto activated carbon is favorable except pH 8 (Table 3). Furthermore, calculated R_L values are very close to the zero for pH 2 and 4, which indicates the irreversible Cr(VI) adsorption, and consequently quite strong interaction between hydrogen chromate anions and activated carbon surface.

3.7. Adsorption kinetics

The rate at which dissolved heavy metal ions are removed from aqueous solution by solid adsorbents is a significant factor for application in water quality control. The rapid adsorption processes are quite useful for practical use, due to the need of short contact time in the actual process. When stoichiometric amount of activated carbon dosage (0.4 g/L) was used, more than 24 h (not shown in figure) were required in order to remove all chromium (Figure 7). However, almost all Cr(VI) was adsorbed within 1.5 h for the case of twice of stoichiometric carbon dosage.

There are different equilibrium times from 60 min to 24 h for Cr(VI) adsorption onto activated carbon in the literature [15, 18, 22, 25, 27, 29–31]. However, all of the given results did not reflect the same conditions as used in this study; it can be concluded that 1.5 h equilibrium time is applicable.

It is essential to evaluate adsorption kinetics as well as adsorption equilibrium using theoretical models in order to design and control the adsorption process units. Pseudo first-order and pseudo second-order kinetic models were used to perform the Cr(VI) adsorption kinetics. Both models include all steps of adsorption such as external film diffusion, adsorption, and internal particle diffusion, so they are pseudo models [41].

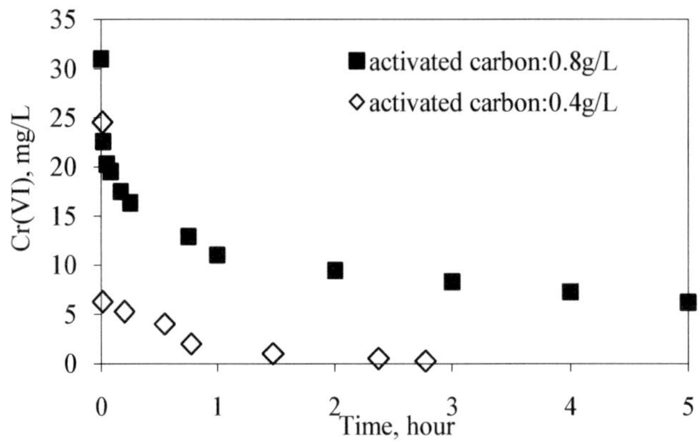

Figure 7. Effect of contact time on Cr(VI) adsorption

The *pseudo first-order kinetic model* can be expressed as follows:

$$\frac{dq_t}{dt} = k_1 \cdot (q_e - q_t) \tag{15}$$

where k_1 is the rate constant of pseudo first-order model (per min) and q_e and q_t denote the adsorption capacity at equilibrium and at time t (mg/g), respectively.

After integration of Eq. 15 by applying boundary conditions, $q_t = 0$ at $t = 0$ and $q_t = q_t$ at $t = t$, becomes

$$\log\left(\frac{q_e}{q_e - q_t}\right) = \frac{k_1 \cdot t}{2.303} \tag{16}$$

A plot of log $(q_{eq} - q_t)$ against t should give a straight line to confirm favorability of kinetic model. The plot of log $(q_{eq} - q_t)$ against t did not give any straight line (R^2 =0.731–0.900) for all kinetic data; consequently, it was concluded that the Cr(VI) adsorption kinetics did not fit the pseudo first-order equation.

The *pseudo second-order kinetic model* suggests that adsorption process involves chemisorption mechanism [41] and can be expressed as follows:

$$\frac{dq_t}{dt} = k_2 \cdot \left(q_e - q_t\right)^2 \tag{17}$$

where k_2 is the rate constant of second-order model. For the same boundary conditions, the integrated form of Eq. 17 becomes

$$\frac{t}{q_t} = \frac{1}{k_2 \cdot q_e^2} + \frac{1}{q_e} \cdot t \tag{18}$$

If the experimental data fit the second-order kinetics, the plot of t/q against t should give linear relationship. q_e and k_2 can be determined from the slope and intercept of this plot. Initial adsorption rates (h, mg/g, min) were also calculated from the data of pseudo second-order kinetic model according to the Eq. (19) [44]:

$$h = k_2 . q_e^{\ 2} \tag{19}$$

Pseudo second-order kinetic model expressed the Cr(VI) adsorption kinetics very well at all the studied conditions (Table 5).

T (°C)	AC (g/L)	C_i (mg/L)	q_e, *exp* (mg/g)	q_e, *calc (mg/g)*	k_2 *(g/mg.h)*	*h, init. rate* (mg/g.min)*	R^2
22	0.4	30	72.93	73.10	0.0290	2.584	0.997
22	0.8	30	33.56	33.57	0.8464	15.894	0.999
32	0.4	30	67.52	67.96	0.0304	2.339	0.998
45	0.4	30	69.56	70.09	0.0319	2.615	0.998

Table 5. Pseudo second-order kinetic model parameters

This situation can be followed with extremely high correlation coefficients ($R^2 = 0.997 - 0.999$). Moreover, there is an excellent correlation between calculated q_e values and experimental q_e values. As can be known that pseudo second-order kinetic model involves chemisorption mechanism, hence it can be suggested that Cr(VI) adsorption onto prepared activated carbon is based on chemical adsorption. Similarly, almost all researchers expressed their kinetic data by using pseudo second-order kinetic model [15, 18, 19, 21, 22, 24, 25, 27, 31]. The rate constant and the initial rate increased with temperature, as expected (Table 5). It is interesting when activated carbon dosage was increased two times, Cr(VI) adsorption rate constant increased about 30 times. Diffusion is an endothermic process; an increase in temperature causes a raise on the diffusion rate of the adsorbate molecules across the external boundary layer and within the pores.

3.8. Regeneration of activated carbon

Regeneration studies were carried out by using different concentrations of acid, alkali, and salt solutions during the five cycles of adsorption–desorption studies. The best regeneration was obtained with 10% NaOH and 10% HCl solutions (Figure 8).

Although it is typical to use salt (NaCl) and caustic-salt (NaOH.NaCl) for chromate recovery from anion exchangers in the treatment of cooling water [45], these solutions were not effective for regeneration of peach stone activated carbon. Although regeneration percentages increased across the cycles, Cr(VI) adsorption capacity decreased about 55% of initial value (Figure 9).

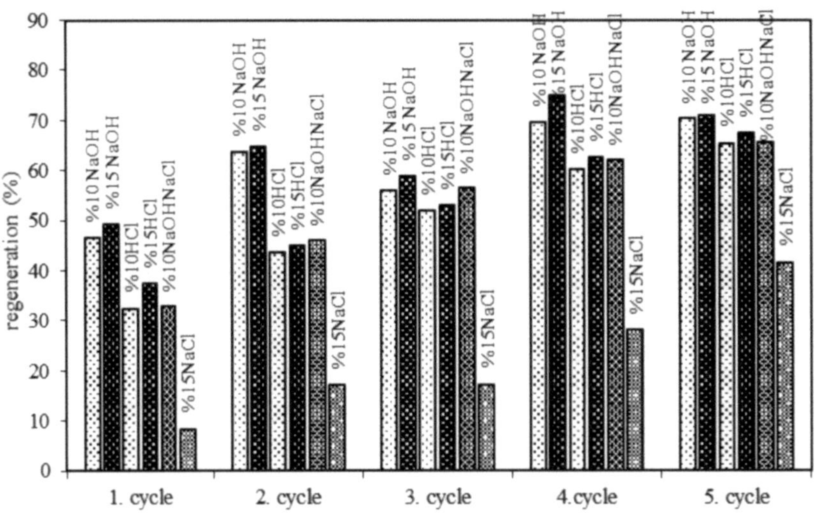

Figure 8. Regeneration percentages during the adsorption–desorption cycles

It can be concluded that Cr(VI) adsorption capacity of activated carbon is considerably high, but regeneration of it is not good. Similarly, irreversible Cr(VI) adsorption onto nitric acid oxidized activated carbon has been indicated in reference [46]. Many researchers reported the regeneration problems of Cr(VI) exhausted activated carbons [3, 19, 24].

3.9. Activation energy

The activation energy, E_A, was determined using the Arrhenius equation

$$k_{ad} = k_0 . e^{\frac{-E_A}{R.T}} \tag{20}$$

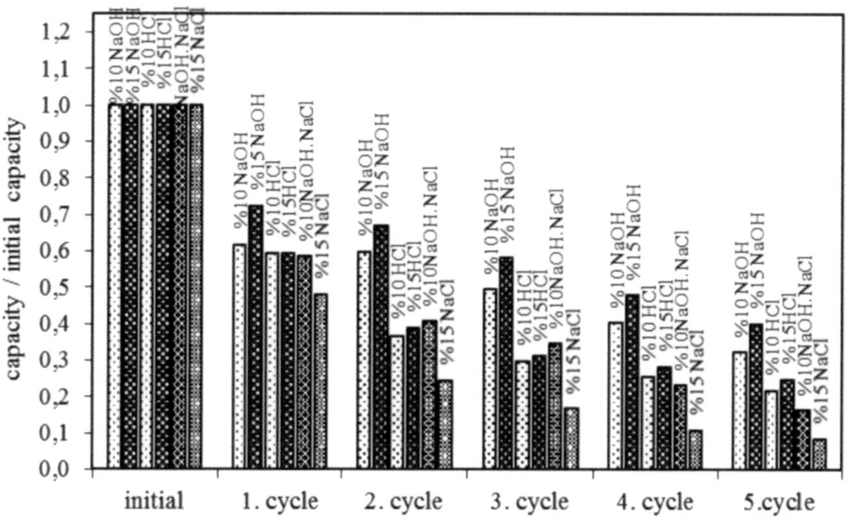

Figure 9. Cr(VI) adsorption capacity variations during the adsorption–desorption cycles

where k_{ad} is the adsorption rate, k_0 is the pre-exponential factor, and R is the gas constant. Linear form of Eq. 20 can be written as in below:

$$\ln k_{ad} = \ln k_0 - \frac{E_A}{R.T} \tag{21}$$

The plot of ln k_{ad} against to $1/T$ was found linear ($R^2 = 0.9991$). The activation energy, E_A, was calculated from the slope as 3.249 kJ/mol.

Because the energy requirement of physical adsorption is very low, activation energy of physical adsorption is usually not more than 4.184 kJ/mol [47]. Although activation energy of chromium adsorption onto peach stone based activated carbon is very low and could be a sign of physical adsorption, irreversible chromium adsorption process gets into mind the chemical interaction; consequently, activation energy is expected to be higher than 4.184 kJ/mol. Two kinds of chemisorption: activated and, less frequently, nonactivated have been defined in reference [47]. Reaction rate varies with the temperature on the base of Arrhenius equation (high E) in case of activated chemisorption. For nonactivated chemisorption, reaction occurs extremely fast, suggesting the activation energy is near zero [48]. Actually, reaction rate constant of chemisorption increases with temperature so the activation energy can be calculated relatively high. On the other hand, if the diffusion to the pores is very low, apparent rate may not be show a significant raise by temperature. Consequently, apparent activation energy can be calculated from the slope of the plot between 1/T and ln K as relatively low [49]. It can be concluded that, activation energy of chromium adsorption onto activated carbon was calculated very low, because of the diffusion resistance of pores. In addition, relatively low activation energy suggested that Cr(VI) adsorption was a diffusion controlled process [48].

3.10. Mean free sorption energy, *E*

The mean free sorption energy, *E*, is defined as free energy change when one mole of adsorbate in aqueous phase is transferred to the solid surface.

It is calculated by the following equation [44]:

$$E = \frac{1}{(2\beta)^{1/2}} \tag{22}$$

β is the constant related to the sorption energy (mol^2/kJ^2) and could be calculated by using Dubinin–Radushkevich (D–R) equation. Linearized D-R equation can be written as follows:

$$\ln q_e = \ln q_m - \beta \cdot \varepsilon^2 \tag{23}$$

where q_e and q_m are equilibrium and maximum adsorption capacity (mol/g), respectively, ε is the Polanyi potential. In order to obtain Polanyi potentials, experimental adsorption data should be applied to Eq. (24).

$$\varepsilon = R \cdot T \cdot \ln\left(1 + \frac{1}{C_e}\right) \tag{24}$$

A linear plot of ln q_e against ε^2 gives the values of β and q_m. Thus the mean free energy of adsorption, $E_,$ can be calculated by using Eq. (22).

The mean free sorption energy for ion exchange processes is between 8 and 16 kJ/mol, and activation energy of physical adsorption is not higher than 4.184 kJ/mol [47]. In this study, mean free sorption energy values were calculated >22 kJ/mol at pH 2 (Table 6), which indicates chemical interaction between carbon surface and hydrogen chromate ions. This chemical interaction could consist of reduction reactions of Cr(VI).

pH	*T* (°C)	*E* (kJ/mol)	R^2
2	22	23.63	0.995
2	32	22.81	0.985
2	45	26.12	0.987
4	22	20.98	0.993
6	22	15.68	0.998
8	22	2.10	0.956

Table 6. Mean free adsorption energy of Cr(VI) adsorption (C_i:30 mg/L)

http://dx.doi.org/10.5772/60206

3.11. Adsorption thermodynamics

Thermodynamic parameters such as Gibbs free energy ($\Delta G°$), enthalpy ($\Delta H°$), and entropy ($\Delta S°$) changes can be determined from the variation of thermodynamic equilibrium constant, *K*, at different temperatures. *K* values were determined by using the following equation:

$$K = \frac{C_{Ae}}{C_e} \tag{25}$$

where C_{Ae} (mg/g) and C_e (mg/g) indicate the Cr(VI) amounts on adsorbent phase and solution phase, respectively. The standard enthalpy change of adsorption ($\Delta H°$) and the standard entropy change of adsorption ($\Delta S°$) could be calculated from the slope and intercept of Van't Hoff plot (ln *K* versus 1/T), respectively.

$$\ln K = -\frac{\Delta H°}{RT} + \frac{\Delta S°}{R} \tag{26}$$

R (8.314 kJ/(mol.K)) refers to the universal gas constant and *T* (K) denotes the absolute temperature. The standard free energy changes of adsorption ($\Delta G°$) is determined according to Eq. (27).

$$\Delta G° = -R.T.\ln K \tag{27}$$

T (°C)	*K* (10^5)	ΔG° (kJ/mol)
22	4.022	-31.650
32	0.348	-26.515
45	1.787	-31.973

Table 7. Thermodynamic parameters of Cr(VI) adsorption (at pH 2; C_i:30 mg/L)

Thermodynamic data obtained for 30 mg/L initial Cr(VI) concentration at pH 2 were given in Table 7. The Gibbs free energy change values were found as negative, which indicated the feasibility and spontaneity of the adsorption.

3.12. Cr(VI) adsorption mechanism

Investigation with X-ray photoelectron spectroscopy (XPS) could provide valuable information in order to examine adsorption mechanism. Elemental analysis of carbon, oxygen, and other elements can be done by using XPS technique. In addition, by determining the shift in binding energies, XPS analysis provides valuable information about the chemical state of adsorbed ions onto activated carbon [50].

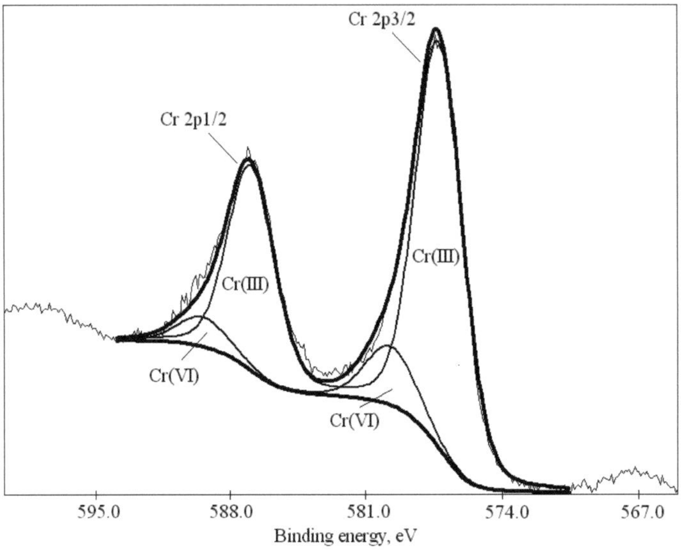

Figure 10. XPS spectrum(Cr $2p^{1/2}$ and Cr $2p^{3/2}$ orbitals) of Cr(VI) exhausted activated carbon (at pH 2 and C_i:30 mg/L)

According to the XPS results corresponding $Cr2p^{1/2}$ and $Cr2p^{3/2}$ orbitals, there are two separate bands for both orbitals (Figure 10). Significant bands appeared at binding energies of 577.0–578.0 eV for Cr2p3/2 and at binding energies of 586.0–588.0 eV for Cr2p1/2 orbital. They correspond to trivalent chromium. CrO_3 was characterized by higher binding energies; 580.0–580.5 eV and 589.0–590.0 eV, since the hexavalent form draws electrons more strongly than the trivalent form [51].

XPS showed that 86% of adsorbed chromium is in Cr(III) form. In other words, some of Cr(VI) adsorbed directly while a significant amount of Cr(VI) adsorbed to the surface after reducing the trivalent form during the adsorption process. Although original carbon surface is not suitable to adsorb cations, carbon surface is oxidized as a result of reduction reaction, and oxidized carbon surface is more attractive for Cr(III) cations than the former situation. Nevertheless, a little amount of Cr(III) was released into the aqueous phase due to electrostatic repulsion between the positively charged groups and Cr(III) ions at pH 2.

4. Conclusion

Adsorption of Cr(VI) onto activated carbon derived from different biomass waste materials, have been reviewed and the factors that influenced the Cr(VI) adsorption were discussed by using both literature and experimental data. Peach stone, one of the biomass waste materials, was used as an activated carbon precursor. Prepared activated carbons showed high Cr(VI) adsorption capacity. Cr(VI) adsorption was strongly pH-dependent and superior at low pH. Freundlich model well expressed Cr(VI) adsorption. Adsorption kinetics followed the pseudo second-order kinetic model which consists of chemisorption.

Low activation energy was interpreted with nonactivated chemisorption. High mean free sorption energy was assigned with chemical interaction between carbon surface and hydrogen chromate. XPS results showed that 86% of adsorbed chromium is in its trivalent form. Possible adsorption mechanism was suggested in three steps: (i) Cr(VI) adsorption onto positively charged groups, (ii) reduction to Cr(III), (iii) adsorption of Cr(III) while a little amount of it released to the aqueous phase. Peach stone based activated carbon could be evaluated as a potential adsorbent to remove Cr(VI) from aqueous media effectively.

Author details

Dilek Duranoğlu* and Ulker Beker

*Address all correspondence to: dduran@yildiz.edu.tr

Chemical Engineering Department, Yıldız Technical University, Istanbul, Turkey

References

[1] EPA. In situ treatment of soil and groundwater contaminated with chromium. Technical Report:EPA/625/R-00/005. Washington; 2000.

[2] Weiner ER. Applications of Environmental Chemistry, A Practical Guide For Environmental Professionals. Florida: CRC Press; 2000.

[3] Mohan D, Pittman U. Activated carbons and low cost adsorbents for remediation of tri- and hexavalent chromium from water. J Hazard Mater 2006;B137:762–811.

[4] Irwin RJ. Environmental Contaminants Encyclopedia chromium VI (Hexavalent Chromium) Entry. National Park Service Water Resources Divisions. Colorado; 1997.

[5] Kozlowski CA, Walkowiak W. Removal of chromium(VI) from aqueous solutions by polymer inclusion membranes. Water Res 2002;36:4870–76.

[6] Lazaridis NK, Bakoyannakis DN, Deliyanni EA. chromium(VI) sorptive removal from aqueous solutions by nanocrystalline akaganeite. Chemosphere 2005;58:, 65–73.

[7] Khezami L, Capart R. Removal of chromium(VI) from aqueous solution by activated carbons: kinetic and equilibrium studies. J Hazard Mater 2005;B123:223–31.

[8] Hu J, Lo IMC, Chen G. Comparative study of various magnetic nanoparticles for Cr(VI) removal. Sep Purif Tech 2007;56:249-56.

[9] Anirudhan TS, Jalajamony S, Suchithra PS. Improved performance of a cellulose-based anion exchanger with tertiary amine functionality for the adsorption of chro-

mium(VI) from aqueous solutions. Coll Surf A: Physicochem Eng Asp 2009;335:107–13.

[10] Lalvani SB, Wiltowski T, Hübner AH, Weston A, Mandich N. Removal of hexavalent chromium and metal cations by a selective and novel carbon adsorbent. Carbon 1998;36:1219–26.

[11] Kobya M. Removal of Cr(VI) from aqueous solutions by adsorption onto hazelnut shell activated carbon: kinetic and equilibrium studies. Bioresour Technol 2004;91:317–21.

[12] Sulaymon AH, Abid BA, Al-Najar JA. Removal of lead copper chromium and cobalt ions onto granular activated carbon in batch and fixed-bed adsorbers. Chem Eng J 2009;155:647–53.

[13] Gosh PK. Hexavalent chromium [Cr(VI)] removal by acid modified waste activated carbons. J Hazard Mater 2009;171:116–22.

[14] Dubey SP, Gopal K. Adsorption of chromium(VI) on low cost adsorbents derived from agricultural waste material: a comparative study. J Hazard Mater 2007;145:465–70.

[15] Demiral H, Demiral I, Tümsek F, Karabacakoglu B. Adsorption of chromium(VI) from aqueous solution by activated carbon derived from olive bagasse and applicability of different adsorption models. Chem Eng J 2008;144:188–96.

[16] Duranoglu D, Trochimczuk AW, Beker U. Kinetics and thermodynamics of hexavalent chromium adsorption onto activated carbon derived from acrylonitrile-divinylbenzene copolymer. Chem Eng J, 2012;187:193–202.

[17] Demirbas E, Kobya M, Konukman AES. Error analysis of equilibrium studies for the almond shell activated carbon adsorption of Cr(VI) from aqueous solutions. J Hazard Mater 2008;154:787–94.

[18] El Nemr A, Khaled A, Abdelwahab O, El-Sikaily A. Treatment of wastewater containing toxic chromium using new activated carbon developed from date palm seed. J Hazard Mater 2008;152:263–75.

[19] Isa MH, Ibrahim N, Aziz HA, Adlan MN, Sabiani NHM, Zinatizadeh AAL, Kutty SRM. Removal of chromium(VI) from aqueous solution using treated oil palm fibre. J Hazard Mater 2008;152:662–68.

[20] Giraldo-Gutiérrez L, Moreno-Piraján JC. Pb(II) and Cr(VI) adsorption from aqueous solution on activated carbons obtained from sugar cane husk and sawdust. J Anal Appl Pyrolysis 2008;81:278–84.

[21] Acharya J, Sahu JN, Sahoo BK, Mohanty CR, Meikap BC. Removal of chromium(VI) from wastewater by activated carbon developed from tamarind wood activated with zinc chloride. Chem Eng J 2009;150:25–39.

[22] Bansal M, Singh D, Garg VK. A comparative study for the removal of hexavalent chromium from aqueous solution by agriculture wastes carbons. J Hazard Mater 2009;171:83–92.

[23] Hsu NH, Wang SL, Liao YH, Huang ST, Tzou YM, Huang YM. Removal of hexavalent chromium from acidic aqueous solutions using rice straw-derived carbon. J Hazard Mater 2009;171:1066–70.

[24] Liu W, Zhang J, Zhang C, Wang Y, Li Y. Adsorptive removal of Cr(VI) by Fe-modified activated carbon prepared from Trapa natans husk. Chem Eng J 2010;162:677–84.

[25] Muthukumaran K, Beulah S. Removal of chromium(VI) from wastewater using chemically activated Syzygium jambolanum nut carbon by batch studies. Proc Environ Sci 2010;4:266–80.

[26] Owlad M, Aroua MK, Daud WMAW. Hexavalent chromium adsorption on impregnated palm shell activated carbon with polyethyleneimine. Bioresour Technol 2010;01:5098–103.

[27] Zhang H, Tang Y, Cai D, Liu X, Wang X, Huang Q, Yu Z. Hexavalent chromium removal from aqueous solution by algal bloom residue derived activated carbon: equilibrium and kinetic studies. J Hazard Mater 2010;181:801–08.

[28] Cronje KJ, Chetty K, Carsky M, Sahu JN, Meikap BC. Optimization of chromium(VI) sorption potential using developed activated carbon from sugarcane bagasse with chemical activation by zinc chloride. Desalination 2011;275:276–84.

[29] Özdemir E, Duranoğlu D, Beker Ü, Avcı AÖ. Process optimization for Cr(VI) adsorption onto activated carbons by experimental design. Chem Eng J 2011;172:207– 18.

[30] Giri AK, Patel R, Mandal S. Removal of Cr(VI) from aqueous solution by Eichhornia crassipes root biomass-derived activated carbon. Chem Eng J, 2012;185–186:71–81.

[31] AL-Othman ZA, Ali R, Naushad M. Hexavalent chromium removal from aqueous medium by activated carbon prepared from peanut shell: adsorption kinetics, equilibrium and thermodynamic studies. Chem Eng J 2012;184:238–47.

[32] Boehm HP. Advances in Catalyst. New York: Academic Press; 1966.

[33] Helfferich F. Ion Exchange. New York: Dover Publications; 1995.

[34] Boehm HP. Surface oxides on carbon and their analysis: a critical assessment. Carbon 2002;40:145–49.

[35] Marsh H, Rodriguez-Reinoso F. Activated Carbon. Oxford: Elsevier Science & Technology Books; 2006.

[36] Strelko V Jr, Malik DJ, Streat M. Characterization of the surface of oxidized carbon adsorbents. Carbon 2002;40:95–104.

[37] Duranoglu D, Trochimczuk AW, Beker U. A comparison study of peach stone and acrylonitrile-divinylbenzene copolymer based activated carbons as chromium(VI) sorbents. Chem Eng J, 2012;165:56–63.

[38] Selomulya C, Meeyoo V, Amal R. Mechanisms of Cr(VI) removal from water by various types of activated carbons. J Chem Technol Biotechnol 1999;74:111–22.

[39] Tsuruhara T, Takehara K, Yoshimura K, Matsuoka S, Saputro S, Aizawa J. Kinetic study on Cr(VI) reduction in natural water by means of flow injection-solid phase spectrometry (FI-SPS): Proc 4th Int Conf Ion Exchange, 15–19 October 2007, Chiba, Japan.

[40] Malkoc E, Nuhoglu Y. Potential of tea factory waste for chromium(VI) removal from aqueous solutions: thermodynamic and kinetic studies. Sep Purif Technol 2007;54:291–8.

[41] Aksu Z, Balibek E. Chromium(VI) biosorption by dried Rhizopus arrhizus: effect of salt (NaCl) concentration on equilibrium and kinetic parameters. J Hazard Mater 2007;145:210–20.

[42] Woodard F. Industrial Waste Treatment Handbook. USA: Butterworth-Heinemann; 2001.

[43] Hall KR, Eagletow LC, Acrivos A, Vermeulen T. Pore and solid kinetics in fixed-bed adsorption under constant-pattern condition. Ind Eng Chem Fund 1966;5:212–23.

[44] El-Kamash AM, Zaki AA, Abed El Geleel M. Modeling batch kinetics and thermodynamics of zinc and cadmium ions removal from waste solutions using synthetic zeolite A. J Hazard Mater 2005;B127:211–20.

[45] Letterman RD. Water Quality and Treatment. New York: McGraw-Hill; 1999.

[46] Dobrowolski R, Otto M. Study of chromium(VI) adsorption onto modified activated carbons with respect to analytical application. Adsorption 2010;16:279–86.

[47] Sag Y, Kutsal T. Determination of the biosorption activation energies of heavy metal ions on Zoogloea ramigera and Rhizopus arrhizus. Process Biochem 2000;35:801–07.

[48] Raji C, Anirudhan ST. Batch Cr(VI) removal by polyacrylamide-grafted sawdust: kinetics and thermodynamics. Water Res 1998;32(12):3772–80.

[49] Levenspiel O. Chemical Reactor Omnibook. Oregon: OSU Book Store; 1989.

[50] Zhao N, Wei N, Li J, Qiao Z, Cui J, He F. Surface properties of chemically modified activated carbons for adsorption rate of Cr(VI). Chem Eng J 2005;115:133–38.

[51] Park D, Yun YS, Park JM. XAS and XPS studies on chromium-binding groups of biomaterial during Cr(VI) biosorption. J Colloid Interface Sci 2008;317:54–61.